Research Series on the Chinese Dream and China's Development Path

Drawing on a large body of empirical studies done over the last two decades, this Series provides its readers with in-depth analyses of the past and present and forecasts for the future course of China's development. It contains the latest research results made by members of the Chinese Academy of Social Sciences. This series is an invaluable companion to every researcher who is trying to gain a deeper understanding of the development model, path and experience unique to China. Thanks to the adoption of Socialism with Chinese characteristics, and the implementation of comprehensive reform and opening-up, China has made tremendous achievements in areas such as political reform, economic development, and social construction, and is making great strides towards the realization of the Chinese dream of national rejuvenation. In addition to presenting a detailed account of many of these achievements, the authors also discuss what lessons other countries can learn from China's experience.

More information about this series at http://www.springer.com/series/13571

Jiahua Pan · Manhong Shen
Editors

Chinese Dream and Practice in Zhejiang – Ecology

SSAP

Springer

Editors
Jiahua Pan
Institute for Urban and Environmental Studies
Chinese Academy of Social Sciences
Beijing, China

Manhong Shen
Ningbo University
Ningbo, Zhejiang, China

Published with support of Zhejiang People's Publishing House

ISSN 2363-6866 ISSN 2363-6874 (electronic)
Research Series on the Chinese Dream and China's Development Path
ISBN 978-981-13-7208-7 ISBN 978-981-13-7209-4 (eBook)
https://doi.org/10.1007/978-981-13-7209-4

Jointly published with Social Sciences Academic Press, Beijing, China
The print edition is not for sale in China Mainland. Customers from China Mainland please order the print book from: Social Sciences Academic Press.

Library of Congress Control Number: 2019935833

This Springer imprint is published by the registered company Springer Nature Singapore Pte Ltd.
The registered company address is: 152 Beach Road, #21-01/04 Gateway East, Singapore 189721, Singapore

Series Preface

Since China's reform and opening began in 1978, the country has come a long way on the path of Socialism with Chinese Characteristics, under the leadership of the Communist Party of China. Over thirty years of reform, efforts and sustained spectacular economic growth have turned China into the world's second-largest economy and wrought many profound changes in the Chinese society. These historically significant developments have been garnering increasing attention from scholars, governments, and the general public alike around the world since the 1990s, when the newest wave of China studies began to gather steam. Some of the hottest topics have included the so-called China miracle, Chinese phenomenon, Chinese experience, Chinese path, and the Chinese model. Homegrown researchers have soon followed suit. Already hugely productive, this vibrant field is putting out a large number of books each year, with Social Sciences Academic Press alone having published hundreds of titles on a wide range of subjects.

Because most of these books have been written and published in Chinese; however, readership has been limited outside China—even among many who study China—for whom English is still the lingua franca. This language barrier has been an impediment to efforts by academia, business communities, and policy-makers in other countries to form a thorough understanding of contemporary China, of what is distinct about China's past and present may mean not only for her future but also for the future of the world. The need to remove such an impediment is both real and urgent, and the *Research Series on the Chinese Dream and China's Development Path* is my answer to the call.

This series features some of the most notable achievements from the last 20 years by scholars in China in a variety of research topics related to reform and opening. They include both theoretical explorations and empirical studies and cover economy, society, politics, law, culture, and ecology; the six areas in which reform and opening policies have had the deepest impact and farthest-reaching consequences for the country. Authors for the series have also tried to articulate their visions of the "Chinese Dream" and how the country can realize it in these fields and beyond.

All of the editors and authors for the *Research Series on the Chinese Dream and China's Development Path* are both longtime students of reform and opening and recognized authorities in their respective academic fields. Their credentials and expertise lend credibility to these books, each of which has been subjected to a rigorous peer review process for inclusion in the series. As part of the Reform and Development Program under the State Administration of Press, Publication, Radio, Film, and Television of the People's Republic of China, the series is published by Springer, a Germany-based academic publisher of international repute, and distributed overseas. I am confident that it will help fill a lacuna in studies of China in the era of reform and opening.

Xie Shouguang

Contents

Chapter 1
The Outlook on Ecological Civilization Leads the Building of a Beautiful Zhejiang

Manhong Chen

In the process of the market-oriented reform Zhejiang has always carried out concrete work to stay ahead and has made tremendous achievements, which have been hailed as the Zhejiang Model, the Zhejiang Experience and the Zhejiang Miracle. Zhejiang pushes forward the drive towards modernization under the condition that Zhejiang is small in terms of land area, natural resources and environmental capacity. Therefore, Zhejiang has also become the first to encounter the growing pains and thus initiate ecological development. Under the guidance of the outlook on ecological civilization that claims that equal emphasis should be placed on the "gold and silver mountains" and clean, clear waters and lush mountains, Zhejiang has upgraded the strategic goal of ecological development from a green Zhejiang to a beautiful Zhejiang, and so a modern Zhejiang which is beautiful and also presents a good life is emerging.

1.1 Equal Emphasis Should Be Placed on the "Gold and Silver Mountains" and Clean, Clear Waters and Lush Mountains

When serving as the Secretary of the Party Committee of Zhejiang Province, Xi Jinping comprehensively expounded the theory of "two mountains"—clean, clear waters and lush mountains, the "gold and silver mountains". Since the 18th National Congress of the Communist Party of China, the theory has been expanded systematically. Xi Jinping criticized the practice of only seeking the "gold and silver mountains" in disregard of clean, clear waters and lush mountains, he required that equal emphasis should be placed on the "gold and silver mountains" and clean, clear waters and lush mountains, he predicted that the idea of clean, clear waters

M. Chen (✉)
Ecological Culture Research Center, Ningbo University, Ningbo, Zhejiang, China
e-mail: 13263169197@163.com

J. Pan and M. Shen (eds.), *Chinese Dream and Practice in Zhejiang – Ecology*,
Research Series on the Chinese Dream and China's Development Path,
https://doi.org/10.1007/978-981-13-7209-4_1

and lush mountains was the "gold and silver mountains", and he advised that the people should prefer clean, clear waters and lush mountains to the "gold and silver mountains", while that criticism, requirement, prediction and advice constitute the complete theory of the "two mountains". Learning and carrying out the theory of "two mountains" is realistically targeted for building a beautiful Zhejiang and is also of great significance as a guide for deeply promoting ecological development aimed at building a beautiful China.

1.1.1 The Background of Developing the Theory of "Two Mountains"

Zhejiang enjoys breathtaking scenery with mountains and rivers, but Zhejiang has fewer natural resources. The geographical feature that there are 70% mountains, 10% water and 20% land determines that Zhejiang is a typical small province in terms of land area, natural resources and environmental capacity. How does a province with a large population survive subject to great environmental and resource constraints? How can it survive better? This is a great issue for the people of Zhejiang.

At the early stage of reform and opening-up, Zhejiang was relatively backward compared with the rest of the country in economic development. As a Chinese saying goes, poverty gives rise to the desire for change. With the Zhejiang Spirit, characterized by ceaseless self-improvement, grittiness, great courage in innovation and pursuit of substantial results, the people of Zhejiang have developed the Wenzhou Model and the Yiwu Model of market-oriented reform which are worth widely applying, and have more extensively brought forth the Zhejiang Model which can offer valuable experience for the rest of the country.

Such market-oriented reform not only has created the Zhejiang Phenomenon, the Zhejiang Miracle and the Zhejiang Mystery, but it also presents extraordinary institutional externality; it provides the Zhejiang Sample for a market-oriented reform in the rest of the country, and stimulates the formation of the China Model, the China Miracle, the China Mystery and the China Experience. At present, the market force has evolved the role of the market mechanism in resource allocation in China from the basic role to the decisive role.

Subject to special conditions in Zhejiang, Zhejiang has become the first to meet with the growing pains in economic development. The contradictions between unlimited resource demand and limited resource supply, inefficient resource utilization, between increasing environmental capacity demand and decreasing environmental capacity supply, low environmental productivity, between the increasing demand of the residents for ecological and environmental quality and less ideal supply of ecological and environmental quality from the government have become very acute. These contrasts have, to varying degrees, caused mass incidents arising out of resource environmental problems.

Against such a background, the Party Committee and the People's Government of Zhejiang Province vowed to improve environmental protection with great resolve, and to carry out change and upgrading better by vacating the cage to change birds. Xi Jinping put forward the theory of "two birds": In order to promote strategic adjustment of the economic structure and a fundamental change in the growth mode, it is necessary to raise "two birds" better—achieving the rebirth of the phoenix amidst fire, vacating the cage to change birds. The rebirth of the phoenix amidst fire means that actions are taken with great courage to break away from dependence upon extensive growth, while vacating the cage to change birds means that the stamina of the people of Zhejiang for making adventures everywhere is carried forward to go out of Zhejiang to develop Zhejiang.[1]

Therefore, the people of Zhejiang have unswervingly carried out the "811" action plan for environmental improvement and the "811" action plan for ecological development. The people of Zhejiang have resolutely shut down outdated production facilities, firmly rejected the industries that pollute, and unequivocally promoted green, circular and low-carbon development. As an important part of the Chinese Dream, the ecological development aimed at building a beautiful China has first come true in Zhejiang.

Zhejiang's transformation and upgrading is underway, but Anji County became the first ecological county in China early on, the first chemical enterprise with zero emissions in China has emerged in Ningbo City, and the construction of the national demonstration area for the circular economy and demonstration area for ecological development is in full swing. Zhejiang has stayed ahead in market-oriented reform, and remains at the forefront in transformation, upgrading and ecological development.

1.1.2 Basic Viewpoints About the Theory of "Two Mountains"

A series of addresses, speeches of Xi Jinping and articles published in newspapers and periodicals by him have systematically expounded the theory of "two mountains".

On August 8, 2003, Xi Jinping expounded the relationship between the "gold and silver mountains" and clean, clear waters and lush mountains from the perspective of the theory of knowledge. He said, "The people only seek the 'gold and silver mountains', having no regard for clean, clear waters and lush mountains; the people only pursue economic interests and development disregarding the environment and without long-term considerations; they exhaust the resources left by their ancestors and do not consider the needs of future generations; this is the first stage of understanding; although the people realize the importance of the environment, they only consider their own small environment and small homes without paying attention to

[1] Xi Jinping, *Carrying out Concrete Work to Stay Ahead—Line of Thought and Practice in Promoting New Development in Zhejiang*, The Party School of the CPC Central Committee Press, 2006, p. 128.

others, and some people even seek economic their own economic interests by doing harm to the environment of others; this is the second stage of understanding; the people truly realize that the ecological issue is borderless and there is only one planet for mankind, the planet is the common homeland for mankind, protecting the environment is the shared responsibility of all of mankind, and thus promoting ecological progress becomes the conscious action; this is the third stage of understanding."[2]

On March 23, 2006, Xi Jinping further made a more complete and more vigorous explanation from the perspective of the unity of opposites between the "gold and silver mountains" and clean, clear waters and lush mountains: the people have experienced three stages in understanding the relationships between "two mountains" in practice: during the first stage, the people exchange clean, clear waters and lush mountains for the "gold and silver mountains" without considering or with less considerations given to the bearing capacity of the environment, and relentlessly exploit its resources; during the second stage, the people endeavor to keep clean, clear waters and lush mountains while seeking the "gold and silver mountains", the contradiction between economic development and a shortage of resources, environmental deterioration has began to loom large, the people realize that the environment is fundamental to their survival and development, and only when clean, clear waters and lush mountains are available can development be fostered; during the third stage, the people realize that clean, clear waters and lush mountains can endlessly bring the "gold and silver mountains", and clean, clear waters and lush mountains themselves are the "gold and silver mountains", evergreen plants are the source of money, ecological advantages become economic advantages, a harmonious integral whole is formed; this stage represents a higher realm and reflects the requirements of scientific development, the philosophy of developing a circular economy and building a resource-conserving and environmentally-friendly society.[3]

Based on the correct understanding of the relationships between the "two mountains", Xi Jinping seriously criticized the line of thought and practices that go against the Scientific Outlook on Development. He said, "Further adoption of the old extensive operational path with high input, high consumption and heavy pollution is not accepted by the national policy, the resource environment or the people."[4] He warned before the governments, leaders at various levels, enterprises and the entire population that a government which did not value ecology was a government which lacked vision; a leader who did not value ecology was an incompetent leader; an enterprise which did not value ecology was a futureless enterprise; a citizen who did not value ecology was a citizen who did not have any awareness about modern civilization.[5] Given the distorted outlook on productive forces and on the performance of officials, Xi Jinping pointed out, "Destroying the ecological environment disrupts the pro-

[2]Xi Jinping, *Zhijiang Xinyu*, Zhejiang People's Publishing House, 2007, p. 13.

[3]Xi Jinping, *Zhijiang Xinyu*, Zhejiang People's Publishing House, 2007, pp. 186–187.

[4]Xi Jinping, *Carrying out Concrete Work to Stay Ahead—Line of Thought and Practice in Promoting New Development in Zhejiang*, The Party School of the CPC Central Committee Press, 2006, p. 23.

[5]Xi Jinping, *Carrying out Concrete Work to Stay Ahead—Line of Thought and Practice in Promoting New Development in Zhejiang*, The Party School of the CPC Central Committee Press, 2006, p. 186.

ductive forces; protecting the ecological environment protects the productive forces; improving the ecological environment develops the productive forces; both economic growth and environmental protection reflect the performance of officials."[6]

When taking charge of drafting the report to be delivered during the 18th National Congress of the Communist Party of China, Xi Jinping laid great importance on ecological development, systematically expounded "vigorously promoting ecological development" by taking it as a separate part, and based on the theory of the "two mountains", he put forward the ambitious goal of building a beautiful China and achieving ever-lasting development of the Chinese nation. On the morning of May 24, 2013, the Political Bureau of the Central Committee of the Communist Party of China carried out the sixth collective learning meeting concerning vigorously promoting ecological development. When presiding over this meeting, Xi Jinping stressed, "What we are doing today to protect the ecological environment will benefit generations to come; it is necessary to be keenly aware that it is urgent and arduous to protect the ecological environment and control environmental pollution, and that it is important and essential to strengthen ecological development, it is necessary to develop a strong sense of responsibility for the people and the future generations, resolutely control environmental pollution and build a better ecological environment, ushering in a new era of socialist ecological civilization and creating a good environment for the production and the lives of the people.[7]

On September 7, 2013, Xi Jinping delivered an important speech "Carrying Forward the People's Friendship, Jointly Creating a Bright Future" at Nazarbayev University, Kazakhstan. When answering the questions about environmental protection raised by students at the university, he pointed out, "China has clearly given higher priority to ecological and environmental protection; we place equal emphasis on clean, clear waters and lush mountains and the 'gold and silver mountains'; we prefer clean, clear waters and lush mountains to the 'gold and silver mountains'; clean, clear waters and lush mountains are the 'gold and silver mountains'. We cannot seek temporary economic development at the expense of the ecological environment. We have put forward the strategic task of fostering ecological development and building a beautiful China, and we have endeavored to leave a beautiful homeland with a blue sky, green land and clean, clear water to the future generations."[8] This answer further expanded the theory of the "two mountains" and also vehemently refuted the wrong outlook on development.

When delivering the speech during the 3rd Plenary Session of the 18th Central Committee of the Communist Party of China, Xi Jinping further elevated the theory of the "two mountains" to a theory of systems. He pointed out, "Mountains, waters, forests, farmland and lakes constitute a community of life, the people's lifeline lies

[6]Xi Jinping, *Carrying out Concrete Work to Stay Ahead—Line of Thought and Practice in Promoting New Development in Zhejiang*, The Party School of the CPC Central Committee Press, 2006, p. 186.

[7]Xinhuanet, Upholding the Basic National Policy of Saving Resources and Protecting the Environment, Moving towards a New Era of Socialist Ecological Civilization, *People's Daily*, May 25, 2013, Page 1.

[8]Wei Jianhua, Zhou Liang, Xi Jinping: Preferring Clean, Clear Waters and Lush Mountains to the "gold and silver mountains", http://www.youth.cn, last accessed: September 7, 2013.

in the farmland, the lifeline of the farmland lies in the water, that of the water lies in the mountain, that of the mountain lies in the soil, and that of the soil lies in the trees; usage control and ecological restoration must follow the law of nature; if those who plant trees only pay attention to planting trees, those who govern water only pay attention to water governance, those who protect the farmland only pay attention to protecting the farmland, it is very easy to only care for particular things and lose other things, so that in the end, the ecological environment is damaged systematically."[9]

As we learn the basic viewpoints about the theory of the "two mountains", we can find that the theory has the following prominent characteristics.

First, it is realistically targeted at the problems. Xi Jinping put forward the theory of the "two mountains" on the basis of reflecting on the traditional developmental mode with high input, high consumption, heavy pollution and high growth, and seriously criticizing the wrong outlook on development which leads to exhaustion of the resources left by ancestors and a lack of consideration of the needs of future generations. Reform and opening-up is an unprecedented cause, a great many problems have loomed large in this process. Therefore, the top-level design must be problem-oriented. If the wrong outlook on development is not criticized, it will be impossible to foster a correct outlook on development. The "Eight-Eight Strategies", put forward when Xi Jinping took charge of Zhejiang, they are the vivid embodiment of the Scientific Outlook on Development in Zhejiang; the 5th strategy among the "Eight-Eight Strategies" stresses efforts to further leverage Zhejiang's ecological advantages, and build an ecological province and a green Zhejiang. The "Eight-Eight Strategies" are the guiding thought for Zhejiang in promoting scientific development in more than ten years and will continue to guide Zhejiang's scientific development.

Second, the theory is logically progressive. The theory of the "two mountains" is progressive from a logical perspective. At the first level, it denies the one-sided view of only seeking the "gold and silver mountains" with no regard for clean, clear waters and lush mountains; at the second level, it presents the view of placing equal emphasis on the "gold and silver mountains" and clean, clear waters and lush mountains; at the third level, it shows the Scientific Outlook on Development in which clean, clear waters and lush mountains are the "gold and silver mountains"; at the fourth level, it takes an extreme approach to stress the view of preferring clean, clear waters and lush mountains to the "gold and silver mountains". As the theory is logically progressive, the theory can be elevated to a strategic level; Xi Jinping incorporated ecological development into the "Eight-Eight Strategies" when taking charge of Zhejiang, and he has incorporated ecological development into the five-sphere integrated plan since he served as the General Secretary of the Central Committee of the Communist Party of China.

Third, it is highly popular in language. The theory of the "two mountains" does not serve the theory itself; instead, it guides practice. A theory for guiding the practice in scientific development and ecological development must be acceptable to, understood

---

[9]Xi Jinping, Explanatory Notes for the Decision of the Central Committee of the Communist Party of China on Some Major Issues Concerning Comprehensively Intensifying the Reform, *People's Daily*, November 12, 2013.

and welcomed by the people. The theory of the "two mountains" is easy to understand and catchy; it represents the people's voice and is greatly accessible to the people; it can be easily converted into the people's conscious practice. There are many views like the theory of the "two mountains"; however, they are not as direct and kind as the theory of the "two mountains"; for instance, "you should value both fronts" is too academic; "economic growth and environmental protection are equally important" is too straightforward. The theory of the "two mountains" is profound in thought and popular in language.

1.1.3 The Compliance of the Theory of the "Two Mountains" with the Objective Law

According to the theory of the "two mountains", the "gold and silver mountains" represents economic growth, and clean, clear waters and lush mountains represents the ecological environment. Xi Jinping put forward the theory of the "two mountains" on the basis of mastering the objective law, thus it obviously follows the objective law.

1. The theory of the "two mountains" conforms to the basic law of development which governs ecological development

Ecological civilization refers to all the efforts and all the achievements made by mankind to achieve sustainable economic and social development by following the natural law of development, the economic law of development, the social law of development and the humanistic law of development, actively improving and optimizing the relationship between the people and nature, among the people, between the people and the society in economic and social activities.[10] The prerequisite for ecological development is compliance with objective laws; the acts which run counter to objective laws will certainly deliver fewer outcomes with more efforts, and might even meet with retaliation from objective laws. In fact, the theory of the "two mountains" requires us to follow the economic law of development and the natural law of development. Moreover, when the economic law of development conflicts with the natural law of development, the natural law of development must prevail. We cannot agree with the mechanical view of denying, conquering and transforming nature; we should uphold the philosophy of ecological development of respecting, conforming to and protecting nature. If clean, clear waters and lush mountains and the ecological environment are destroyed, the basic conditions for economic development and the foundation for the "gold and silver mountains" will disappear. As a Chinese saying goes, as long as there are lush mountains, there is firewood. As long as clean, clear waters and lush mountains are available, there is the foundation for ever-lasting development and the ecological advantage can be turned into economic advantage.

[10] Shen Manhong, Cheng Hua, Lu Genyao et al., *A Study of the Strategy of Ecological Development and Coordinated Regional Economic Development*, Science Press, 2012, p. 4.

2. The theory of the "two mountains" conforms to the law of coordinated ecological and economic development

The theory of the "two mountains" is fully consistent with the law of coordinated ecological and economic development. From the perspective of the unity of the "two mountains", it is necessary to maximize the comprehensive benefits including the ecological, economic and social benefits, give overall considerations to and properly protect the ecosystem while developing the economy to increase the ecological capital, and improve the economic system while protecting the ecological environment to enhance the economic quality, ultimately achieving benign interaction and coordinated development between the ecological system and the economic system. Given the opposites between the "two mountains", it is necessary to proceed from the overall, long-term and fundamental interests, prefer ecological benefits to economic benefits which are generated at the expense of the ecology, and give priority to delivering ecological benefits, and then turn ecological benefits into economic benefits. The ultimate goal of ecological development is to increase the index of the people's happiness, but that index does not merely depend upon the level of income. On the one hand, the index of happiness is directly proportional to per capita income; on the other hand, the index of happiness is inversely proportional to environmental pollution. If the economic level increases but the environmental quality declines, the index of happiness does not necessarily rise. Only the level of a green income free of damage from environmental pollution is closely associated with the index of happiness. For the sake of the people's happiness, the Party Committee of Zhejiang Province arranged "three renovations and one demolition"—renovating old residential areas, old factory areas in urban areas and the villages in cities, demolishing illegal buildings—in 2013, and governed "five waters" in tandem—treating sewage, controlling floods, discharging the accumulated water, guaranteeing a supply of water and saving water—in 2014. These measures have pushed forward the actual action plan for building a beautiful Zhejiang; the people of Zhejiang have a stronger sense of happiness, and the index of happiness has been on the increase. This is the ultimate pursuit of the people's government!

3. The theory of the "two mountains" conforms to the law of the increment of ecological needs

The high-quality ecological environment and ecological products meet high-level consumption needs. Where food and clothing are inadequate, the people first hope that their basic needs, including food, clothing, housing and transportation, are met and do not pursue the high-quality ecological environment and ecological products. At the initial stage of industrialization, various countries generally experienced the stage of exercising control after pollution or only seeking the "gold and silver mountains" disregarding clean, clear waters and lush mountains. Why? This is determined by the structure and the hierarchy of the needs; for instance, in the 1980s, some township enterprises discharged wastewater like ink, but the residents did not boycott them. The people held that they would rather be poisoned to death than be starved to death. This is because the main conflict at that time consisted in the problem of food and clothing. In a moderately prosperous society, even an affluent society, with an increasing level of per capita income, the people have more needs for the high-quality ecological environment and ecological products as well as other high-grade products because the high-quality ecological environment and ecological products mean an improvement in the quality of their life. Therefore, the law of the increment in ecological needs can be identified as follows: with an increasing level of income, consumers' ecological needs are on the rise.[11] Against the special background that currently mountains are not lush, water is not clean and clear, the sky is not blue and land is not clean in China, given the trend of an increment in ecological needs, the Party and the government are duty-bound to actively increase the supply of the ecological environment and ecological products to meet the ever-growing ecological needs of the people. Zhejiang's efforts in promoting ecological development respond to the people's appeals and meet their needs as if the Party Committee of Zhejiang Province had always respected the people's innovations and creations at the initial stage of reform and opening-up.

4. The theory of the "two mountains" conforms to the law of the increment in ecological value

The ecological environment is not free goods which are not priced; instead, it is the priced economic resource, so the compensated use and trading of the ecological environment conforms to the economic law; with economic and social development, the ecological and environmental resources tend to become increasingly scarce, the ecological value increases; therefore, the people can make ecological investments like economic investments to increase the value of the ecological capital.[12] This is the law of the increment in ecological value. "Clean, clear waters and lush mountains are the 'gold and silver mountains'" vividly mirrors the law of the increment of ecological value. How can we ensure that "clean, clear waters and lush mountains are the 'gold and silver mountains'"? First, it is essential to strictly protect clean, clear waters and lush mountains and resolutely oppose the act of destroying them. Only when there are lush mountains can firewood be available. Clean, clear waters and lush mountains are the fundamental guarantee for ever-lasting development. Second, it is necessary to stop uncompensated use of the ecological environment, and ensure compensated use of it. As clean, clear waters and lush mountains are scarce, but precious resources, where the property rights in the ecological environment can be defined, the scarcity of the ecological environment can be demonstrated through price signals to realize the compensated use of the ecological environment. Third, it is necessary to actively invest in the ecological environment to ensure an increase in the value of ecological capital. The ecological environment is no longer used on an uncompensated basis, the government should invest in the ecological environment as public goods; the ecological environment should be used on a paid basis, the government and enterprises may invest in the ecological environment. The practice in different areas shows that following the law of the increment in ecological value is

[11] Shen Manhong, *Ecological Economics*, China Environmental Science Press, 2008, p. 32.

[12] Shen Manhong, *Ecological Economics*, China Environmental Science Press, 2008, p. 33.

conducive to not only ecological protection but also to economic development. For instance, the system of compensated use of the initial emission permit is adopted to deal with the environmental capacity resources in an economic way; the system of trading water usage rights is carried out to achieve optimal allocation of natural resources; the incentives for investment in ecological protection are provided to preserve and increase the value of natural capital. Therefore, the 3rd Plenary Session of the 18th Central Committee of the Communist Party of China decided to dedicate special provisions to expound the system of compensated use of resources and the ecological compensation system and pointed out that it was necessary to accelerate the pricing reform for natural resources and their products to give full expression to their market supply and demand, the extent of resource scarcity, ecological and environmental damage costs and restoration benefits. The plenary session stressed that it was essential to uphold the principle that whoever uses the resources must pay for them, and that whoever pollutes the environment and damages the ecology must pay compensation, and gradually expand resource taxes to cover the possession and use of all natural and ecological resources; they determined that it was necessary to stabilize and expand the areas where farmland is converted back to forests and grassland, adjust the usage of farmland in heavily polluted areas or areas where underground water has been overused, and implement an orderly rehabilitating system for farmland, rivers and lakes; it is also essential to establish an effective comparative pricing mechanism to regulate industrial land and residential land prices, and raise the price of land for industrial use, persist in the principle that he who profits should compensate, improve the ecological compensation mechanism for key ecological functional areas, and promote the establishment of a horizontal compensation system between regions; moreover, it is also necessary to develop the environmental protection market, implement a trading system for energy conservation, carbon emission, waste discharge and water usage rights, establish a market-oriented mechanism to attract private capital to ecological and environmental protection, and implement third-party treatment of environmental pollution.[13] This is the specific policy for realizing the ecological value and following the law of the increment in ecological value.

1.2 From a Green Zhejiang to a Beautiful Zhejiang

The theory of the "two mountains" is an important guiding thought for ecological development aimed at building a beautiful China. A beautiful China is a special concept and a goal with Zhejiang's elements; it is realistically targeted for building a beautiful Zhejiang. Therefore, the successive Party Committees and the People's Governments of Zhejiang Province have spared no efforts in this endeavor. Zhejiang's strategic goal of ecological development can be broadly summarized as four stages: a green Zhejiang → an ecological province → an ecological Zhejiang → a beautiful

[13]Decision of the Central Committee of the Communist Party of China on Some Major Issues Concerning Comprehensively Intensifying the Reform, *People's Daily*, November 16, 2013.

Zhejiang. The evolution of that strategic goal of ecological development shows the consistency of the policy in carrying through one blueprint on the one hand and the progressiveness in connotation and level on the other. This is a model of innovation amidst inheritance.

1.2.1 The Trajectory of Evolution from a Green Zhejiang to a Beautiful Zhejiang

1.2.1.1 The Evolution of Zhejiang's Strategic Goal of Ecological Development

1. From ecological and environmental development to the building of a green Zhejiang

Ecological development and environmental protection constitute a systematic project and must be promoted in a systematic way. The 11th Party Congress of Zhejiang Province, held in 2002, completed the evolution from simply ecological and environmental development to comprehensive building of a green Zhejiang. When delivering the report in the Party Congress, the then Secretary of the Party Committee of Zhejiang Province, Zhang Dejiang, pointed out, "The building of a green Zhejiang is a great event for achieving sustainable development in Zhejiang; it is essential to proceed from the overall interests and long-term development, give higher priority to developing green industries, strengthening environmental protection and ecological development; it is also necessary to speed up the development of ecological agriculture, ecological industry, ecotourism and the environmental protection industry, actively promote clean production, strictly control and actively deal with environmental pollution, improve the environmental quality in urban and rural areas, better develop the non-commercial ecological forest, strengthen the comprehensive governance of river basins and establish a mechanism of compensation for ecological protection, bring about beautiful hills and waters; finally, it is necessary to rationally develop, utilize and protect the land, the water, minerals, forests and other natural resources and build a resource-conserving society."

The strategic goal of building a green Zhejiang has the following basic characteristics: First, the foundation for building a green Zhejiang is ecological development, environmental protection and resource saving; second, the building of a green Zhejiang focuses on developing the ecological sectors including ecological agriculture, ecological industry and ecological service industry; third, the building of a green Zhejiang no longer merely stresses environmental protection, instead, it coordinates environmental protection and economic growth.

2. From the building of a green Zhejiang to the building of an ecological province

The report, delivered by the then Secretary of the Party Committee of Zhejiang Province Xi Jinping on behalf of the Party Committee of Zhejiang Province during the 4th Enlarged Session of the 11th Party Committee of Zhejiang Province held in Hangzhou in 2003, put forward the "Eight-Eight Strategies", and stressed the commitment to leveraging eight advantages and pushing forward eight measures: one, further leveraging Zhejiang's advantages regarding institutions and mechanisms, earnestly stimulating diverse forms of ownership to develop side by side with public ownership remaining dominant, continuously improving the socialist market economic system; two, further drawing on Zhejiang's locational advantage, proactively keeping in line with Shanghai, actively engaging in cooperation and communication with the Yangtze River Delta Region, constantly increasing the level of opening-up internally and externally; three, further leveraging the advantages of massive characteristic industries in Zhejiang, speeding up the building of advanced manufacturing bases, taking the path of new industrialization; four, further giving play to Zhejiang's advantages in coordinated urban and rural development, quickening urban-rural integration; five, further utilizing Zhejiang's ecological advantage, turning Zhejiang into an ecological province, building a green Zhejiang; six, further exploiting the advantages of mountain and marine resources in Zhejiang, energetically developing the marine economy, boosting leapfrog development of the underdeveloped areas, and turning the development of the marine economy and the underdeveloped areas into the new economic growth point of Zhejiang Province; seven, further utilizing Zhejiang's environmental advantages, actively pushing forward principal construction mainly including five major ten-billion projects, practically strengthening legal construction, credit construction and improving the efficiency of organs; and eight, further leveraging Zhejiang's humanistic advantage, actively rejuvenating the province through science, technology and talents, stepping up efforts to turn Zhejiang into a culturally large province.[14]

The "Eight-Eight Strategies" are the main strategic thoughts applied in the period during which Xi Jinping took charge of Zhejiang. One of the main parts of the "Eight-Eight Strategies" lies in further utilizing Zhejiang's ecological advantage, turning Zhejiang into an ecological province, and building a green Zhejiang. The building of an ecological province will become the keynote and main theme of Zhejiang's ecological development in more than ten years, or even a longer period. The strategic goal of building an ecological province has the following salient features: First, it is necessary to dialectically view Zhejiang's conditions, be aware of the environmental problems arising out of Zhejiang's rapid economic development and also identify Zhejiang's advantages in ecological development and environmental protection; second, it is necessary to comprehensively push forward the building of an ecological province—that building is more inclusive than the building of a green Zhejiang and covers the economic, political, social, cultural, ecological and other fields; third, the building of an ecological province needs a strong organizational guarantee—the then Secretary of the Party Committee of Zhejiang Province, Xi Jinping,

[14]Xi Jinping, *Carrying out Concrete Work to Stay Ahead—Line of Thought and Practice in Promoting New Development in Zhejiang*, The Party School of the CPC Central Committee Press, 2006, pp. 71–73.

served as the leader of the leading group for building an ecological province so as to ensure real execution of the work on various fronts.

3. From the building of an ecological province to the building of an ecological Zhejiang

The 7th Plenary Session of the 12th Party Committee of Zhejiang Province specially studied ecological development. The *Decision of the Party Committee of Zhejiang Province on Promoting Ecological Development*, adopted in this plenary session, is a landmark document. According to the decision, under the guidance of the Deng Xiaoping Theory and the Important Thought of Three Represents, Zhejiang will deeply carry out the Scientific Outlook on Development, comprehensively implement the "Eight-Eight Strategies" and the overall strategy of making the people rich by starting business and of building a strong province through innovation, uphold the strategy of building an ecological province, take the path of building the province through ecological development, vigorously develop the ecological economy, continuously optimize the ecological environment, value the development of the ecological culture, improve systems and mechanisms, more rapidly shape an industrial structure, growth mode and consumption mode conducive to conserving energy resources and protecting the ecological environment, build an ecological Zhejiang which is affluent, beautiful, harmonious and peaceful, achieve sustainable economic and social development and continuously enhance the quality of the lives of the people of Zhejiang.[15]

The decision contains the following new ideas: First, as the Central Committee of the Communist Party of China makes the unified arrangements for the strategy of ecological development, Zhejiang becomes the first to upgrade from comprehensive building of an ecological province to the building of an ecological Zhejiang; second, subject to continuing to uphold the strategy of building an ecological province, Zhejiang vows to build the province through ecological development and further strengthens the extreme importance of ecological development; third, Zhejiang combines ecological development with the people's well-being, while the objective of building an ecological Zhejiang which is affluent, beautiful, harmonious and peaceful is to enhance the quality of the lives of the people of Zhejiang.

4. From the building of an ecological Zhejiang to the building of a beautiful Zhejiang

The report of the 18th National Congress of the Communist Party of China put forward the line of thought of ecological development aimed at building a beautiful China. In a sense, the objective of China's ecological development—building a beautiful China—is associated with Zhejiang, a beautiful land, and also guides Zhejiang's ecological development aimed at building a beautiful Zhejiang. Since Xi Jinping served as the General Secretary, he has still shown great care for Zhejiang's ecological development. In early 2013, when exchanging views with Huang Kunming, the former Secretary of the Party Committee of Hangzhou City, Xi Jinping

[15]The Party Committee of Zhejiang Province: The Decision of the Party Committee of Zhejiang Province on Promoting Ecological Development, *Zhejiang Today*, 2010(13).

pointed out, "I hope that all of you can more solidly push forward ecological development to turn Hangzhou into a model for building a beautiful China." Zhejiang has naturally become the forerunner in building a beautiful China. Therefore, the 5th Plenary Session of the 13th Party Committee of Zhejiang Province, held in 2014, specially studied ecological development and made the *Decision of the Party Committee of Zhejiang Province on Building a Beautiful Zhejiang and Creating a Good Life*.

According to the decision, the 18th National Congress of the Communist Party of China incorporated ecological development into the overall plan for the socialist cause with Chinese characteristics, vowed to build a beautiful China and move towards a new era of socialist ecological civilization, achieving ever-lasting development of the Chinese nation. The 3rd Plenary Session of the 18th Central Committee of the Communist Party of China identified accelerating institutional development of an ecological civilization as an important part of efforts to comprehensively intensify reforms and stressed that it was essential to build a systematic and complete ecological institutional system and adopt institutions to protect the ecological environment. Xi Jinping emphasized that moving towards a new era of ecological development and building a beautiful China is an important part of the Chinese Dream of national rejuvenation. He also put forward a number of new thoughts, new views and new requirements, including "mountains, waters, forests, farmland and lakes constitute a community of life", "clean, clear waters and lush mountains are the 'gold and silver mountains'" and "the people's aspirations to live a good life are our goal". This shows the Party's further understanding of the law of socialism with Chinese characteristics and the strong will and determination of the Party on upholding the five-sphere integrated plan and strengthening ecological development. The decision further pointed out that building a beautiful Zhejiang and creating a good life was Zhejiang's practice of building a beautiful China, also inherited and upgraded the strategic goals set by the successive Party Committees of Zhejiang Province for building a green Zhejiang, an ecological province and the national demonstration area for ecological development. In recent years, in the practice of ecological development, Zhejiang has always taken the "Eight-Eight Strategies" as the overarching principle, further leveraged Zhejiang's ecological advantages, firmly upheld the developmental line of thought by which clean, clear waters and lush mountains are the "gold and silver mountains", ensured that the successive Party Committees and the People's Governments worked hard and carried through one blueprint and have given priority to ecological development; it has sought development amidst protection and kept protection during development, focused its tasks on developing the ecological economy and improving the ecological environment; it has upheld overall coordinated development and given prominence to key points, worked on solving the salient environmental problems affecting sustainable development and harming the people's health; it has also exercised strict supervision, optimized services, taken guaranteeing the ecological and environmental safety and safeguarding social harmony and stability as the basic requirements; it has persisted in the pattern in which the Party plays the leading role with social participation, sought important guarantees by innovating systems and mechanisms and advocating co-building and sharing; as a result,

Zhejiang has made significant and active progress in ecological development, laying a solid foundation for building a beautiful Zhejiang and creating a good life.[16]

The decision brings about the following new ideas: First, given that ecological development is holistic, the systematic line of thought that mountains, waters, forests, farmland and lakes constitute a community of life is turned into the guiding thought for ecological development; second, given the rapid growth of the people's ecological needs, the objective of ecological development is upgraded to the building of a beautiful Zhejiang; third, given the Party's fundamental purpose of wholeheartedly serving the people, creating a good life is taken as the ultimate goal of building a beautiful Zhejiang.

1.2.1.2 The Main Characteristics of the Evolution of the Strategy for Ecological Development

1. Always upholding the idea that the people's aspirations to live a good life are our goal

The Party's fundamental purpose is to wholeheartedly serve the people. The people's satisfaction, pleasure and recognition are the fundamental criteria for testing economic and social development; in the final analysis, it depends upon the people's happiness. The higher the degree of realizing a good life for the people, the higher the index of the happiness of the people. The people's aspirations to live a good life vary with different historical stages. When the problem of food and clothing has not yet been solved, the people's greatest aspiration lies in solving the problem of food and clothing; after the problem of food and clothing is solved, the people's greatest aspiration lies in becoming well-off; after the people become well-off, their greatest aspiration lies in becoming affluent. Moreover, the people's aspirations are not a single goal; instead, they are a combination of multiple goals. For instance, as Zhejiang's per capita GDP has reached USD 13,000, the aspiration of the people of Zhejiang lies in being materially affluent, politically democratic, culturally prosperous, socially harmonious and environmentally beautiful. Besides the pursuit of being materially affluent, the people also highly yearn for a beautiful Zhejiang with clear waters, green hills, a blue sky and clean land. Therefore, the decision made by the Party Committee of Zhejiang Province on building a beautiful Zhejiang and creating a good life fully embodies the people's aspirations.

2. Always upholding the spirit of carrying through one blueprint

The building of a green Zhejiang, an ecological province, an ecological Zhejiang and a beautiful Zhejiang epitomizes Zhejiang's goals of ecological development in different periods. These goals are expressed in different ways, but the main line

[16]The Party Committee of Zhejiang Province: The Decision of the Party Committee of Zhejiang Province on Building a Beautiful Zhejiang and Creating a Good Life, *Zhejiang Today*, May 29, 2014.

is consistent: upholding the Scientific Outlook on Development and the Outlook on Ecological Civilization, properly dealing with the relationship between the "gold and silver mountains" and clean, clear waters and lush mountains, placing equal emphasis on the "gold and silver mountains" and clean, clear waters and lush mountains, achieving sustainable economic and social development, ensuring the people's happiness and well-being. Ecological development is a beneficial project which entails great efforts in unifying the line of thought and massive investments and slowly produces effects, so it is very difficult to carry out the project for a long time if actions are taken to seek instant benefits in a short-sighted way. The admirable spirit of the successive Party Committees and the People's Governments of Zhejiang Province consists in tenaciously persevering in keeping lush mountains, successively working hard and carrying through one blueprint. This is a valuable experience from the building of a beautiful Zhejiang and deserves to be promoted in an all-round way.

3. Always carrying forward the spirit of making painstaking efforts

As there are correct goals and strategic arrangements, the key lies in implementation. Xi Jinping placed great emphasis on the spirit of working hard. The "811" action plan embodies that spirit. In the past, "811" was a code; at present, it has become an appealing brand in Zhejiang's drive towards environmental protection, thus it has served as an effective carrier for promoting Zhejiang's ecological development, it has lasted for a decade. In October, 2004, a province-wide sweeping action for environmental pollution control—the "811" action for environmental pollution control—was initiated in Zhejiang. The "8" refers to 8 major water systems in Zhejiang, the "11" means 11 cities divided into districts in Zhejiang, which are the 11 provincial-level key environmental supervision areas with a particularly salient regional and structural pollution identified by the People's Government of Zhejiang Province. The first round of the three-year "811" environmental pollution control action strongly put a check on the environmental pollution and ecological damage throughout the province. The second round of the three-year "811" environmental protection action was initiated in 2008. At that time, the "8" became the goals in 8 respects and the main tasks in 8 respects in the environmental protection field; the "11" meant the policy measures in 11 respects put forward at that time and 11 key environmental problems identified by the People's Government of Zhejiang Province. With two rounds of "811" action in six years, Zhejiang entered the period with the most input, the greatest intensity of input and the greatest number of significant achievements in the field of ecological and environmental protection in Zhejiang. In June, 2010, the Party Committee of Zhejiang Province made the *Decision of the Party Committee of Zhejiang Province on Promoting Ecological Development*, calling for further carrying out the "811" action plan for promoting ecological development (2011–2015) and taking five years to basically ensure that economic and social development is commensurate with the bearing capacity of the resource environment and that the environmental quality is commensurate with the improvement in the people's well-being, as well as ensuring that Zhejiang continues to stay ahead nationwide in the building of an ecological province and ecological development. Thus the "811" action has shifted its focus from comprehensively promoting environmental protection to pushing forward ecological development in a multidimensional way.

1.2.1.3 The Overall Framework of Ecological Development Aimed at Building a Beautiful Zhejiang

1. Adopting the systematic line of thought to grasp the building of a beautiful Zhejiang

First, strengthening the building of an ecological environment with beautiful mountains and rivers, mainly covering ecological development and environmental governance: (1) intensifying the natural ecological protection in mountains, rivers and the sea. Making more efforts to protect the important functional ecological areas, the ecological and environmental sensitive areas and vulnerable areas, ensuring that primitive ecology is maintained in the areas where the main river basins, including the Qiantang River, the Ou'jiang River and Taihu Lake, originate and the marine ecological functional areas; focusing on natural restoration supplemented by artificial restoration, returning cultivated land to forests, closing hillsides to facilitate afforestation, conducting stock enhancement and releasing, imposing a fishing ban and moratorium to ensure regeneration of the ecosystem, carrying out engineering restoration of the ecosystem which cannot restore itself; reinforcing the building of a green ecological protective screen, completely carrying out the project of stimulating the farmers on high and remote mountains to descend from those mountains and migrate to other areas, intensifying the protection of forest resources, comprehensively promoting the greening of the plains, forest coverage and quality enhancement, improving the quality of forest stands and increasing the volume of forest stock, enhancing the functions of the forest ecosystem; (2) pushing forward comprehensive environmental governance focusing on water governance. Governing "five waters" in tandem to make water clearer, curbing the haze to make the sky bluer, purifying the soil to make the land cleaner.

Second, speeding up the development of the ecological economy characterized by conservation and environmental protection. More rapidly developing the ecological economy, shaping a modern industrial system conducive to saving resources and protecting the environment. More quickly shutting down outdated production facilities with high energy consumption and high emissions, actively developing new energy such as solar and wind energy and renewable energy. Strictly exercising the total quality management of water usage, accelerating the building of a water-conserving society. Actively promoting the development of the circular economy, energetically pushing forward renovations for making parks circular, comprehensively increasing the level of comprehensive utilization of renewable resources. More rapidly establishing and promoting a modern ecological circular agricultural mode, earnestly developing pollution-free agricultural products, green food and organic products. Developing a modern forest economy to make forest workers in mountainous areas rich and increase their income. Pushing forward the ecological renovation of industrial parks, comprehensively promoting clean production audits. Encouraging enterprises to develop green low-carbon products, developing and implementing a green purchasing consumption policy. Actively bringing about a low-carbon economic developmental mode based on low energy consumption, low pollution and low emissions.

Third, actively promoting the cultivation of the ecological culture, which features unity of the people and nature. Attaching importance to the ecological philosophy and thought in Zhejiang's traditional culture, cultivating and arousing the awareness of all citizens about building a beautiful Zhejiang and creating a good life, vigorously carrying forward the philosophy of respecting, conforming to and protecting nature, proactively drawing upon the advanced philosophies, effective practices and concrete systems of the developed countries regarding ecological civilization, enhancing the awareness about ecological ethics, ecological morality and ecological value, shaping a social action system of interaction among the government, enterprises and the general public.

2. From the planning of spatial structure to the layout of building a beautiful Zhejiang

The building of a beautiful Zhejiang should be carried out in the concrete spaces, so it is necessary to optimize the urban-rural regional spatial layout for ever-lasting development. This mainly includes: following the requirements of keeping coordination among the population, the economy, the bearing capacity of the resource environment and of positioning the main functional areas to innovatively develop the provincial master plan, promote the integrated development of a plan for economic and social development, urban and rural planning, land use planning, underground spatial planning, environmental functional zoning, marine development plan and river basin plan, create a spatial planning system with clear positioning and strict control to ensure strict execution of the plans; carrying out the *Plan of Zhejiang Province for the Main Functional Areas* to build the marine economic development demonstration areas, the modern pattern of agricultural development and the large industrial agglomeration platforms in Zhejiang, improve the strategic pattern of new-type urbanization, build an ecological security system, gradually shape the overall pattern of spatial development with coordination among the population, the economy, resources and the environment; thoroughly carrying out the *Plan of Zhejiang Province for Extensively Promoting New-type Urbanization* and the implementation opinions on extensively promoting new-type urbanization to boost industrialization, IT application, urbanization and agricultural modernization at the same time and more rapidly form the framework for the layout of urban-rural spaces and the structure of the urban-rural system.

The building of a beautiful Zhejiang should be carried out in the living environment, thus it is necessary to optimize the living environment which features marvelous scenery south of the Yangtze River; this mainly covers two aspects: (1) speeding up the planning and building of beautiful cities. Considering the bearing capacity of the environment and the population, the land resources available for development and the level of economic and social development to further improve the plans for a provincial urban system. Conducting coordinated provincial planning, promoting coordinated planning and development of metropolitan areas, central regional cities, county areas and central towns, stimulating high-end factors to metropolitan areas, providing classified guidance for the development of central regional cities, turning county areas, small cities and central towns into the strategic nodes for urban and rural development. Considering the characteristics of natural resources and humanistic characteristics to scientifically design the urban living environment, landscape and architectural color, strengthen the protection and construction of the urban ecological landscape, promote pilot construction of small ecological and humanistic cities, build a number of small towns with a fascinating style and landscape south of the Yangtze River, showing the unique charm of that area. Upholding the "boundary" of urban development, promoting the development of green cities, smart cities and humanistic cities. Scientifically developing and utilizing urban underground spaces, controlling urban light pollution. (2) Increasing the level of the construction of a beautiful countryside. Carrying out the *Regulations on the Construction of a Beautiful Countryside* to increase standards, optimize the layout and enhance the characteristics so that the people can get access to lush mountains, clean, clear waters and local lingering charm. Thoroughly carrying out the project of "1000-village demonstrations and 10,000-village improvements", promoting the organic renewal of villages in the ecological field. Stepping up contiguous comprehensive improvement of the rural environment, establishing the long-term mechanism of management. Vigorously creating green towns and ecological demonstration villages, protecting the local natural landscape and characteristic cultural villages. Strengthening village planning and development, rural house design services, giving expression to the characteristics of rural houses south of the Yangtze River. Improving the work on renovating the rural houses and dilapidated houses, meticulously building a batch of residential houses with Zhejiang's characteristics.

3. Better conducting a design of institutions and mechanisms in order to guarantee the building of a beautiful Zhejiang

The *Decision of the Party Committee of Zhejiang Province on Building a Beautiful Zhejiang and Creating a Good Life*, made on May 23, 2014, systematically expounded the institutional and mechanism guarantee for the building of a beautiful Zhejiang.

First, establishing and improving the systems and mechanisms for exercising strict control at the source. This mainly includes: exploring and establishing the system of property rights of natural resource assets and the environmental spatial control system; identifying three red lines including the baseline of guaranteeing the ecological functions, the baseline of the safety of environmental quality and the upper line of natural resource utilization; carrying out the system of environmental access involving space, total quantity and project; linking environmental capacity with total regional quantity, environmental quality and project-related environmental impact assessment; exercising strict control in environmental access; further strengthening the control over total energy consumption and the intensity of energy consumption;

improving the total quantity control system involving the main pollutants, promoting the control over pollutants of the total industrial and regional quantity.

Second, establishing and improving the systems and mechanisms for exercising strict control over the process. This mainly includes: establishing and improving the system of the management of environmental protection covering strict monitoring and supervision of all pollutant emissions, independent environmental monitoring and supervision and administrative law enforcement; improving the comprehensive evaluation system involving the yield per mu (unit of area), establishing a mechanism of incentives and restraints for differentiated use of resource factors, the incentive mechanism for the exit of inefficient enterprises and the mechanism of choosing the excellent new projects; exploring and building the market system of the efficient flow of resource factors; speeding up the pricing reform for natural resources and their products, improving the mechanism for the formation of environmental prices of resources for fully reflecting the extent of resource scarcity, ecological and environmental governance and restoration costs; establishing the mechanism of joint defense and joint control in sea areas by local governments and the departments of the central government, promoting the management mode of joint defense and joint control in areas, river basins and offshore areas, improving the regional coordination and inter-regional comprehensive governance mechanism for environmental protection.

Third, establishing and improving the systems and mechanism for severe punishment of evil consequences. (1) Establishing a lifelong accountability system for environmental damage, such as, showing no tolerance towards the acts of polluting the environment and destroying the ecology, strengthening administrative law enforcement and criminal punishment, intensifying investigations into the accountability and punishment of law breakers to really protect citizens' environmental rights and interests, and to safeguard social fairness and justice. (2) Establishing the system of punishment for environmental damage, such as, building the environmental pollution accountability system based on compensation for environmental damage, carrying out the strict compensation system for those responsible for ecological and environmental damage, intensifying administrative enforcement and administrative punishment, investigating any criminal responsibilities according to laws. Exploring and establishing a fund for compensation responsibility risks for damage due to environmental pollution, encouraging the adoption of the liability insurance system for damage due to environmental pollution, subjecting the high-risk enterprises to acquiring compulsory liability insurance for environmental pollution.

1.2.2 The Strategic Content from the Building of a Green Zhejiang to the Building of a Beautiful Zhejiang

1. Implementing the strategy of the ecological development of industries to promote transformation and upgrading of the industrial economy

The development of the ecological economy is the focus of the building of a green Zhejiang. Zhejiang has spared no efforts in promoting transformation from black development to green development, from linear development to circular development, from high-carbon development to low-carbon development, and it has made staged achievements in transforming the developmental mode and the industrial structure.

The rigid constraint means for pollution and emission reduction have been adopted and the mode of phasing out a batch, transferring a batch and upgrading a batch has been introduced to promote the reduction in total quantity, improve the quality and optimize the development. Zhejiang has started with resource conservation and environmental protection to make structural adjustments, and has carried out ten major projects for energy conservation and the reduction of consumption; during the period of the 11th Five-Year Plan, Zhejiang shut down 5.31 million KW small thermal power units, 2.31 million tons of outdated steelmaking facilities, 23.97 million tons of outdated cement facilities, 578,000t of outdated papermaking facilities, 2100 inefficient industrial boilers and 2547 clay brick and tile kilns. Zhejiang has continuously carried out compulsory clean production in key industries and key enterprises, and the enterprises' sense of social responsibility for environmental protection has improved significantly.

Zhejiang has earnestly developed its industries in an ecological way and has regarded the ecological renovation of industrial parks as the priority of transformation and upgrading. In 2006, Zhejiang issued the *Guiding Opinions on Further Promoting the Ecological Development and Renovation of Development Zones (Industrial Parks) in Zhejiang Province*. Zhejiang has made staged achievements in key projects, including ecological industrial demonstration parks, enterprise clean production demonstration projects, wind power generation and other renewable energy and high-efficiency energy-saving technology demonstration projects.

Zhejiang has actively developed ecological circular agriculture, extensively carried out the development of modern forest parks, and has promoted healthy aquiculture across the board by means of modern fishery parks and the projects of conducting fish farming and stable grain production for increasing income. Zhejiang has developed the ecological service industry, concretely pushed forward the development of ecotourism areas, the ecological management of Grade A tourist attractions and the improvement of their surrounding areas, and further standardized the supervision over agritainment infrastructures and environments.

2. Carrying out the green consumption strategy to fundamentally transform the manner of consumption

Both ends of consumption behavior connect nature: exploiting resources from nature, remitting pollutants. Excessive exploitation will waste resources and destroy biodiversity; excessive emission will pollute the environment and destroy the ecology. The contradiction between the people's unlimited demand for consumption and the limited supply of goods shows that it is essential to strengthen the management of the demand and adjust the people's manner of consuming, enable the change from a linear manner of consuming to a green manner. Green consumption means the relations between the people and nature are properly dealt with, that an environmentally friendly consumption awareness, consumption mode and consumption habit gradually take shape. Implementing the strategy of making consumption ecological and promoting green consumption is the core of ecological development.

The establishment of green organizations internalizes the philosophy of green consumption. Zhejiang has given great importance to establishing green organizations. Since the appraisal and selection of green schools was initiated in 2002, Zhejiang has been home to 49 national green schools and 1095 provincial green schools. The establishment of green schools has made the ecological cultural philosophy profoundly rooted among the teachers and students in various types of schools at various levels. The theme activity "Low-carbon Home · Fashion Life" carried out by the Women's Federation of Zhejiang Province, the activity "Refuse Classification Starts in the Households, Adding Lustre to the Building of Culturally Advanced Cities" organized by the Women's Federation of Hangzhou City, the activity of appraising and selecting the "Top 10 Green Households" conducted by Cixi City have produced good effects. As of 2011, there were 27 national green communities, 702 provincial green communities, 22 national green households and 1688 provincial green households in Zhejiang.

Zhejiang has fully leveraged the Ecology Day in an effort to publicize and guide. On the important occasions, including World Environment Day, World Earth Day, China Water Week, National Land Day and China Tree Planting Day, Zhejiang has promoted a low-carbon life and encouraged green consumption. This has also fostered the atmosphere for the Ecology Day with Zhejiang's characteristics. The first county-level ecology day in China was established in Anji County in 2003, which was a successful experience. In 2009, Zhejiang established the first provincial ecology day in China, and designated each June 30 as Zhejiang Ecology Day. The Secretary of the Party Committee of Zhejiang Province and the Governor of Zhejiang Province have participated in ecological development together with the people. In Deqing County, the first ecological consumption day in China was established on March 15, 2011, the Ecological Consumption Education Forum was founded in 2012, and one activity of ecological consumption education is carried out each March, promoting ecological consumption education at various levels to enhance the people's awareness of ecological consumption and to foster an environment of ecological consumption.

Zhejiang has adopted the two-way green incentive fiscal policy for restricting non-green products and encouraging green products to guide residents to consciously uphold green consumption. Zhejiang has introduced the restrictive policy for large-displacement vehicles and the subsidy policy for battery cars as well as the policies for exercising total quantity control over fossil energy, encouraging the development of renewable energy, strictly regulating the polluted food and ensuring high quality and preferential prices of green food. The consumption behaviors of Zhejiang's residents are changing.

3. Carrying out the strategy of resource conservation to promote circular utilization of resources and energy

The contradiction between rapid expansion of economic aggregate and limited supply of natural resources, low productivity of natural resources is a salient problem at present. Thus, we cannot continue adopting the extensive resource development and utilization mode. Resource conservation means that it is necessary to intensively develop and utilize natural resources, improve the structure of resource productivity and resource input, combine resource development and utilization with ecological and environmental protection to ensure sustainable development and utilization of resources.

Given that Zhejiang is a small province in area and resources, Zhejiang has given great importance to resource conservation. As early as 2005, Zhejiang issued the *Plan of Zhejiang Province for the Development of the Circular Economy*, calling for implementing the "991" action plan for the circular economy, namely, developing 9 key fields of the circular economy, carrying out demonstration projects involving 9 batches and 100 key projects. In the meantime, Zhejiang has comprehensively carried out the "4121" and the "733" project for the industrial circular economy, and actively promoted work on rating the demonstration parks and enterprises for their industrial circular economy, green enterprises and the enterprises for comprehensive resource utilization. Zhejiang has carried out the "2115" demonstration project for ecological circular agriculture, initiated and established 18 provincial demonstration counties and 39 demonstration districts for ecological circular agriculture, identified and recognized 21 provincial demonstration enterprises for ecological circular agriculture and arranged 100 provincial projects for ecological circular agriculture. The cycle at "points" represented by circular utilization of chlorine within Wynca Chemical in Jiande City, the cycle on "lines" represented by the chemical circular economic park in Zhenhai District and the cycle on "planes" represented by the circular utilization of bamboo resources in Anji County constitute a three-dimensional circular economic pattern.

4. Implementing the strategy of environmental cleanliness to make the ecological environment beautiful and pleasant

Ecological and environmental protection is the foundation for building a beautiful Zhejiang. Zhejiang has made three breakthroughs in ecological development: it has developed a non-commercial ecological forest and natural reserves, so the ecological functions have improved significantly; as a resource-conserving society is built, the output per unit of land resource, energy resource and water resource has increased markedly, putting Zhejiang in the lead of the rest of the country in this regard; the "811" action plan for environmental protection and the "811" action plan for ecological development have curbed all-round degradation of environmental quality, showing the trend of gradual improvement.

The construction of a beautiful countryside initiates a "beautiful" cause. The first national ecological county in China, Anji County, creatively focused the construction of a new socialist countryside on the construction of a beautiful countryside in 2008, bringing about the Anji Model for ecological development. Based on the successful practice in Anji County, Zhejiang developed the *Action Plan for the Construction of a Beautiful Countryside (2011–2015)*, thus the construction of a beautiful countryside was carried out in an all-round way and the Zhejiang Experience preliminarily took shape. Afterwards, the construction of a beautiful countryside was rapidly promoted in southern, northern and western China, and it became a scenery in China's construction of a new countryside. Based on rich practice in the construction of a beautiful countryside, the report delivered at the 18th National Congress of the Communist Party of China declared the building of a beautiful China. The policy of building a beautiful China contains Zhejiang's elements, and the people of Zhejiang have made important contributions to the building of a beautiful China.

"Three renovations and one demolition" have achieved multiple goals. The People's Government of Zhejiang Province decided to take the three-year action of renovating old residential areas, old factory areas in urban areas and the villages in cities, demolishing illegal buildings ("three renovations and one demolition") from 2013 to 2015. The action is an important measure for carrying out the principles adopted during the 18th National Congress of the Communist Party of China and accelerating the building of a materially affluent and culturally advanced modern Zhejiang; it is urgently needed for promoting new-type urbanization, improving the appearance of urban and rural areas, optimizing the living environment and building a beautiful Zhejiang; it is an effective way to strengthen urban and rural planning and construction, to push forward the intensive use of land and speed up the transformation of the developmental mode; it is objectively required to strengthen and innovate social management and build a harmonious society. "Three renovations and one demolition" have basically achieved the goals of promoting development, expanding spaces, optimizing the environment, maintaining stability and benefiting the people.

The action of governing "five waters" in tandem has restored the scenic waterscape in the south of the Yangtze River. Zhejiang becomes famous, prosperous and beautiful thanks to water. Governing "five waters" in tandem—treating sewage, controlling floods, discharging the accumulated water, guaranteeing a supply of water and saving water—is determined by the objective law of development, a specific stage of development and the goal of scientific development. Water is fundamental to production. Production mode and industrial structure determine the quality of water in the bodies of water. Water governance is conducive to change. Water is essential for ecology, so making air and soil clean entails the efforts to make water clean, and water governance is vital for ecology; water is the source of life, the people use water to wash their faces, they drink water every day, irrigation needs water, so water governance has a vital bearing on the people's livelihood. Thus, governing "five waters" in tandem is a multi-purpose measure; it expands investments and promotes change, and also improves the environment and benefits the people.

5. Implementing the strategy of ecological economization to promote the optimal allocation of environmental resources

As the province with the earliest market-oriented reform and the highest degree of marketization, Zhejiang is the first province to introduce the ecological compensation system in China. As early as 2005, the People's Government of Zhejiang Province issued the *Several Opinions on Further Improving the Ecological Compensation*

Mechanism. The development of non-commercial ecological forests and the protection of the protected areas of water sources reflect the basic spirit that ecological protection is protection of productive forces, and give rise to a historic change from uncompensated ecological protection to paid ecological protection. During the period 2007–2011, 5.1 billion yuan of an ecological compensation fund was allocated from the fiscal transfer payment from the provincial finances. Based on the practice for many years, Zhejiang has also continuously expanded the ecological compensation mechanism: (1) expanding the single ecological compensation mechanism into the system of combining ecological protection compensation with the compensation for environmental damage, and determining compensation or indemnity on the basis of the results of the monitoring of the quality of water concerning trans-boundary rivers; (2) expanding intra-regional ecological compensation into inter-regional ecological compensation, officially starting the implementation of the *Pilot Implementation Plan for Water Environmental Compensation for the Xin'an River Basin*.

As Zhejiang is the national pilot province for compensated use and trading of emission rights, the People's Government of Zhejiang Province has issued a number of regulation and policy documents, including the *Interim Measures of Zhejiang Province for Pilot Work on Compensated Use and Trading of Emission Rights*, while relevant departments have also introduced a series of supporting policies. There have been 11 provincial policy documents and 68 local policy documents, the framework for the regulation and policy system involving the compensated use and trading of emission rights has basically taken shape. With five years of pilot practice, Zhejiang has seen 3863 emission rights transactions, 9537 transactions concerning the compensated use of emission rights, the compensated use and trading of emission rights amounted to 2.5 billion yuan, placing Zhejiang No. 1 nationwide, accounting for more than 2/3 in the country. As of June, 2014, the pilot work on emission rights trading had been conducted in 11 cities divided into districts and 60 counties in Zhejiang, the coverage in cities and counties reached 100% and 75%, respectively, and the coverage in Zhejiang was the widest in the country. The transformation from uncompensated use of emission rights to compensated use of emission rights and from non-tradable emission rights to tradable ones is a revolution. That revolution has resulted in evolving the system involving compensated use and trading of emission rights into the mechanism for the attraction and selection of investments.

Ecological economization is embodied in not only the economization of ecological and environmental capacity and resources but also in the industrialization of ecological advantages. Under the principle of ecological economization and economic ecologization, with a focus on cultivating the health preservation (old-age care) service industry, Lishui City, reputed as the first city with an ecological environment in China, has, under the theme "With Gorgeous Mountains, Lishui, a Paradise for Health Preservation", fully leveraged its advantages and started with building two major platforms, including the scenic spot for ecological leisure tourism and the health preservation (old-age care) base (leisure health preservation and old-age care city), to promote the all-round development of the industries relating to ecological leisure health preservation (old-age care) and built China's paradise for health preservation and old-age care which is the largest in China and internationally well-known;

this has helped realize Lishui's goal of achieving green development, making the people rich through ecological development and bringing about leapfrog scientific development.

Both the mechanism for ecological compensation and the system for the compensated use of emission rights embody the philosophy that the ecological environment is a scarce resource which needs to be optimally allocated. Zhejiang has stayed ahead nationwide in ecological economization.

1.2.3 The Important Inspirations from the Building of a Green Zhejiang to the Building of a Beautiful Zhejiang

1. Speeding up mindset innovation, adopting ideas to guide practices

Mindset innovation must be guided by the Scientific Outlook on Development. In mindset innovation concerning ecological development, it is essential to first answer these questions: For whom is development pursued?—the overwhelming majority of the people; Who is relied upon in development?—the overwhelming majority of the people; What is the objective of development?—the happiness of the overwhelming majority of the people. An ecological environment is the basic condition for the survival of the people; the ecological and environmental rights and interests are the people's basic rights and interests; ecological and environmental quality has a direct bearing on the quality of the lives of the people.

Mindset innovation must be guided by materialist dialectics. The view, in materialist dialectics, about a universal connection can guide the people to understand the relations among the ecosystem, the economic system and the eco-economic system; the view about eternal development can guide the people to understand the law of economic, social and ecological development; the view about the law of the unity of opposites can guide the people to understand the contradiction and conflict between economic development and ecological protection and their conversion; the view about the law of mutual change of quality and quantity can guide the people to understand the finiteness of the ecological and environmental capacity and the reversibility of the ecological and environmental quality; the view about the law of negation of the negation can guide the people to understand the laws of ecological change and ecological innovation including the fact that the people cannot beat the heavens $\rightarrow$ the unity of the heavens and the people $\rightarrow$ the people will conquer nature $\rightarrow$ harmony between the heavens and the people.

2. Accelerating scientific and technological innovation, promoting change and upgrading through green technology

Science and technology determine environmental cost and benefit, while environmental cost and benefit determine the economic developmental mode. The scenarios including "ecological but uneconomical", "circular but uneconomical" and "low-carbon but uneconomical" are caused by scientific and technological underdevelopment. The developmental mode which is both "ecological and economical", "circular and economical" and "low-carbon and economical" must rely on ecological scientific and technological innovation.

The upgrading and transformation of the traditional industries is an important task, so it is essential to actively push forward ecological technology research, development and promotion in the traditional industries. (1) Researching, developing and promoting the technologies for energy conservation and emissions reduction to ensure that the task of energy conservation and emissions reduction is completed; (2) researching, developing and promoting the technologies for circular utilization, actively promoting circular science and technology at the enterprise, industrial and social levels; (3) researching, developing and promoting the technologies for the development of agglomerations, following the requirement of park-based development and agglomeration to promptly research, develop and promote the suitable technologies consistent with the characteristics of the massive economy.

If a country or territory lacks high-end technologies and continues to be at the end of the industrial chain in the new round of industrial division, it will be difficult for it to obtain generous profits. Therefore, Zhejiang must make great efforts to press ahead with research, development and promotion of new and high technologies, occupy the commanding heights in some fields of green, circular and low-carbon technologies, and turn them into the new growth points for Zhejiang's industrial development.

Innovation in the environmental testing technology is an important part of the innovation in green science and technology. Zhejiang has remained ahead nationwide from online monitoring to swiping cards for pollution emissions and from a single test to the building of the testing system. In the first half of 2014, Zhejiang reached the evaluation requirements of the Ministry of Environmental Protection in all of the three indicators concerning the building of the system for the monitoring of emission reduction, including the effective rate involving the transmission of automatic monitoring data on pollution sources, the proportion of the automatic monitoring results released and the proportion of supervision-oriented monitoring results released. As shown by the ranking in these three indicators and the number of the national key enterprises subject to control over pollutant emissions, Zhejiang ranked no.1 among the six provinces and municipalities which passed the evaluation concerning these three indicators across the country.

3. Speeding up institutional innovation, enabling coordination among the main players to promote harmonious development

Ecological development calls for synergy among the government, enterprises and the general public, and for coordination among the government mechanism, the market mechanism and the social mechanism, and for a scenario in which these three forces work together and complement each other's advantages to bring about joint forces.

Innovation in the government mechanism has been promoted. Zhejiang's important mechanism for promoting ecological development is the system for the evaluation of ecological development, which incorporates environmental protection as the constraint indicator. However, merely adopting the comprehensive evaluation system is not enough. The correct way is as follows: adopting the differentiated modes for the evaluation of officials' performance according to the functional orientation of different players in different areas, making comparisons among similar items, thus effectively stimulating the supply of public goods including ecological development. In the meantime, it is necessary to develop an institutional system for ecological development, introduce a mix of the compulsory policies including total quantity control, the selective policies for weighing the advantages and disadvantages, including the trading of emission rights, the moral education-focused guiding policies, including environmental protection education. In governing the "five waters" in tandem, Zhejiang has established the river chief system covering the provincial, municipal, county and town levels to ensure that each river, each river segment is in the charge of a river chief; the river chief system has effectively put in place the responsibility for governing waters.

Innovation in the market mechanism has been promoted. Zhejiang has gained successful experience in carrying out the system for ecological protection compensation, the water usage rights trading system, the emission rights trading system, the fishing ban system, the system involving the compensated use of mining rights and the third-party governance mechanism. Zhejiang should promptly promote this successful experience; in particular, Zhejiang should comprehensively promote the system involving the compensated use and trading of emission rights, the mechanism of full compensation for ecological protection and the mechanism of compensation and insurance for environmental damage, so that they can better serve ecological development.

Innovation in the social mechanism has been promoted. Environmental democracy is the quintessence and soul of the rule of law in environmental resources. The popularization of the philosophy of environmental democracy will certainly democratize the laws of environmental resources and legally institutionalize the environmental democratic means. Therefore, it is necessary to, based on public participation in supervision over environmental protection, build the most extensive united front in environmental protection. It is necessary to give play to the role of scientific research institutions, the institutions of higher learning, experts and scholars as think tanks so that necessary intellectual and technical support is provided for environmental protection; it is essential to give scope to the roles of non-governmental organizations for environmental protection and volunteers to push forward the work on the safeguarding of rights in environmental protection, environmental protection publicity and education and environmental improvement; play should be given to the roles of letters and visits relating to environmental complaints, prize-based reporting of offences, appraisal and review of industrial tendencies, thus ensuring better public participation in environmental protection and supervision over environmental protection, converting the people's willingness, passion and wisdom into concrete actions for joint ecological development and the sharing of an ecological civilization.

Chapter 2
The Development of a Conservation-Oriented and Environmentally Friendly Ecological Economy

Zhibin Li, Kaiwen Lin and Hongyan Tian

The building of a beautiful Zhejiang and the creation of a good life need to be guaranteed by a beautiful economy including beautiful ecological industries and beautiful green consumption. As a province which met with the growing pains at the earliest, Zhejiang has the deepest understanding of the important significance of the transformation of the developmental mode and economic transformation and upgrading, so under the guidance of the developmental outlook through which clean, clear waters and lush mountains are the "gold and silver mountains", Zhejiang has vigorously promoted the transformation from a black economy to a green economy, from a linear economy to a circular economy, from a high-carbon economy to a low-carbon economy, and it has made gratifying achievements in this regard, preliminarily demonstrating the charm of a conservation-oriented and environmentally friendly economy and providing the model for national economic transformation and upgrading.

2.1 Clean, Clear Waters and Lush Mountains Are the "Gold and Silver Mountains"

Xi Jinping pointed out, "The people have experienced three stages in understanding the relationships between 'two mountains'—clean, clear waters and lush mountains, the 'gold and silver mountains'—in practice: During the first stage, the people exchange clean, clear waters and lush mountains for the 'gold and silver mountains' without considering or with less considerations given to the bearing capacity of the environment, the people relentlessly exploit the resources; during the second stage, the people endeavor to keep the clean, clear waters and lush mountains while seeking

Z. Li (✉) · K. Lin · H. Tian
The Zhejiang Research Center for Ecological Development, Zhejiang Sci-Tech University, Hangzhou, China

J. Pan and M. Shen (eds.), *Chinese Dream and Practice in Zhejiang – Ecology*, Research Series on the Chinese Dream and China's Development Path,
https://doi.org/10.1007/978-981-13-7209-4_2

the 'gold and silver mountains', the contradiction between economic development and a shortage of resources, environmental deterioration has begun to loom large, the people realize that the environment is fundamental to the people's survival and development, only when clean, clear waters and lush mountains are available can development be fostered; during the third stage, the people realize that clean, clear waters and lush mountains can endlessly bring the 'gold and silver mountains', and clean, clear waters and lush mountains themselves are the 'gold and silver mountains', evergreen plants are the source of money, ecological advantages become economic advantages, a harmonious integral whole is formed; this stage represents a higher realm and reflects the requirements of scientific development, the philosophy of developing the circular economy and building a resource-conserving and environmentally friendly society. The above three stages are a process in which the economic growth mode is transformed, continuous progress is made in the development of the mindset, the relations between the people and nature are incessantly adjusted and tend to be harmonious."[1]

2.1.1 The Evolution of the Strategy of Economic Development for Mountainous Areas

With more than 20 years of reform and development, in 2001, Zhejiang achieved, in advance, a leap from the state of ample food and clothing to the state of being well-off, but economic development was unbalanced among regions, in particular, the level of economic and social development in Quzhou City, Lishui City, their counties (county-level cities), Taishun County and Wencheng County was much lower than the province's average level. Accelerating the development of the underdeveloped areas has a vital bearing on achieving coordinated and sustainable economic and social development throughout the province and the basic realization of modernization across the province ahead of schedule. Therefore, the Party Committee and the People's Government of Zhejiang Province issued the *Several Opinions on Accelerating the Economic and Social Development of the Underdeveloped Areas*, initiating the path of coordinated regional development. In 2001, with the goal of eliminating the towns with absolute poverty, Zhejiang officially carried out the 100-Town Poverty Alleviation Program. In 2002, Zhejiang became the first province without poor towns in the country; in the same year, Zhejiang started to implement the Program for Boosting Development through Mountain-Sea Cooperation, through which Zhejiang adopted the new path of focusing on project cooperation, transfer of the industrial gradient and rational allocation of factors to promote the transfer of industrial gradient from the developed areas to the underdeveloped areas, ultimately achieving interactive and win-win development between the underdeveloped and developed areas. As of June, 2012, Zhejiang had carried out 7485 character-

[1] Xi Jinping, *Carrying out Concrete Work to Stay Ahead—Line of Thought and Practice in Promoting New Development in Zhejiang*, The Party School of the CPC Central Committee Press, 2006, p. 198.

istic industrial projects for mountain-sea cooperation with an actual investment of 194.6 billion yuan. According to an estimation, the economic value added from the mountain-sea cooperation projects in Quzhou City accounted for 40% of the city's GDP, the mountain-sea cooperation projects in Lishui City contributed more than 40% to the growth of the industries above the designated size throughout the city. In 2003, the project for turning the underdeveloped villages and towns into the well-off ones was launched, injecting new vigor into the poverty alleviation project.

With poverty alleviation through more than 30 years of development, Zhejiang has achieved two leaps from eliminating absolute structural poverty to eliminating regional absolute poverty and then relative regional poverty. In 2008, the 12th Party Congress of Zhejiang Province unveiled the *Action Plan for Increasing the Income of the Low-income Groups (2008–2012)*, and carried out the Project for Making the Low-income Rural Households Become Well-to-do. The efforts at poverty alleviation shifted from regional underdeveloped towns to class-induced low-income rural households, from making the underdeveloped towns well-to-do to making the low-income households well-to-do; with a focus on improving the capacity of the low-income people for getting rich and increasing the level of their income, Zhejiang combined coordinated urban and rural development with coordinated regional development to make the low-income people more rapidly increase their income, become rich and well-to-do. In October, 2008, the 4th Plenary Session of the 12th Party Committee of Zhejiang Province vowed to speed up the development of the underdeveloped areas while more rapidly transforming the economic developmental mode and promoting economic transformation and upgrading, and to build a "Zhejiang on the Mountains".

In 2011, the growth rates of the main economic indicators in the underdeveloped areas were higher than the province's average levels, but the absolute gap between the underdeveloped and developed areas continued to widen. If no special policies were adopted, Zhejiang might fail to finish building a moderately prosperous society in all respects on schedule during the period of the 12th Five-Year Plan. Therefore, Zhejiang introduced the special support policy for the key underdeveloped counties and initiated a new round of projects for promoting the development of the underdeveloped areas, and actively established provincial experimental areas for rural reform. In Zhejiang, 63.4% of the low-income rural households had a per capita net income above 4000 yuan, up 22 percentage points compared with the previous year. According to Zhejiang's plan, by 2020, the per capita GDP in Zhejiang's mountainous areas would hit 90,000 yuan; the industrial structure will be further optimized; new-type urbanization and the construction of a beautiful countryside would be further promoted; the level of urbanization would further increase; the income of urban and rural residents would continue to grow rapidly, the disposal income of urban residents and the per capita net income of rural residents would reach 61,000 yuan and 27,000 yuan; new mountainous areas which become developed in production, affluent in life, excellent in ecology and vibrant would be built in an all-round way.

2.1.2 Vigorous Development of the Ecological Economy in Mountainous Areas

2.1.2.1 Making Great Efforts to Develop Ecological Agriculture

1. Improving the quality of agricultural products and increasing their level of safety

Zhejiang has attached great importance to grain safety, established and improved the quality standard system, the testing system and the certification system for agricultural products, established and improved the system for the management of the quality of safety monitoring for agricultural products. Zhejiang has extensively promoted international standards and green production, paid a great deal of attention to solving the problem of "table pollution" and endeavored to exercise monitoring in the whole process. Zhejiang has energetically carried out agricultural standardization, more rapidly promoted standardized agricultural production, established and improved the agricultural standardization system. Zhejiang has enhanced the comprehensive capacity for grain production, and strengthened the development of the grain market system and the grain storage system to ensure grain security.

2. Intensifying efforts to cultivate fine-quality, high-value agricultural products under well-known brands

Zhejiang has vigorously pushed forward agricultural development by applying scientific and technological advances, speeded up scientific and technological innovation in the agricultural field, carried out the seed and seedling project; Zhejiang has reinforced the system for the promotion of agricultural technology covering resource conservation, green development, safety and value added processing; Zhejiang has dedicated a great deal of energy to developing well-known, high-quality and novel agricultural products, expedited the cultivation of advantageous industries and brand products, increased the scientific and technological content and added value of agricultural products, and improved market competitiveness.

3. Sharpening the competitive edge of the leading agricultural enterprises and making them more able to stimulate others' development

With a focus on cultivating the leading enterprises, Zhejiang has worked on improving enterprise competitiveness by building various types of agricultural parks and bases. Zhejiang has made a major push to develop large specialized households and specialized cooperatives, enhance the market competitiveness of the leading agricultural enterprises and enable them to have higher capability for driving the development of rural households. Zhejiang has generously supported the leading agricultural enterprises and specialized cooperative organizations to develop the agricultural product processing industry and organize the sales of agricultural products. In order to support the development of the leading agricultural enterprises in the underdeveloped areas, each year Zhejiang has allocated to these enterprises not less than 1/3 of the provincial special funds dedicated to supporting the leading agricultural enterprises and the leading enterprises engaged in the exportation of agricultural products.

4. Speeding up the building of agricultural demonstration parks

Zhejiang has strengthened the building of the development-oriented agricultural bases. During the period of the 10th Five-Year Plan, each year Zhejiang allocated from 20 million yuan in special funds from the provincial finance to support the building of the development-oriented agricultural bases, pollution-free green food and organic agricultural bases. Zhejiang has made efforts to expand the scale of green agricultural product bases and improve the popularity of brand agricultural products. In order to support the building of the agricultural demonstration parks and the seed and seedling projects in the underdeveloped areas, each year Zhejiang has allocated to these parks and projects not less than 1/3 of the special funds dedicated to supporting the hew and high agricultural technology demonstration parks, the high-quality, high-efficiency agricultural demonstration bases and the seed and seedling projects.

In the meantime, Zhejiang has a number of high-efficiency ecological agricultural bases. With a focus on the development of the leading sectors in agriculture, Zhejiang has intensified the construction of infrastructure including water conservation, promoted the building of grain production functional areas and modern agricultural parks, increased the level of agricultural science, technology and management, supported and developed a number of characteristic bases for planting and raising crops and bases for the planting and breeding of new varieties. By 2017, there will be 70 modern agricultural comprehensive areas above 20,000 mu, 140 leading industrial demonstration areas and 350 characteristic parks for high-quality agricultural products.

Moreover, Zhejiang has given full play to the resource advantages of mountainous areas, vigorously developed characteristic advantageous agriculture, more rapidly pushed forward a number of key characteristic parks for high-quality agricultural products. Zhejiang has fully leveraged the natural resource advantages of mountainous areas, and has, under the principle of being green, efficient, distinctive and making the people rich through ecological development, focused on promoting the building of 23 characteristic parks for high-quality agricultural products, including the Longyou Selenium-rich Industrial Park and the Qingyuan High-quality Edible Fungus Park. They will be turned into the main platforms for high-efficiency ecological agriculture in mountainous areas by conducting land circulation, increasing the scientific and technological level and developing fine and deep processing.

2.1.2.2 Actively Cultivating Characteristic Industries

Zhejiang has energetically cultivated characteristic pillar industries and key enterprises. During the period of the 10th Five-Year Plan, Zhejiang established a special interest subsidy fund for industrial development in the underdeveloped areas, each year Zhejiang allocated 30 million yuan to support the underdeveloped areas for relying on local resources to actively develop small and medium-sized technology enterprises, characteristic advantageous enterprises and key enterprises. In the meantime, Zhejiang has further broadened financing channels, increased credit input, developed direct financing to support the eligible enterprises to go public or issue bonds, and more quickly cultivated characteristic pillar industries and the regional massive economy.

Zhejiang has proactively promoted the development of provincial-level industrial parks. For the underdeveloped areas with a certain industrial foundation, relevant provincial departments have appropriately lowered thresholds to support the building of provincial-level characteristic industrial parks or specialized industrial areas in provincial-level towns, have introduced relevant preferential policies for them and provided them with support in land use.

Zhejiang has propelled the transfer of industrial gradients. Zhejiang has supported the underdeveloped areas to adopt preferential measures for attracting enterprises within and outside the province to make investments and build factories, or participate in the restructuring of state-owned and collective enterprises by holding shares, mergers and acquisitions and the transfer of technology. Zhejiang has further encouraged the development of individual private enterprises.

Zhejiang has speeded up the development of an open economy. During the period of the 10th Five-Year Plan, the provincial development fund for foreign trade exportation provided interest subsidies relating to export commodities according to the standard of 0.02 yuan per USD 1 to the underdeveloped areas in the field of general trade exportation, and the Zhejiang Investment and Trade Symposium provided the underdeveloped areas with exhibition zones covering a certain area free of charge.

In 2012, Zhejiang vowed to give priority to developing the advanced manufacturing industry in the coastal areas in eastern Zhejiang. The factor and resource advantages in coastal areas, including land and talents, should be fully leveraged to transform and upgrade the traditional advantageous industries, including textiles and apparel, metal products and auto and motorcycle parts, expand the industrial chain and improve the value chain, efforts should be made to transform and upgrade the massive economy, including the tie sector in Shengzhou and the pump valve sector in Yongjia, to industrial clusters. Actions should be taken to more rapidly develop such industries as electronic information, biomedicine and advanced equipment manufacturing and make the advanced manufacturing industry large and strong. In the mountainous areas with hills and basins in central and northern Zhejiang, priority should be given to developing the advanced manufacturing industry. With a focus on transformation and upgrading of the traditional industries, priority should be given to developing the advanced manufacturing industry covering full vehicle and key parts and components, new-type textiles and apparel and hardware manufacturing; great efforts should be made to transform and upgrade the massive economy, including auto and motorcycle parts in Yongkang and Wuyi in Jinhua City, stockings in Datang, Zhuji City, hardware in Yongkang, Wuyi and Jinyun, to industrial clusters, with active measures taken to build some important advanced manufacturing bases in China. In the inland mountainous areas in southwestern Zhejiang, priority should be given to developing the characteristic ecological industries. Steps should be taken to actively develop the advanced manufacturing industry focusing on aerodynamic equipment, equipment for high-voltage power transmission and distribution, engineering machinery, the electronic information industry focusing on sensing devices, new-type display devices, semiconductor lighting devices, and the new material industry focusing on organic silicon materials, fluorosilicone composite materials and special fibres; great efforts should be made to build a resource-conserving and environment-friendly low-carbon industrial system, more quickly transform and upgrade such a massive economy as fluorosilicone in Quzhou and automobile and air-conditioning parts in Longquan to industrial clusters.

2.1.3 The Successful Mode of Industrial Development in Mountainous Areas

Five representative developmental modes have taken shape in the mountainous areas in Zhejiang.

2.1.3.1 The Mode of High-Quality Agriculture

1. The Xianju Sample

Xianju County has become the first nationwide to build the "3 + x" new agricultural service system to develop China's top-end agriculture. Under the system, based on agricultural public services including the promotion of agricultural technology, the prevention and control of animal and plant epidemic diseases, supervision over the quality of the safety of agricultural products, in light of the actual needs in different areas and of farmers, the park-level, village-level and other service workstations have been built, and service-oriented cooperatives and leading enterprises are being cultivated to provide public services and socialized services.

Xianju's environment for agricultural investment has been optimized continuously. In late 2007, Xianju became the first in Zhejiang to establish a specialized market for green agricultural products, with green agricultural products exclusively managed, made, supplied and sold. Initially, the market was based on investments from the government and jointly controlled by the supply and marketing cooperatives at the national, provincial, municipal and county levels, with a total investment of 16 million yuan. At present, 20 leading agricultural enterprises, 35 specialized cooperatives and 18 large planting households have set up direct sale stalls in the market, more than 450 varieties of green agricultural products are dealt with in the market.

The manner of supervising the safety of the quality of agricultural products has been built. The network of supervision of the quality of agricultural products covering the county, towns, villages and bases in Xianju County has basically ensured that there is a network, personnel, a mechanism, equipment and a guarantee, with an investment of 11 million yuan; a county-level agricultural product testing center with the provincial qualification of metrology accreditation has been built, thus really exercising supervision over the quality of the safety of agricultural products from the field to the table.

So far, the county has been home to 300,000-mu of bases for the production of green agricultural products and a number of modern agricultural parks, it has become the national base for the production of organic food and the national demonstration base for the processing of green agricultural products; actions are being taken to actively build Xianju's green agricultural product brand and promote the transformation from traditional agriculture to modern agriculture.

2. The connotation of the mode

The mode of high-quality agriculture is aimed at accelerating the strategic transformation of the traditional extensive agricultural growth mode, stimulating mountainous areas to deliver more high-quality ecological agricultural and sideline products to the whole society; it fully embodies the strategic requirement of increasing the value of mountainous areas.

3. Priorities of development

The manner of high-quality agriculture has the following basic characteristics: First, relying on the ecological environment to enhance and develop the manner of high-quality agriculture, basing the development of high-quality agriculture on the improvement in the ecological and environmental quality; second, acting in an organized way and on a large scale to accelerate the development of high-quality agriculture, encouraging and promoting the circulation of farmland and forest land to increase the degree of large-scale agricultural operations; third, leveraging capital and technologies to innovate and transform the manner of high-quality agriculture, attracting industrial and commercial capital to the agricultural industry; fourth, introducing policies to support and help the manner of high-quality agriculture to become large and strong, making agriculture large and strong through measures of support and safeguarding.

4. Experience and inspirations

The manner of high-quality agriculture can vary with different mountainous areas: in the economically developed mountainous areas, actions can be taken to actively develop factory-based high-quality agriculture, energetically attract industrial and commercial capital to actively develop protected agriculture, enhance the quality and economic performance of agricultural products, make agricultural development more capital- and technology-intensive; in the underdeveloped mountainous areas, efforts can be made to actively develop high-quality, ecological agriculture, rely on a good ecological environment, intensify the capital, scientific and technological input in agriculture and increase the level of agricultural organization; in the areas surrounding central cities, the sightseeing and leisure-oriented high-quality agriculture can be actively developed, convenient transportation, the beautiful natural environment and good agricultural industrial foundation can be fully relied upon and guidance in planning and more facility input can be provided.

2.1.3.2 The Manner of the Small County and the Large City

1. The Yunhe Sample

Yunhe County has given play to the leading role of the government in resource allocation; with the county town as the growth pole, measures are being adopted to break the urban-rural boundary, make the county town large and strong, produce the effect of stimulating rural development through urban areas and boosting agricultural development through the industry, thus promoting urban-rural integration and driving the holistic development of the county in the economic and social fields.

Agglomeration has been promoted to make the county town large and strong. In Yunhe County, the urban planning area has been expanded from the original 8.8 to 22 km^2, the development and construction has been accelerated in the southern part of county town, the built-up area of the county town has been expanded to 5.2 km^2. Actions have been taken to speed up the construction of transportation, information, energy networks and municipal infrastructure and improve the environment of urban development. Low hills and gentle slopes have been developed to extend the area of the Yunhe Industrial Park from 0.17 km^2 in 2001 to 3.2 km^2 and it has been turned into the typical provincial-level industrial park and the demonstration area for the mountain-sea cooperation project.

Steps have been taken to make farmers rich and benefit them, build a new socialist countryside. Yunhe's natural resource advantages and traditional industrial foundation have been relied upon to vigorously develop the leading agricultural industries dominated by an edible fungus, the Yunhe pear, organic fish and ecological tea. Yunhe County annually makes 80 million bags of edible fungus, it has stayed ahead nationwide in the quality of its planting per household and income per household. The county has pushed forward a number of projects for the benefit of the people, including village demonstration improvement, broad roads in the countryside and drinking water for farmers, and it has improved the production and living conditions for the farmers who stay behind. The county has speeded up internal agglomeration and external migration, and has adopted, in the central towns and central villages where conditions permit, various means, including self-built houses, apartments and financial aid, to guide the relatively poor people to descend from the mountains and migrate to the nearby areas.

The county has expedited integration and promoted coordinated urban and rural development. The county has strengthened vocational skill training among the transferred farmers, carried out the project for quality training thoroughly among 10,000 people as surplus labor force from rural areas to improve farmers' employment skills and overall quality. The county has pushed forward the reform of the household registration system, removed the unreasonable restrictions and various registration items relating to the employment of rural migrant workers in urban areas, and gradually exercised the management based on the temporary residence permit. The county has actively pressed ahead with institutional reforms involving family planning, social security and benefits for veterans, progressively made sure that urban and rural residents get equal access to treatments and tangible benefits. The county has treated rural migrant workers as citizens and has provided more than 30,000 jobs while ensuring rural migrant workers' access to housing. The county has endeavored to make rural migrant workers enjoy the same benefits as citizens and turn them into citizens.

2. The connotation of the mode

The manner of the small city and the large city refers to a way of enabling coordinated allocation of factors in urban and rural areas, actively guiding the agglomeration of the population, industries, capital and construction land in the county town to make the central county town, as the growth pole, large and strong, increasing the level of intensive development and stimulating holistic leapfrog development of the county in the economic and social fields.

3. Priorities of development

The manner of the small city and the large city has the following basic characteristics: First, enhancing the agglomeration of industries, the population and social programs, developing industries and programs to intensify employment support for population agglomeration, moving forward the urbanization drive moderately ahead of schedule to enhance industrialization and urban service functions; second, concurrently making the county town large and promoting the construction of a new socialist countryside, impelling the coordinated development of urban and rural areas; third, further innovating the mechanism for the circulation of such factors as land to inject more vigor into development.

4. Experience and inspirations

The manner of the small city and the large city is an important mode of development for transforming the urban-rural pattern in the underdeveloped mountainous areas; it is mainly suitable for the sparsely populated areas with the following characteristics: population outflow from the county town continuously occurs, the towns except the county town within a county can provide less support and the central county town has contiguous spaces for development, such as Taishun, Wencheng, Jingning, Pan'an, Kaihua and Changshan.

2.1.3.3 The Manner of Land-Sea Coordination

1. The Linhai Sample

Linhai City has persisted in land-sea coordination and combined terrestrial advantages with marine advantages to achieve coordinated development of terrestrial and marine industries, coordinated construction of infrastructures, coordinated allocation of resource factors and coordinated protection of the ecological environment.

Land and sea transportation has been strengthened. Linhai has proceeded from the whole province and the whole Taizhou to develop three extensive plans for the construction of transportation, including an extensive plan for a network of roads—building two railways, four highways, three ring roads, six connections and three extensions; an extensive plan for ports—building one coastal port area, one navigation channel and one line; an extensive logistics plan—building five stations, two parks and one center, to establish a half-hour economic transportation circle in the city; Linhai has actively built a three-dimensional network of coastal transportation and an integrated logistics network to turn itself into an important transportation hub in the central coastal part of Zhejiang, a port city and a logistics base.

The cooperation platform has been built. With a focus on building a circular economic industrial agglomeration area at Taizhou Bay and a marine economic development platform has been built. Actions have been taken to speed up the construction of the Toumen port area in east Linhai, leverage its excellent deepwater navigation channel, berths and 150 km^2 of land reclamation space from the sea to stimulate the development of the industries in the hinterland of central Zhejiang, turn Toumen Port into an industrial port, effectively promote Linhai's industrial transformation and upgrading, make Linhai leap from a closed type of inland development to a kind of development based on land-sea coordination.

Industrial development has been arranged. In the eastern block, several main industries, including pharmaceutical chemicals, glasses, stainless steel and mechanical industries, have preliminarily taken shape, great efforts have been made to build the pharmaceutical chemical base, the medicinal chemical base and the petrochemical base, and with emphasis placed on cultivating emerging marine industries and upgrading the traditional manufacturing industry; moreover, efforts are being made to develop a massive economy into industrial clusters and breakthroughs have been made in port-based industrial development and in industrial transformation and upgrading.

2. The connotation of the mode

The mode of land-sea coordination is a way of further leveraging the capital, talents, industrial and other advantages of developed coastal areas, giving play to the role of the marine economy in stimulating the development on other fronts, guiding the extension of infrastructures, industrial, commercial and human capital from coastal areas to mountainous areas, reinforcing the role of mountainous areas in supporting the development of coastal areas, and thus achieving coordinated development of coastal areas and inland mountainous areas.

3. Priorities of development

The manner of land-sea coordination has the following basic characteristics: first, strengthening coordination among the governments at various levels, achieving industrial and infrastructure connectivity between mountainous areas and coastal areas, enhancing the pattern of integrated development; second, promoting the two-way flow of resource factors to shape the pattern of two-way flow of factor resources in which the people come out of mountains and capital enters the mountains.

4. Experience and inspirations

The mode of land-sea coordination is divided into two types; the scope of application varies with different types. At the provincial level, the mode of land-sea coordination focuses on 26 underdeveloped mountainous areas across the province, especially 12 key underdeveloped mountainous counties, it intensifies paired assistance and land-sea coordination, stimulates capital, industries, talents and other resources to flow from the developed coastal areas to the underdeveloped mountainous areas, thus achieving balanced development of the whole province. At the county level, the mode of land-sea coordination proceeds from 11 coastal cities and mountainous counties, including Yuyao City, Fenghua City, Ninghai County, Xiangshan County, Yueqing City, Yongjia County, Shaoxing County, Linhai City, Sanmen County, Pingyang County and Cangn'an County, and fully leverages the sea and land resources to promote the integrated development of mountainous towns and coastal towns.

2.1.3.4 The Mode of Non-local Development

1. The Pan'an Sample

As one of the sources of the four main water systems, including the Qiantang River, the Cao'e River, the Ou'jiang River and the Lingjiang River, Pan'an County is one of the most important ecologically sensitive areas in Zhejiang. As the space for economic development is small and transportation is not convenient in Pan'an County, Pan'an County is economically underdeveloped. In order to support Pan'an County in fostering non-local development of the industrial economy, the Party Committee and the People's Government of Jinhua City have planned 3.8 km^2 of land in the Jinhua Economic and Technological Development Zone to establish the Jinpan Poverty Alleviation Economic Development Zone. In 18 years since its inception, the Jinpan Development Zone realized 19.5 billion yuan in industrial sales value, 7.9 billion yuan in foreign trade exportations, more than 2 billion yuan in tax payments and other non-tax revenue, and 8 billion yuan in fixed assets and infrastructure investments; it accounted for more than 1/4 of the county's quantity in the main economic indicators.

The Jinpan Development Zone has become the main industrial base and important economic pillar for Pan'an County and is also the provincial-level demonstration area for mountain-sea cooperation. The successful practice of Pan'an County in non-local development has promoted innovations in the poverty alleviation system for the underdeveloped areas, the inter-regional ecological compensation mechanism and the management system for non-local development zones.

2. The connotation of the mode

The mode of non-local development is aimed at leveraging the coordination and cooperation advantages of the governments at the provincial, municipal and county levels to support some mountainous areas with a shortage of land and capital, a sensitive ecological environment and an urgent need for improving the people's life, and establishing industrial development zones on a non-local basis to develop an "enclave" economy and achieve the non-local development of the mountainous areas.

3. Priorities in development

The mode of non-local development has the following basic characteristics: first, coordinating the interest relations among multiple stakeholders in the mode of non-local development to ensure that multiple stakeholders obtain benefits and that win-win development is achieved; second, arousing the enthusiasm of the players in the mode of non-local development, ensuring that the governments at higher levels introduce certain preferential policies in ecological and environmental protection, tax revenue distribution, transfer payment, energy conservation and emission reduction and other fields to enhance the enthusiasm of the underdeveloped areas regarding the mode of non-local development.

4. Experience and inspirations

The mode of non-local development is dominated by the mode of non-local development across the county-level administrative divisions within the province. Specifically, it can be divided into three types with different scopes of application.

The mode of non-local development across the county-level administrative divisions within the province is mainly suitable for the key underdeveloped mountainous counties and focuses on inter-regional non-local development within the prefecture-level city where the county government is located. Besides Pan'an County in Jinhua and Jingning County in Lishui, Zhejiang can focus on exploring and promoting non-local development of Wencheng County and Taishun County in Wenzhou City, and that of Qingyuan County in Lishui City and that of Kaihua County in Quzhou City.

Under the mode of non-local development across the town-level administrative divisions within the county, the mountainous counties across the province can, in light of their developmental reality, carry out non-local development across administrative villages and across towns within their administrative divisions. In Xinchang County, the joint construction of property is conducted on a non-local basis in rural areas, certain industrial land in the county's industrial development zone and industrial functional areas is designated for some administrative villages with a weak collective economy in the county with a view to building new factory buildings through the funds jointly raised by village collectives and generating income to develop the collective economy.

The mode of non-local development outside the province is mainly suitable for the areas with a developed market economy and great financial strength; these areas can rely on market players to organize massive private capital for non-local development outside the province. In other words, enterprises rent or purchase land outside the province or in foreign countries to carry out large-scale agricultural development, the building of raw material bases and industrial parks.

2.1.3.5 The Mode of Ecotourism

1. The Suichang Sample

Suichang County has built the Suichang Sample for the scientific development of mountainous areas in which rural leisure tourism leads the way, and the primary, secondary and tertiary industries develop in a coordinated way; it has successfully become one of the strong counties with an economy based on tourism and it has been included in the first batch of pilot counties for comprehensive tourism reform in Zhejiang.

The government plays the leading role. The plan for development for the tourist industry has been developed scientifically, and the developmental strategy for the tourist industry has been carried out. The county has intensified its guidance regarding policies and infrastructure input; the county has promoted tourism projects on the basis of high quality to actively seek financial support. The county government has allocated 600,000 yuan to support gold mine tourism; in August, 2006, its gold mine park was included by the Ministry of Land and Resources into the first batch of national gold mine parks and received 10 million yuan in special financial subsidies.

Investors have been attracted to participate in the development of the tourism resource. The county has guided well-known local enterprises to invest in the tourist industry, it has supported and guided local leading enterprises–private capital represented by the Yuanli Group and the KAN Group—to invest in the tourist industry. The county has attracted capital from outside the county to make investments. The Party Committee and the People's Government of Suichang County have attached great importance to investment attraction and promotion relating to tourism projects and have actively introduced tourism projects. With agritainment as the carrier, the county has built a platform for investment in tourism for rural private capital, thus greatly broadening the channels for rural private capital to participate in developing the tourist industry.

The strategy for high-quality tourism and the holistic marketing strategy have been implemented. The county has incorporated more cultural connotations in tourist products so as to upgrade tourism. The county has tapped resource advantages to make the ecological factors in tourist products appealing. The design of tourist products stresses the full utilization of ecological factors and to the greatest extent embodies Suichang's style of breathtaking landscape and unconstraint.

2. The connotation of the mode

Under the mode of ecotourism, based on natural scenery and humanistic resources, actions are being taken to actively promote comprehensive development of resources, deliver more economic benefits to the local people, enhance local enthusiasm about the comprehensive protection of the natural environment and humanistic resources, and achieve win-win outcomes—development and protection.

3. Priorities in development

The mode of ecotourism has the following basic characteristics: first, pushing forward the construction of various facilities including roads, hotels, inns and scenic spots, encouraging and attracting private capital to invest in the tourist industry, increasing the level of facility support under the mode of ecotourism; second, properly dealing with the relations between development and protection under the mode of ecotourism, giving priority to ecology, scientifically determining the total number of vacationists who can be received and the scale of development and construction, intensifying environmental protection and governance, advocating ecological vacations and ensuring sustainable development.

4. Experience and inspirations

In the mountainous counties with comparative advantages in ecotourism across the province, ecotourism can be taken as the main or one of the important modes for accelerating local development. In northern Zhejiang, with a focus on the tourists from central cities, lake and mountain leisure vacations and rural ecological leisure tourism are being actively developed; in eastern Zhejiang, the mountain and sea advantages and the industrial advantages from industrial and commercial prosperity are leveraged to promote the transformation and upgrading of traditional scenic spots, energetically develop ecological leisure tourism based on both the mountains and the sea; in central Zhejiang, the circle of ecotourism dominated by Jinhua and Yiwu has been built, the hot spring leisure and health preservation vacation industry in Wuyi has been developed and enhanced; in southwestern Zhejiang, the advantage of holistic ecotourism resources is relied upon, investment in construction has been accelerated, health preservation and old-age care leisure vocation tourism is being actively developed.

2.2 "Better Raising Two Birds" to Promote the Transformation and Upgrading of the Economic Structure

Xi Jinping pointed out, "It is necessary to better raise two birds—achieving the rebirth of the phoenix amidst fire and vacating the cage to change birds; 'achieving the rebirth of the phoenix amidst fire' means that it is essential to act with great resolve to break away from dependence on extensive growth, actively enhance the capability for independent innovation, build a scientifically and technologically strong province and a large province of brands, stimulate industrialization through IT application, foster bases for advanced manufacturing, develop the modern service industry, change from manufacturing to creation, from OEM to brand creation, make industries and enterprises 'born of fire' and 'born again'; 'vacating the cage to change birds' means that it is necessary to carry forward the spirit of the people of Zhejiang of blazing new trails everywhere, to go out of Zhejiang to develop Zhejiang, follow the requirement of coordinated regional development, actively participate in the country's regional cooperation and communication, make room for a high degree of industrial development in Zhejiang; combine 'going global' with 'bringing in', introduce high-quality foreign and domestic capital to promote industrial structural adjustment and cover the shortage in the industrial chain, maintain the alignment with the international market, thus cultivating and introducing "superior birds which eat less, lay more eggs and fly high". 'Achieving the rebirth of the phoenix amidst fire' and 'vacating the cage to change birds' are the objective trends and inevitable choices for a high level of industrial development."[2]

2.2.1 The Evolution of the Strategy of Economic Structural Transformation and Upgrading

With the practice of reform and opening up for more than 30 years, Zhejiang has carved out a path of economic and social development with regional characteristics and has achieved a historic leap in economic and social development, Zhejiang's strategy of economic structural transformation and upgrading has gradually shifted its focus from promoting economic transformation and upgrading to boosting the transformation of its manner of economic development.

Based on in-depth surveys and studies, in July, 2003, the 4th Plenary Session of the 11th Party Committee of Zhejiang Province put forward the "Eight-Eight Strategies"; afterwards, Zhejiang vowed, on many occasions, to unremittingly promote economic structural adjustment and industrial optimization and upgrading with the courage of "achieving the rebirth of the phoenix amidst fire", the measure of "vacating the cage to change birds" and the mettle of being "born of fire". In 2006, the 11th Five-Year Plan called for changing the manner of economic growth, promoting IT application in the national economy and the society, really venturing down the path of new-type industrialization, optimizing the industrial structure and increasing the industrial level.

The 12th Party Committee of Zhejiang Province completely carried out the "Eight-Eight Strategies" and the overall strategy of making the people rich by starting businesses and of building a strong province through innovation, and a new pattern of more rapidly transforming the developmental mode and promoting economic transformation and upgrading took shape.[3] In December, 2007, the economic work conference of Zhejiang Province put forward the major strategic task of accelerating economic transformation and upgrading for the first time. On September 26, 2008, the 4th Plenary Session of the 12th Party Committee of Zhejiang Province vowed to

[2]Xi Jinping, *Carrying out Concrete Work to Stay Ahead—Line of Thought and Practice in Promoting New Development in Zhejiang*, The Party School of the CPC Central Committee Press, 2006, pp. 128–129.

[3]Luo Yiqun, Zhejiang Speeds Up "Vacating the Cage to Change Birds" to Promote Economic Transformation and Upgrading, *Zhejiang Daily*, July 16, 2013.

more rapidly establish four new great patterns, including modern industrial development, intensive innovative economical development, coordinated stimulation by consumption, investment and exportation, and coordinated urban and rural regional development.

In March, 2010, Lu Zushan put forward three major philosophies of development—including sustainable and sound development, development that is more beneficial to the people and socially harmonious and stable development—and called for optimizing the industrial structure, the structure of factor support, the demand structure and the structure of spatial layout, strengthening the guarantee of mechanism and policies and organizational leadership.

In May, 2010, Zhao Hongzhu called for correctly handling the relations between "large" and "small", seeking new breakthroughs in fostering new advantages, properly dealing with the relations between "new" and "old", breaking new ground in industrial structural adjustment, correctly addressing the relations between "introduction" and "research and development", making new breakthroughs in scientific and technological innovations, properly handling the relations between "mountain" and "sea", achieving new breakthroughs in cultivating a new economic growth point, correctly dealing with the relations between "break" and "make", seeking new breakthroughs in deepening reforms, correctly addressing the relations between "internal" and "external", bringing about new breakthroughs in inter-regional opening and international opening, correctly handling the relations between "urban" and "rural", breaking new ground in coordinated urban and rural development, correctly addressing the relations between "utilization" and "protection", making new breakthroughs in ecological development and properly handling the relations between "co-building" and "sharing", making new breakthroughs in comprehensively improving the people's wellbeing.

The 13th Party Congress of Zhejiang Province set the overall goal of building a modern Zhejiang which is materially affluent and culturally advanced and raised more urgent and higher requirements for more rapidly "vacating the cage to change birds". The Party Committee and the People's Government of Zhejiang Province further strengthened strategic coordination and put forward the goals, requirements and policy measures for promoting "vacating the cage to change birds", and stressed that it was necessary to ensure the leading role of market regulation, adopt economic, legal and administrative means, integrate the measure of imposing reverse pressure, an incentive measure and a service measure, carry out law-based phase-out, local transformation and transfer the gradient, establish a long-term working mechanism for "vacating the cage to change birds", push forward the change in the industrial gradient, structural optimization and improvement in competitiveness, as well as strive to greatly enhance the efficiency of resource utilization and rapidly increase effective investments by the end of the period of the 12th Five-Year Plan.

Accelerating the policy of "vacating the cage to change birds" has a vital bearing on Zhejiang' future economic and social development. In September, 2012, the General Office of the People's Government of Zhejiang Province issued the *Implementation Opinions of the General Office of the People's Government of Zhejiang Province on Accelerating "Vacating the Cage to Change Birds"*, further specifying work tasks and measures in five respects including enhanced work coordination, policy coordination, institutional and mechanism innovation, pilot work and demonstration, guidance and incentives, and resulting in establishing the provincial coordinating group for the work on "vacating the cage to change birds" headed by the vice-governor in charge, including an office, four working groups for shutting down outdated production facilities, optimal allocation of energy, intensive use of land and environmental pollution control. The cities, counties (county-level cities, districts) also established a mechanism for coordinating the work on "vacating the cage to change birds". The mechanism for promoting the policy of "vacating the cage to change birds" which features coordination between the provincial and municipal levels and among departments, joint promotion and systematic efforts has been preliminarily established in Zhejiang and has produced positive effects.

Subject to bottlenecks and restrictions from factor supply and environmental bearing capacity, the constraints imposed on sustainable development by the extensive growth mode have become increasingly salient. Zhejiang has adopted the strategy of optimizing agriculture, making the industry strong and rendering the tertiary industry prosperous, and it has developed ecological agriculture and quality agriculture to replace traditional agriculture, transformed and upgraded traditional industries through the new-type manufacturing industry. In the meantime, Zhejiang has fostered innovation-driven development, and has changed from cost to technology in securing the leading position, brought about "vacating the cage to change birds", removed the projects and enterprises involved in low-level and redundant development to introduce the strategic emerging industries, new mechanisms and new growth modes.

2.2.2 The Strategic Measures for the Transformation and Upgrading of the Economic Structure

2.2.2.1 Earnestly Promoting Independent Innovations in Enterprises

Given the complex and changing economic situation, the Party Committee and the People's Government of Zhejiang Province developed the overall line of thought of addressing both the symptoms and the root cause, maintaining stability and promoting adjustment, endeavored to overcome developmental difficulties, encouraged enterprises to make independent innovations, and it became the first to undergo transformation and upgrading for excellence and higher capability. In 2006, the Party Committee and the People's Government of Zhejiang Province specially issued the *Several Opinions on More Rapidly Enhancing the Capability for Independent Innovation and Building an Innovative Province and a Scientifically and Technologically Strong Province*. In 2008, the Zhejiang Administration for Industry and Commerce issued the *Circular Concerning Carrying out the Policy of Maintaining Stability and Promoting Adjustment to Further Mitigate the Burden on the Enterprises*, and the

Zhejiang Provincial Administration of the Supervision of Quality and Technology issued the *Circular Concerning Reducing and Exempting Charges to Mitigate the Burden on Enterprises*, supporting and encouraging enterprises' independent innovation in an all-round way.

1. Guiding enterprises to increase scientific and technological input

Zhejiang has given full scope to the role of enterprises as the main players in research and development input, technical innovation activities and the application of innovation achievements, and it has guided enterprises to increase their scientific and technological input, especially that for research and development, and to continuously develop new technologies, products and processes. Where enterprises are identified and recognized as the new and hi-tech enterprises and small and medium-sized technology enterprises above the provincial level, their research and development expenses must not be less than 5 and 3% of the annual sales revenue. The new and hi-tech enterprises above the provincial level and large and medium-sized enterprises are encouraged and supported to establish research and development bodies. Small and medium-sized enterprises are encouraged and supported to engage in cooperative research and development by joint investment, joint commissioning and other means to build up reliable technical support. Zhejiang has actively cultivated innovative enterprises, supported and developed a number of national and provincial-level innovative enterprises with continuous capability for innovation, proprietary intellectual property rights and well-known brands. Zhejiang has encouraged scientific and technical personnel and private capital to establish small and medium-sized technology enterprises through various channels. The provincial-level small and medium-sized technology enterprises can enjoy relevant preferential policies for new and hi-tech enterprises.

2. Enhancing scientific and technological cooperation and communication among enterprises

Efforts have been made to enhance scientific and technological cooperation in the Yangtze River Delta Region, to jointly build an integrated and open system of innovation. Zhejiang has thoroughly carried out the strategy of introducing large institutions and well-known universities, encouraged local authorities to work with large domestic and foreign institutions and well-known universities for jointly building various forms of innovation carriers, including research and development bodies, mainly through enterprises by focusing on introducing teams, high-level talents and core technologies. The areas where conditions permit are supported to establish the bases which will become home to research and development bodies and are designed to facilitate scientific and technological innovations, achievement conversion and industrialization. The enterprises where conditions permit are encouraged and supported to set up research and development centers in foreign countries and overseas to fully leverage foreign and overseas science and technology, talents and resources for improving the capability for independent innovation. Local authorities have given priority support to various types of research and development bodies regarding construction funds, land expropriation, talent introduction, scientific and technological projects, and to provide policy support to the enterprises which set up foreign and overseas research and development centers.

3. Strengthening the protection of intellectual property

Actions have been taken to establish and improve a modern intellectual property institution with clear ownership, well-defined rights and responsibilities, standardized management and smooth transfer to enhance vitality in invention and creation, strengthen the protection of intellectual property and to push forward industrialization of patent achievements. Enterprises are encouraged and guided to own patented technologies, technical secrets and software copyrights, especially invention patents; invention patents are taken as important factors for approving and initiating scientific and technological projects and performance evaluation. Zhejiang has pushed forward the project for building patent demonstration enterprises, and it has nurtured and developed a number of enterprises with advantages in intellectual property rights. Zhejiang has intensified the enforcement of intellectual property rights laws, and fostered a good environment for respecting and protecting intellectual property rights. Zhejiang has further reinforced the protection and management of intellectual property, given full play to the role of industrial associations in promoting self-discipline in the protection of intellectual property, continuously improved the system for the protection of intellectual property with government supervision, enterprise self-discipline, supervision by public opinion and the people's participation.

4. Solidly pushing forward the project for identifying and recognizing well-known trademarks, big-name trademarks and famous trade names

Enterprises are encouraged to go down the path of independent innovation and brand development. Enterprises are supported in bringing about technical innovations to improve the quality of products and upgrade products, and to create their self-owned brands. Efforts are being made to cultivate and develop well-known trademarks, famous trademarks and enterprise brands. Small and medium-sized enterprises are stimulated to jointly create brands and build regional brands. Export enterprises are encouraged and supported to register trademarks and seek quality certification overseas, turn domestic brands into international brands and to foster a number of world-famous brands which are internationally influential and competitive. Enterprises are encouraged to establish the self-regulatory organizations for trademarks, trade names and brands, carry out innovations to, protection of and publicity for brands. The project for identifying and recognizing well-known trademarks, famous trademarks and famous trade names is being implemented to improve the enterprises' intangible assets.

5. Supporting enterprises to vigorously cultivate and attract innovative talents

Actions are being taken to intensify the reform of the enterprise distribution system and further carry out the policy of allowing technical factors to participate in distribution. Visiting research posts for enterprises' innovative talents have been set up at the institutions of higher learning and scientific research institutions, while senior experts from enterprises are appointed to concurrently serve as professors or research fellows. The teachers at the institutions of higher learning and the research fellows at scientific research institutions are encouraged to concurrently work at large and medium-sized leading enterprises and at new and hi-tech enterprises. Postdoctoral research stations have been established at enterprises with a view to attracting outstanding postdoctoral research fellows to engage in scientific and technical innovations at enterprises; enterprises are encouraged to build bases for student internships for the institutions of higher learning and scientific research institutions. Where postgraduate and undergraduate students at the institutions of higher learning establish private technology enterprises, their names are retained on the school roll with the approval of schools.

6. Increasing financial support

Various types of financial institutions have been stimulated to engage in close cooperation with scientific and technological projects and venture capital; active experiments are underway in extending loans for scientific and technical development to support technical innovations in enterprises. Financial institutions have intensified credit support for the new and hi-tech enterprises which have production capacity and enjoy a great potential for growth. Priority has been given to arranging loans, including working capital loans, for the enterprises which export the products based on independent innovation and are able to repay the loans. Experiments are being made regarding pilot pledge loans relating to intangible assets, including intellectual property rights. The new and hi-tech enterprises and small and medium-sized technology enterprises where conditions permit are being supported in going public on the domestic main board, SME board and overseas. The new and hi-tech enterprises where conditions permit are being supported in issuing corporate bonds. The credit guarantee system for small and medium-sized technology enterprises has been established and improved; guarantee institutions are encouraged to further support the technical innovation activities in technology enterprises. All credit guarantee institutions which have provided a credit guarantee to new and hi-tech enterprises and private technology enterprises in an amount above 70% of the cumulative amount guaranteed can enjoy the preferential policy for new and hi-tech enterprises according to the regulations. Insurance companies are being supported in developing such insurances as enterprise property insurance, product liability insurance, export credit insurance and business interruption insurance, and they provide insurance services to new and hi-tech enterprises. The province and the cities and counties where conditions permit establish special funds dedicated to offering loan interest subsidies for scientific and technical development, achievement conversation and industrialization projects. Commercial banks are being stimulated and supported in establishing a stable relationship with small and medium-sized technology enterprises so as to support the enterprises with a great vitality for innovation. Actions are being taken to more rapidly establish an enterprise and individual credit system to promote credit development and support commercial banks in improving their financial services for technology enterprises.

2.2.2.2 Pushing Forward an Economical and Intensive Use of Land

1. Carrying out economical use of and reduction in the newly-added construction land

First, the structure of land supply has been optimized. Comprehensively strengthening the management of the land use plan, strictly controlling the total scale of the newly-added construction land. Giving prominence to the orientation of regional industrial development, strengthening the management of access, and building the joint examination system for access involving industrial projects. Exploring the differentiated land supply policy, giving priority to guaranteeing the land for the emerging strategic industries, major industries, advanced manufacturing industries, scientific and technological entrepreneurial and innovation projects. Strictly executing the *Catalogue of Projects for Which Land Use Is Restricted*, the *Catalogue of Projects for Which Land Use Is Prohibited* and the *Control Indicators Concerning Construction Land for Industrial Projects* to curb, at the source, land use for the projects with an excess capacity for production and redundant projects. Second, land use standards have been strictly adopted. Revising the *Control Indicators of Zhejiang Province Concerning Construction Land for Industrial and Other Projects*, gradually raising the access threshold for the newly-added land for construction projects. Strictly executing land use standards, exploring and establishing a system for the evaluation and expert review of project land saving for the special projects for which national and local land use standards and construction standards have not yet been issued, so as to increase the intensity of investment in and the efficiency of use of the newly-added construction land. Third, behavior regarding land supply have been standardized. Strictly executing the system for the transfer of clean land—the land for which demolition and leveling efforts have been made—and profit-oriented land through bidding, auction and listing, comprehensively carrying out the requirement for the announcement and disclosure of land supply, timely uploading the written decision on transfer and on the contract of land transfer to the system of dynamic monitoring and supervision of the land market, conducting online management of the entry of information and its filing, and truly strengthening supervision and regulation of the land market.

2. More rapidly tapping the potential of the available construction land to enhance efficiency

First, activities are underway to tap the potential of industrial land for improving efficiency. Actively pushing ahead with the policy of "exchanging space for land", encouraging enterprises to, subject to meeting the requirements of urban and rural planning, quality and fire protection and not changing the usage of land, making technical renovations and reducing the land for green fields and supporting facilities in increasing their land used for production activities; adding storeys to factory buildings, utilizing underground spaces and adopting other means to make technical renovation based on zero increase in land use, increasing investments, enhancing the intensity of investments in industrial land and the efficiency of industrial land use. Second, industrial enterprises have been stimulated to use industrial land space for developing the tertiary industry. The enterprises where conditions permit are encouraged to use their available land and factory buildings to engage in the service industry—including commercial service, e-commerce, research and development design, cultural creative services, goods storage and transportation, sales of fresh and live agricultural products—according to laws and regulations. Third, the adjustment of the policy concerning the tax on urban land use has been piloted. Giving play to the role of taxes as a lever in economic regulation to promote the intensive use of land.

3. Accelerating the reduction of the approved but unused land

First, the "dual linking" system has been strictly implemented. Developing concrete measures for the use of the approved but unused land and the allocation of quotas in the plan of the newly-added construction land, and linking them with the evaluation based on the system for the responsibility of the goal of cultivated land protection. Intensifying the execution of the "dual linking" system, speeding up the effective utilization of the land approved but not supplied or the supplied but unused land. Second, the crucial activity of using the approved but unused land has been carried out. The land that has been approved but not supplied before 2005 was basically reduced in late 2014. The Department of Land and Resources of Zhejiang Province has strengthened dynamic monitoring of the total scale of the land approved but not supplied, and has established the irregular reporting system, urged local authorities to more quickly remove and use the land approved but not supplied. Third, the post-approval supervision over construction land has been strengthened. Improving the system of post-approval supervision over construction land, reinforcing supervision over the execution of the land transfer contract, increasing the proportion of the construction land where construction begins on schedule after supply; strictly executing the *Measures for the Disposal of Idle Land*, intensifying its disposal and preventing land resources from becoming idle and being wasted.

4. Redeveloping the inefficiently used land in urban areas

First, relevant work has been carried out to check the inefficiently-used land. With a focus on the built-up areas in cities and towns, efforts should be organized to check the inefficiently-used construction land in urban areas, establishing a database for that land, thus laying a good foundation for redeveloping it. Second, the redevelopment of the inefficiently-used land in urban areas has been piloted. According to the requirements specified in the *Circular of the Ministry of Land and Resources on Printing and Circulation of the Guiding Opinions on Piloting the Redevelopment of the Inefficiently Used Land in Urban Areas*, a plan for that work in Zhejiang should be developed. Third, the inefficiently used land in urban areas has been disposed of in a classified way. In coordination with the three-year plan for "three renovations and one demolition", the redevelopment of the inefficiently-used land in urban areas should be accelerated, the mechanism for the distribution of incremental benefits from land should be improved, and its disposal should be promoted in an orderly fashion and there should be an efficient reuse of that land.

5. Pushing forward intensive and efficient land use in development zones

First, activities have been carried out to establish the demonstration areas for economical and intensive land use in development zones. An evaluation system has been established for economical and intensive land use in development zones, a notice of commendation has been circulated involving the good performers reaching the standard for economical and intensive land use; an activity of appraising and selecting the demonstration areas for economical and intensive land use in development zones has been organized and conducted, a number of the demonstration areas have been built for economical and intensive land use in development zones across the province. Second, intensive use of the land in development zones has been promoted by giving prominence to the industrial orientation in development zones, raising the standard for control indicators concerning the land for construction projects, strengthening the management of the land supply, pushing forward the integration and use of the existing construction land, and increasing the rate of land use. Third, the supervision over land use in development zones has been reinforced. Regularly surveying the changes in the current use of the land in development zones, updating the results of the evaluation. Dynamic supervision has been intensified over the land supply, its development and use in the development zones, and the idle and inefficiently-used land has been discovered and disposed of.

2.2.2.3 Actively Shutting Down the Outdated Production Facilities

Shutting down the outdated production facilities is the objective requirement for changing the manner of economic development and promoting industrial transformation and upgrading; it is also an important measure for expanding the space for development and increasing the potential of economic development. The Party Committee and the People's Government of Zhejiang Province have given great importance to the shutdown of outdated production facilities and they have resolutely shut them down in the industries with high energy consumption and heavy pollution, making room for industrial transformation and upgrading.

1. Strengthening organizational leadership and the building of a mechanism

In 2010, based on national requirements and the needs for making progress, Zhejiang set up a coordinating group for the work on shutting down outdated production facilities; in 2012, Zhejiang established the provincial coordinating group for the work on "vacating the cage to change birds", including an office containing four working groups for shutting down those facilities, the optimal allocation of energy, the intensive use of land and environmental pollution control, while local governments at various levels established the corresponding mechanisms, built and improved a number of mechanisms for promoting the relevant work, evaluating pilot work, ensuring long-term incentives and imposing reverse pressure through vacation, covering the division of responsibilities, program application, task decomposition, inspection and acceptance, announcement and disclosure, evaluation and incentives as well as pilot work and demonstrations.

2. Strengthening the guidance regarding policies and standards

In 2010 and 2011, the People's Government of Zhejiang Province issued the policy opinions on more rapidly shutting down outdated production facilities; in 2012, the People's Government of Zhejiang Province issued the *Several Opinions of the People's Government of Zhejiang Province on More Rapidly "Vacating the Cage to Change Birds" to Promote Economic Transformation and Upgrading (Trial)*, the *Implementation Opinions of the General Office of the People's Government of Zhejiang Province on More Rapidly "Vacating the Cage to Change Birds"*, specifying the goals, requirements and policy measures for the province-wide work on "vacating the cage to change birds".

3. Taking multi-pronged measures to promote relevant work in an all-round way

Zhejiang has adopted the legal, economic, technical and necessary administrative means to strengthen the work on "vacating the cage to change birds" and shutting down outdated production facilities. These means include strictly safeguarding market orders, ensuring orderly competition for selecting the superior and eliminating the inferior, giving full play to the role of the economic means, revising and issuing the interim measures for identifying and recognizing the enterprises subject to the policy of differentiated electricity prices, raising the standards for identifying and recognizing the enterprises subject to shutdown and restriction, intensifying the execution of differentiated electricity prices, earnestly imposing reverse pressure on shutting down outdated production facilities, giving full scope to the role of laws and regulations as constraints and the role of technical standards as a threshold and stepping up law enforcement and punishment and shutting down outdated production facilities according to laws and regulations.

4. Reinforcing coordinated and joint efforts of various departments

According to the documents of the People's Government of Zhejiang Province, the Office of Zhejiang Province for Shutting Down Outdated Production Facilities promptly assigned 29 key tasks under ten main working measures to 23 provincial units, cities, counties (county-level cities, districts), promoting the work on shutting down outdated production facilities in a vertical, horizontal and systematic way. Actions have been taken to strengthen the coordination of the work among provincial departments, joint efforts have been made to drive "vacating the cage to change birds" and the shutdown of outdated production facilities; for instance, the provincial economic and information commission and the environmental protection department have jointly developed a plan for rectifying and improving six industries, including printing and dyeing, papermaking, leather and chemical industries, thus accelerating the shutdown of outdated production facilities through vigorous industrial rectification and improvements.

2.2.2.4 Transforming and Upgrading the Traditional Industries

The traditional industries are the foundation for Zhejiang's national economy and have created the majority of output value, profits and tax revenue as well as job opportunities; they are the basic and pillar industries as well as the industries that are crucial for the people's well-being, so transforming and upgrading the traditional industries is of great significance for Zhejiang's economic development, transformation and upgrading.

1. Strengthening guidance regarding policies guidance, building a system of policies for transforming and upgrading the traditional industries

Actions have been taken to establish and improve an industrial plan and a system to guide industrial policies, improve the mechanism for supporting the advantageous pillar industries, build the mechanisms for promoting the transformation and upgrading of traditional industries and technical renovation in enterprises; actions have also been taken to improve the industrial system of productive services for the massive economy and to establish an effective mechanism for boosting the transformation of the massive economy into modern industrial clusters.

2. Supporting the leading enterprises in industries, cultivating the leading enterprises for transforming and upgrading the traditional industries

In 2010, the General Office of the People's Government of Zhejiang Province issued the *Several Opinions of the People's Government of Zhejiang Province on Supporting the Leading Enterprises in Industries to Accelerate Development*, stressing that the leading enterprises in industries were the forerunners in Zhejiang's innovative industrial development, the main forces in economic transformation and important signs of economic strength and competitiveness. It is necessary to give play to the role of the leading enterprises in stimulating the development of others, drive them to expand the industrial chain, improve the value chain, optimize the enterprises' organizational structure and further increase industrial concentration and the level of coordination among enterprises.

3. Intensifying support, cultivating a number of small and medium-sized growth enterprises amidst the transformation and upgrading of traditional industries

Zhejiang has increased its support for micro, small and medium-sized enterprises and it has guided small and medium-sized enterprises to become specialized, high-precision, distinctive and excellent. In 2011, the Zhejiang Provincial Economic and Information Commission issued the *Circular Concerning the Work on Recommending Small and Medium-sized Growth Enterprises (Industrial Cluster Demonstration Areas) in Zhejiang Province*; the Zhejiang Provincial Administration of Small and Medium-sized Enterprises issued the *Annual Implementation Plan for the 100-1000-10,000 Project of Small and Medium-sized Enterprises*, actively guiding and stimulating Zhejiang to change from a quantitatively large province to a qualitatively strong province in small and medium-sized enterprises, supporting a batch of small and medium-sized enterprises which are consistent with the orientation of industrial development and have outstanding growth potential to broaden their financial channels, and providing them with key support in financing guarantees and the projects under special funds.

4. Giving prominence to demonstration and pilot work, developing the effective carriers for transforming and upgrading the traditional industries

Full scope has been given to the leading role of the massive economy. In 2010, the General Office of the People's Government of Zhejiang Province issued the *Several Opinions of the People's Government of Zhejiang Province on Further Accelerating the Building of Demonstration Areas for Transforming and Upgrading the Massive Economy into Modern Industrial Clusters*, calling for grasping the inherent law of upgrading the massive economy to modern industrial clusters, choosing a number of industrial clusters with better independent innovations as the cultivation priorities and demonstration points in Zhejiang, gradually fostering regional industrial clusters which will lead the way in China, are internationally competitive and highly influential in industries and have a high degree of capability for technical innovations. In the meantime, Zhejiang has piloted the transformation and upgrading of the industrially large counties. In 2012, the General Office of the People's Government of Zhejiang Province released the list of 20 pilot counties (county-level cities, districts) for developing into the industrially strong counties (county-level cities, districts). The pilot counties (county-level cities, districts) will proceed from their local reality to stimulate the transformation and upgrading of the massive economy, cultivate the headquarters, the brands, listing-oriented enterprises, new and hi-tech enterprises and the enterprises led by industrial alliances, and they will become the first to break new ground in building the characteristic bases for industrial design and smart cities.

5. Speeding up the integration of IT application and industrialization, exploring the effective ways to transform and upgrade the traditional industries through information technology

With a focus on ten main traditional industries, including equipment manufacturing, light industries, the textile industry as well as the chemical, pharmaceutical, electric power, building material, metallurgical, automobile and shipbuilding industries, Zhejiang has carried out a plan of action for the deep integration of IT applications and industrialization to transform enterprises from single service applications to integrated applications of multiple services, from single enterprise applications to coordinated applications along the upstream and downstream sections of the industrial chain, to enable full penetration, integration and complete fusion of information technology in the traditional manufacturing industry, promote innovative industrial development, green development and intelligent development, and to increase the intensiveness of industrial production. Zhejiang has popularized and promoted the tools of information intelligence, promoted the application of such new industrial research and development modes as simulation experiments, platform integration, collaborative research and development and it has built a research and development system for collaborative innovation, pushed forward the integrated operations and integration of the enterprise management information system to exercise fine and flexible production management; moreover, it has accelerated digital and network-based development of major equipment and key production equipment, adopted information technology to optimize the functions of equipment and devices, increase the information level of machine tools for industrial production and major industrial equipment, applied information technology to transform traditional processes and production flow, make production flow automatic and intelligent, make improvements in the production efficiency and the rate of comprehensive energy utilization, ensure the safety of industrial production, reduce pollution emissions and promote the integrated recycling of resources; it has also encouraged enterprises to thoroughly develop and utilize the resources of industrial information, establish industry-specific specialized knowledge bases, databases and information banks, and give play to the role of information resources in guiding and supporting the development of small and medium-sized enterprises in industrial clusters.

2.2.2.5 Energetically Developing the Strategic Emerging Industries

More quickly developing the strategic emerging industries is crucial to promoting industrial transformation and upgrading. In 2011, Zhejiang identified nine, including the Internet of Things, high-end equipment manufacturing, new energy, new materials, energy conservation and environmental protection, biology, new energy automobiles, the emerging marine industry and the nuclear power-related industry, as the strategic emerging industries. Zhejiang has followed industrial development plans to identify developmental lines of thought, the goals of development, the main fields, key tasks and policy measures, to develop and carry out annual implementation plans, ascertain annual targets and tasks and the assignment of responsibilities and to speed up the cultivation of and support in boosting industrial development.

1. Continuously pushing forward the 1000-100-10 Cultivation Project

Zhejiang has focused on cultivating 1000 enterprises by taking enterprises as the main players. Zhejiang has completely carried out the work on cultivating 1000 key enterprises, it has improved the repository of key enterprises, actively guided and promoted the transformation and upgrading of a number of the leading enterprises towards the strategic emerging industries, supported the leapfrog development of a number of small and medium-sized innovative enterprises with technical advantages, given scope to the role of key enterprises in guiding and stimulating the development of others and serving as typical models. Zhejiang has exercised dynamic management of the enterprises in the repository, and it has given priority to adopting preferential policies for the enterprises in the repository in the arrangement of special funds from provincial finance, application for relevant special national projects, tax exemption for imported equipment and major technical equipment if the enterprises comply with relevant regulations.

With projects as carriers, Zhejiang has vigorously carried out the 100-project Program. Zhejiang has organized efforts to implement the *Plan of Zhejiang Province for the 100-project Program for Strategic Emerging Industries (2013)*, covering 101 key projects with investments totaling 57.22 billion yuan. Zhejiang has focused on pushing forward a number of major key technologies, major technology industrialization projects and demonstration application projects. About three key fields in each industry have been intensively supported in speeding up the growth of emerging industries, carrying out the projects for research and development, industrialization and demonstration application and thus improving the performance in the use of fiscal funds.

With parks as the main fronts, Zhejiang has built ten industrial bases. The new and hi-tech industrial parks and industrial agglomeration areas with a better industrial foundation are scientifically planned and rationally arranged through efforts led by the government with key enterprises as the main players in cooperation with scientific research institutions, universities, industrial associations and other forces. Emphasis has been placed on building about ten distinctive and domestically influential bases of strategic emerging industries, which have become the new economic growth pole in Zhejiang.

2. Pressing ahead with the building of bases for the demonstration of innovations

In light of the foundation of industrial development in different areas and local cultivation and developmental priorities, a number of bases for the demonstration of innovations of strategic emerging industries should be built through government-led efforts with key enterprises as the core in cooperation with scientific research institutions, universities, industrial associations and other forces, and by relying on the areas of industrial agglomeration, new and hi-tech industrial developmental zones, economic developmental zones (industrial parks), industrial cluster demonstration areas—all of which are above the provincial level, giving prominence to key points and making scientific planning, further improving the propelling mechanism, strengthening factor guarantees and optimizing the market environment. As the provincial-level industrial agglomeration areas and the new and hi-tech industrial development zones (parks) above the provincial level are the main fronts for cultivating and developing the strategic emerging industries, it is required that the proportion of the added value from their strategic emerging industries in the GDP should increase by more than 3 percentage points each year. For the economic developmental zones (parks) and industrial agglomeration demonstration areas above the provincial level, that proportion is required to increase by more than 2 percentage points each year.

3. Increasing fiscal support

Zhejiang has actively sought support from central finance for the key projects of the strategic emerging industries in Zhejiang, and has better utilized the provincial special fiscal funds for those industries to further support their key projects. Actions have been taken to enhance communication and coordination with financial institutions, build a multilevel guarantee system, comprehensively adopt fiscal policies, including risk compensation, to establish bank-enterprise connections and a cooperation platform, actively recommend the key projects of the strategic emerging industries to financial institutions. Efforts have been made to change financing mindsets, broaden financing channels, energetically introduce direct financing means, including private equity funds, to provide a multi-tier fund guarantee for the development of the strategic emerging industries.

4. Promoting standardization

The role of standards in leading technical innovation has been strengthened; emphasis has been placed on supporting the conversion of the invention patents of demonstration enterprises into standards. Steps have been taken to improve the building up of the capacity for organizational support, raise the standard setting and revision level, improve the standard technical system, encourage and support the leading role of the characteristic advantageous enterprises engaged in the strategic emerging industries, scientific research institutions or industrial associations focusing on the strategic emerging industries in setting and revising international, national, industrial and local standards. Relevant work has been carried out to promote popularization and implementation of standards, reinforce the supervision over and inspection of the implementation of standards, really give play to the role of standardization in leading, normalizing and promoting rapid, healthy and orderly development of the strategic emerging industries.

2.2.3 Marked Effects in the Transformation and Upgrading of the Economic Structure

2.2.3.1 The Enterprises' Capability for Independent Innovation Has Improved Significantly

Enterprises' scientific and technological input has grown steadily. In 2013, the scientific and technological expenditure in the industries above the designated size totaled 45,153 million yuan, the expenditure on the purchase of technological achievements was 1177 million yuan. The extent of increase in the scientific and technological input in the new and hi-tech industries was greater than that in the industries above the designated size.

The new and hi-tech enterprises, small and medium-sized technology enterprises have been cultivated and developed. The Department of Science and Technology of Zhejiang Province has extensively carried out the Project for the Cultivation of 10-100-1000-10,000 Enterprises. As of 2013, there were 5160 new and hi-tech enterprises, 10,000 small and medium-sized technology enterprises, 44 national innovative pilot enterprises and 34 national technology enterprise incubators in Zhejiang. A total of 155 well-known and industry-leading enterprises built provincial-level enterprise research institutions, and the new products developed by them delivered more than 250 billion yuan in sales revenue, more than half of the total sales revenue.

Table 2.1 The changes in the intensity of fixed asset investment in the newly-added industrial projects in Zhejiang, 2007–2012, unit: 10,000 yuan/mu, %

Year	2007	2008	2009	2010	2011	2012	Annual growth rate in 2007–2012 on average	Average in 2007–2012
Intensity of fixed asset investment	181	187	185	203	214	233	5.18%	200.5

The enterprises with advantages in intellectual property rights have been cultivated in an orderly way. As of late 2010, Zhejiang had been home to 3 cities for implementing the project aimed at promoting the national intellectual property rights strategy for small and medium-sized enterprises, 2 national patented technology demonstration and trading centers, 84 national demonstration and pilot enterprises and public institutions for intellectual property rights, 2 national enterprise patent work exchange stations, 722 provincial patent demonstration enterprises and 4299 municipal and county-level patent demonstration and pilot enterprises. In 2010, the number of patent applications from enterprises for inventions reached 7691 and 2436 invention patents were granted to enterprises in Zhejiang, up 4.05 times and 5.63 times compared with 2007, accounting for 42.7 and 38% of the total number of invention patents in Zhejiang; Hangzhou H3C ranked No. 6 among single enterprises across the country.

2.2.3.2 The Economical and Intensive Use of Land Has Reached a New Level

Economical use and reduction of the newly-added construction land has been carried out. The average intensity of investments in the newly-added industrial land in Zhejiang increased from 1.81 million yuan/mu in late 2007 to 2.33 million yuan/mu in 2012, an average increase of 5.18% in six years (see Table 2.1). In 2010, the newly-added construction land per unit of fixed asset investment was 0.85 m/10,000 yuan.

The potential of the available construction land has been tapped to improve the efficiency. The net value of fixed assets per unit of industrial land and the fixed asset input per unit of industrial land grew annually by an average of 9.04 and 6.87% in 2007 and 2011. The total amount of profit per unit of industrial land, total profit taxes per unit of industrial land, total output value per unit of industrial land, added value per unit of industrial land and sales revenue per unit of industrial land, which reflect the output level of industrial land, grew annually by an average of 12.93, 13.04, 9.44, 9.89 and 9.19% during the period 2007–2011; these five indicators concerning the output per unit of industrial land grew by an average of 10.90% (see Table 2.2).

Table 2.2 The growth of the input-output efficiency per unit of industrial land in Zhejiang, 2007–2011

Year			2007	2008	2009	2010	2011	Annual growth rate in 2007–2011 on average (%)
The output efficiency of industrial land	Total amount of profit per unit of land	10,000 yuan/mu	4.9	4.2	5.6	8.4	8.0	12.93
	Total profit in taxes per unit of land	10,000 yuan/mu	8.4	7.7	9.6	13.5	13.7	13.04
	Total output value per unit of land	10,000 yuan/mu	133.4	143.9	141.2	169.3	191.4	9.44
	Industrial added value per unit of land	10,000 yuan/mu	25.7	28.3	28.0	33.5	37.4	9.89
	Sales revenue per unit of land	10,000 yuan/mu	97.6	105.0	104.5	132.2	138.8	9.19
The input intensity of industrial land	Net value of fixed assets per unit of land	10,000 yuan/mu	25.3	20.0	34.6	37.3	35.8	9.04
	Fixed asset input in the secondary industry per unit of land	10,000 yuan/mu	10.1	10.6	11.3	12.5	13.2	6.87
	Land consumption by the newly-added fixed assets	mu/100 million yuan	55.2	26.6	27.2	36.2	28.7	−15.07

In the meantime, 981,000 mu of land approved but not supplied was reduced and used, 293,000 mu in available construction land was put to use in Zhejiang during the period 2008–2012.

2.2.3.3 Significant Achievements Have Been Made in Shutting Down Outdated Production Facilities

In 2012, Zhejiang phased out about 900,000t of capacity for backward iron and steel (including stainless steel), 7.71 million tons in cement capacity, 320,000t in papermaking capacity, 7.80 million in standard leather capacity, 1.9 billion meters in printing and dyeing capacity, 600 million meters making capacity, 190,000t in chemical fiber capacity, 2.50 million kVAh in lead storage battery capacity, involving 14 main industries and 2411 enterprises, much exceeding the target and task assigned by the state to Zhejiang for shutting down outdated production facilities. With the shutdown of outdated production facilities, Zhejiang recovered 12,744 mu of land resources, saved energy equivalent to 1.39 million tons of standard coal, reduced sulfur dioxide emissions and COD by about 14,582t and 4809t. In 2013, Zhejiang shut down outdated production facilities in 1658 enterprises in 18 industries, saved energy equivalent to 1.44 million tons of standard coal, reduced nearly 30,000t of main pollutants including COD, overfulfilling the annual task assigned by the state.

2.2.3.4 The Traditional Advantageous Industries Have Been Upgraded in an Accelerated Way

Zhejiang has speeded up the transformation and upgrading of the massive economy towards modern industrial clusters. Zhejiang has continuously pushed forward 42 demonstration and pilot areas for the transformation and upgrading of the massive economy towards industrial clusters, including the Hangzhou Equipment Manufacturing Industrial Cluster.

The leading role of the leading enterprises has been enhanced continually. Zhejiang has cultivated a number of large key advantageous enterprises which are at high technical levels and are highly competitive; a large number of enterprises have pursued international development through mergers, acquisitions and reorganizations, the establishment of overseas production bases and the registration of foreign trademarks.

Zhejiang has transformed and upgraded a number of industries and enterprises through integration of IT applications and informationization, cultivated and developed many excellent solutions for thorough integration of IT applications and informationization, outstanding consulting service agencies and excellent industrial bases for applied electronics; it has built a host of information service platforms for providing such services as product design, quality testing and industrial database sharing. Zhejiang has built a number of industrial demonstration bases for the integration of IT applications and informationization in 20 key industries, including printing and dyeing, papermaking and chemical industries. At the enterprise level, Zhejiang has actively developed enterprise e-commerce, changed the traditional marketing mode, and promoted the transformation from traditional specialized markets to modern commercial and trade logistics, from cash spot transactions to online transactions, and it has cultivated 15 enterprises as the demonstration and pilot enterprises for the integration of IT applications and informationization in headquarters-oriented enterprises.

2.2.3.5 The Strategic Emerging Industries Have Developed Rapidly

In 2013, the strategic emerging industries delivered 274.4 billion yuan added value in Zhejiang, up 8.2% compared with the previous year. The sales value was 1481.6 billion yuan, up 6.8% compared with the previous year. From the industrial perspective, new materials, new energy automobiles and nuclear power-related industries were the top three industries in growth rate, and the added value grew by more than 10%.

The added value of the new material industry was 52.9 billion yuan in 2013, up 12.3% compared with the previous year, ranking it no. 1 among nine industries. The added value of the new energy automobile industry grew by 11.5% compared with the previous year, ranking it no. 2. The total profit of the new material industry grew by 12.2% compared with the previous year, while that of the new energy automobile industry grew by 11.3%, and both figures showed relatively high levels. The added value of the nuclear power-related industry, the energy conservation and environmental protection industry, the high-end equipment manufacturing industry grew by 10.2, 9.0 and 8.8%, each figure of which was higher than the average level of the strategic emerging industries. The total profit of the nuclear power-related industry and the energy conservation and environmental protection industry grew by 12.8 and 12.5%.

Scientific and technological input has increased continuously. The scientific and technological expenditure on the strategic emerging industries in Zhejiang was 26.7 billion, up 13% compared with the previous year, which was at a relatively high level and 6.6 percentage points higher than main business revenue. In the meantime, the scientific and technological expenditure on the strategic emerging industries in Zhejiang accounted for 1.8% of the revenue from the main businesses, 0.5 percentage points higher than the industries above the designated size.

Zhejiang has built a number of new and hi-tech parks and new and hi-tech characteristic industrial bases. The Software Park in the Hangzhou New and Hi-tech Industrial Development Zone has been included by the Ministry of Science and Technology into the first batch of pilot projects for software exportation to Europe and the USA. Zhejiang has been home to 5 national new and hi-tech industrial development zones, 25 provincial new and hi-tech parks, 39 characteristic industrial bases under the National Torch Program and 83 provincial new and hi-tech characteristic industrial bases. They have become the core carriers for developing the strategic emerging industries.

2.3 Energetic Development of the Green Economy

In today's world, the development of a green economy has represented an important trend. Efforts should be made to develop a green economy, overcome the bottleneck from resource shortage, solve the environmental pollution and ecological problems, really build a resource-conserving and environmentally-friendly society, intensify resource conservation and environmental protection; these efforts will be conductive to promoting economic structural adjustment and the transformation of the manner of development to achieve sustainable economic and social development; efforts should also be made in stimulating environmental protection and the development of related industries, fostering new economic growth points and creating jobs, as well as in enhancing the environmental awareness and moral quality of the whole society, guaranteeing the physical health of the people; finally, efforts need to be made in safeguarding the long-term interests of the Chinese nation to leave a good space for the future generations to survive and develop in. In particular, with increasingly acute contrasts between economic development and the resource environment, the development of a green economy is the objective requirement and inevitable choice for overcoming the energy resource bottleneck and achieving modernization.[4]

2.3.1 The Philosophy of the Development of a Green Economy

The green economy is a new economic form which is market-driven and aimed at achieving harmony between the economy and the environment; it is a state of development generated and shown by industrial economy for addressing human needs for environmental protection and health.

On June 12, 2002, the 11th Party Congress of Zhejiang Province officially put forward the strategic goal of building a green Zhejiang and stressed that "the building of a green Zhejiang is a great event for achieving sustainable development in Zhejiang; it is essential to proceeding from the overall interests and long-term development, to giving a higher priority to developing green industries, strengthening environmental protection and ecological development and to speeding up the development of ecological agriculture, ecological industry, ecotourism and the environmental protection industry." A green Zhejiang is Zhejiang's inherent need for, at the new historical stage and with a focus on the well-rounded development of the people, promoting harmony between the people and nature, coordination between material progress and cultural and ethical progress, pursuing a model of sustainable development featuring increased production, higher living standards, and healthy ecosystems.

On January 17, 2005, the then Secretary of the Party Committee of Zhejiang Province, Xi Jinping, pointed out that "the main direction for earnestly developing

[4]Liu Shuo, Developing a Green Economy to Promote Sustainable Development, *Economic Daily*, May 24, 2010.

highly-efficient ecological agriculture, improving the overall capacity of agricultural production and the building of modern agriculture is as follows: extensively promoting a strategic adjustment of the agricultural structure and actively developing highly-efficient ecological agriculture by proceeding from green consumption needs, seeking guidance from the philosophy of agricultural industrialization and economic ecologization, centering on enhancing agricultural market competitiveness and the capability for sustainable development".[5] Energetically developing highly-efficient ecological agriculture, transforming the agricultural growth mode and comprehensively improving the overall capacity of agricultural production are some of the important ways to increase the farmers' income and are also essential for giving full scope to Zhejiang's comparative advantages and speeding up agricultural modernization.

When delivering the report during the 7th Plenary (Enlarged) Session of the 12th Party Committee of Zhejiang Province on June 29, 2010, Zhao Hongzhu called for adapting to the new international trends, earnestly adjusting the economic structure and the energy structure, more rapidly developing the strategic emerging industries and the modern service industry so that the economy becomes green, and winning initiative in international competition; he also called for carefully evaluating the impact of green barriers on Zhejiang's imports and exports, giving full play to the important roles of a green economy and a circular economy in implementing the strategy of "going global", developing green industries, promoting green standards, performing green operations, thus further enhancing international competitiveness; moreover, he advocated earnestly developing highly-efficient ecological agriculture, actively pushing forward the development of ecotourism, more quickly developing a green economy, energetically promoting green production, comprehensively carrying out clean production, cultivating a number of clean production enterprises and green enterprises, vigorously developing pollution-free agricultural products, green food and organic food and making agricultural products of a high-quality and pollution-free. Green development has become an important trend in current international economic development; various countries around the world have established green mechanisms and green standards, and have linked ecological and environmental development with economic and trade issues with a view to maintaining competitive advantages. Zhejiang is an important industrial base in China; only when Zhejiang firmly upholds the philosophy of green development, actively promotes independent innovation, constantly pushes forward the transformation and upgrading of the industrial structure and ventures down the path of green development featuring the guidance of science and technology, resource conservation, clean production, ecological protection and circular economy can Zhejiang seize opportunities and win initiative in a new round of regional competition.

When delivering his report during the 13th Party Congress of Zhejiang Province on June 6, 2012, Zhao Hongzhu vowed to actively develop an ecological economy, strictly carry out the system of environmental access involving space, total quantity and projects, more rapidly shut down outdated production facilities with heavy

[5]Xi Jinping, *Zhijiang Xinyu*, Zhejiang People's Publishing House, 2007, p. 109.

pollution and high energy consumption, energetically develop ecological circular agriculture, green manufacturing, ecological services, ecotourism, leisure and health preservation. The development of a green economy entails long-term painstaking efforts. It is necessary to take developing a green economy as an important means for adjusting the economic structure and transforming the developmental mode, and thus promote rapid and sound economic and social development.

On May 23, 2014, the 5th Plenary Session of the 13th Party Committee of Zhejiang Province adopted the *Decision of the Party Committee of Zhejiang Province on Building a Beautiful Zhejiang and Creating a Good Life*. The decision called for more rapidly establishing and promoting a modern mode of ecological circular agriculture, earnestly developing pollution-free agricultural products, green foods and organic products. It is necessary to develop the modern forest economy to increase the income of forest workers and make them rich in mountainous areas.[6] Only when green development is pursued and great efforts are made to build a resource-conserving and environmentally friendly society and a beautiful Zhejiang with clear waters and green hills can a beautiful, clean and comfortable living environment be created for the people and the future generations.

2.3.2 *Vigorous Development of the Green Industries*

A green economy has an extensive connotation and covers ecological agriculture, ecological industry, ecotourism, eco-forestry and clean energy; it is a benign developmental mode which combines economic reproduction with natural reproduction; it is essential for sustainable development of the human society. To develop a green economy, it is necessary to energetically cultivate the industries with low consumption, slight pollution and high returns, increase the green content in Zhejiang's GDP and thus enhance Zhejiang's comprehensive strength and international competitiveness.

2.3.2.1 Substantially Developing Ecological Agriculture

1. Specifying developmental goals with a great deal of attention from the government

The Party Committee and the People's Government of Zhejiang Province have given great importance to the development of the organic food industry. The *Decision of the Party Committee of Zhejiang Province on Promoting Ecological Development* identified "actively developing pollution-free agricultural products, green foods and organic products, more rapidly building a number of large-scale standardized green food production bases with brands" as one of the important tasks for speeding up the development of an ecological economy. The *Action Plan of Zhejiang Province for*

[6]Shen Zhengxi, The Decision of the Party Committee of Zhejiang Province on Building a Beautiful Zhejiang and Creating a Good Life, *Zhejiang Today*, May 29, 2014.

Developing Ecological Circular Agriculture, released in 2010, stated that by 2015, more than 5500 pollution-free agricultural products, green foods and organic foods would have passed certification, the area of production bases would have exceeded 15 million mu. In 2014, the *Decision of the Party Committee of Zhejiang Province on Building a Beautiful Zhejiang and Creating a Good Life* called for more rapidly establishing and promoting the modern manner of ecological circular agriculture, earnestly developing pollution-free agricultural products, green food and organic products.

2. Enhancing inter-departmental cooperation, making joint efforts to promote development

The development of the organic food industry involves multiple departments and fields. Zhejiang has fully leveraged the evaluation mechanism for the building of an ecological province to incorporate such indicators as "the number of newly-added certified pollution-free agricultural products, the number of certified green foods (including organic foods)" into the evaluation of the annual task for the building of an ecological province, and identify "the planting area involving organic, green and pollution-free products in main agricultural and aquatic products" as the indicator for evaluating the provincial-level ecological county, so as to holistically promote the development of the regional organic food industry. According to the *Several Opinions of the General Office of the People's Government of Zhejiang Province on Actively Promoting the Development of the Organic Food Industry* and the *Developmental Plan of Zhejiang Province for the Organic Food Industry*, the Zhejiang Environmental Protection Bureau, in conjunction with the Zhejiang Provincial Development and Reform Commission, the Department of Agriculture of Zhejiang Province and other relevant departments, strengthens the building of, publicity and training concerning the bases for the production of organic food in order to develop a good pattern with extensive mobilization and strong synergy.

3. Intensifying the process of supervision, enhancing the quality of bases

Actions have been taken to strengthen the management at the source, to survey and monitor the soil environment, to establish a system for the supervision and management of the quality of the soil environment, to reinforce soil pollution prevention and control and to improve the safety of agricultural products. Zhejiang has carried out special work relating to a survey on soil pollution in Zhejiang, a typical survey and risk evaluation of soil pollution in Zhejiang's key areas and the monitoring of such items as soil, water and air in the "vegetable basket" planting bases as well as a survey on the background of bases have been carried out throughout the province. Zhejiang has strictly adopted relevant regulations of the Certification and Accreditation Administration of China to ensure that the certification process is fair and standardized and that it guarantees the healthy development of the organic food undertaking. Measures have been taken to step up supervision over various segments including the production of organic foods, their certification, and market circulation, to strengthen the internal quality control in enterprises to ensure standardized operations and strict management, guarantee the quality and increase the level. In recent years, the percent of passes shown by random inspection of pollution-free agricultural products, green foods and organic agricultural products has always been higher than 98% and no major accidents regarding the quality of the products have occurred in Zhejiang.

4. Consolidating the basic support, guaranteeing healthy development

Zhejiang has vigorously nurtured the enterprises engaged in producing pollution-free agricultural products, green foods and organic foods, increased economic and ecological benefits and consolidated the foundation for market operations. Efforts have been made to support farmers in developing specialized cooperatives of organic agricultural products, give play to the service functions of cooperatives in connecting the leading enterprises, the market and rural households. Zhejiang has built the scientific and technological demonstration parks for the production of organic foods to directly connect the production field and accelerate the conversion of scientific and technological achievements. Steps have been taken to further improve the technical service support system for organic agriculture, speed up the research, development and application of organic food production and processing technologies, establish a network for the promotion of organic agricultural technology, regularly conduct training concerning relevant new technologies and new methods, promptly provide farmers with relevant consulting and other services and offer technical guarantees for the production of organic foods. Actions have been taken to strengthen technical support, enhance cooperation with Zhejiang University, the Zhejiang Academy of Agricultural Sciences and other institutions of higher learning, appoint experts to make scientific and technological breakthroughs, actively encourage scientific and technical personnel to work on the front line and extensively carry out technology promotion, technical training and technical consulting.

2.3.2.2 Promoting Industrial Green Development

1. Boosting green design of products

A mechanism for promoting the ecological design of industrial products which combines policy guidance with market promotion has been built. A number of standards for the green design of products have been made; an evaluation, supervision and management mechanism for the green design of products has been preliminarily built, the green design of products has been piloted, the list of evaluation results concerning the green design of products has been released and a number of nontoxic, harmless or less toxic and harmful raw materials (products) and clean production processes and technologies have been developed, applied and promoted.

2. Widely advocating green production

In 2004, the *Measures of Zhejiang Province for Establishing Green Enterprises (Advanced Enterprises for Clean Production) (Trial)* was issued. In the "733" project for an industrial circular economy, the establishment of more than 80 green enterprises is explicitly required, while this has been identified as one of the criteria for evaluating the completion of the annual task of relevant departments. During the period 2010–2013, Zhejiang also conducted the fourth activity of appraising and selecting the avant-garde enterprises for a green low-carbon economy. With appraisal, enterprises are simulated to optimize and upgrade their industrial structure and pursue green production.

3. Developing the environmental protection industry

Actions have been taken to develop and make famous brand products in the environmental protection industry which are competitive on the market; an industrial system for environmental protection with a rational structure, ready sales and high technical content has taken shape. Emphasis has been placed on promoting the innocent treatment and recycling of domestic garbage, livestock and poultry manure and straw, on prevention and centralized control of pollution from industrial enterprises, comprehensive utilization of wastes, development of water resource recycling technologies and equipment and environmental monitoring instruments and equipment. Zhejiang has built the bases for research, development and testing of key technologies and products for environmental protection, and has developed a number of leading enterprises and groups in the environmental protection industry which deliver more than 100 million yuan in output value.

2.3.2.3 Actively Developing Ecotourism

1. Energetically pushing forward ecotourism projects

Actions have been taken to increase the level of the development of ecotourism in forest parks, natural reserves and scenic spots and to accelerate the development of theme ecotourism, including ancient villages (relics), wetland landscapes, geological parks, religious culture, and leisure and health preservation. Priority has been given to carrying out a number of key projects, including the Baizhangji · Feiyun Lake Ecotourism Resort, the Anji South Ecological Leisure Economic Resort, the China (Dongyang) Paleontological Cultural Industrial Demonstration Area, Quzhou Wetland Park and the Lishui Ecological Leisure and Health Preservation Base; the advantages from abundant high-altitude mountainous resources have been fully leveraged to build a number of high-mountain bases integrating leisure, health preservation, summer resorts and other functions and to enhance the market competitiveness of key ecotourism scenic spots in mountainous areas and their comprehensive capacity for stimulating development on other fronts.

2. Continuously improving the infrastructure of ecotourism

With a focus on the infrastructure in scenic spots, the roads to scenic spots and tourist distribution centers, efforts have been made to intensify the construction of hub-type, functional infrastructures, to improve the integrated layout of the multi-tiered transportation network covering highways, railways, airports and shipping, to promote connection among various means of transport, to build a tourist transportation system that features fast transfer, convenient travel and unimpeded access, to upgrade the roads to the existing and planned high-grade scenic spots, intensify the construction of the roads to connect those scenic spots, and build connections among high-grade scenic spots and among counties.

3. Speeding up the construction of tourism service facilities

Efforts have been made to expedite the construction of the functional infrastructure for tourism, including star hotels, characteristic cultural theme hotels, country hotels and economy chain hotels, as well as to strengthen the construction of tourism information service facilities and to build a smart tourism service system; with the adoption of cloud computing, the Internet of Things and other new technologies, the people can get easy and intelligent access to tourism information through the Internet, mobile Internet and portable devices with Internet access.

2.3.2.4 Vigorously Developing the Forest Industry

1. Improving the investment and financing system

First, innovating the investment and financing mechanism, broadening fund channels, extensively utilizing funds from various sources, including enterprises, credit institutions, individuals and consortiums, to increase financial input, building stable financing channels, gradually establishing a multi-channel investment and financing mechanism in which the government provides guidance, enterprises and forest workers are main players with social participation; second, introducing preferential policies in fund support, optimizing the fund structure, adjusting investment priorities and direction, carrying out the policy of replacing subsidies with rewards, and giving full play to the guiding role of fiscal funds.

2. Enhancing the system for the promotion of technology

First, continuously cultivating fine and new varieties, promoting the cultivation techniques for improved varieties and high quality, transforming the characteristic economic forest industry from a resource-supported one to a science and technology-supported one; second, building a number of bases of production for science and technology demonstration with high scientific and technological content, relying on science and technology to transform the industrial growth mode; third, building and improving a network for the promotion of science and technology, carrying out scientific and technological training among farmers, developing more new-type farmers who can master and apply advanced science and technology in rural areas; fourth, focusing on improving the capacity of technical services of and service standards for industrial associations and specialized economic organizations, enhancing the cohesion among associations and economic organizations, giving shape to the technical service and promotion mode with interaction among multiple parties.

3. Carrying out a brand strategy

First, taking brand building as a strategic measure, intensifying brand innovation, actively carrying out the work on brand identification, appraisal and selection as well as trademark registration relating to the products with potential for market development on the basis of expanding sales and improving the quality, so as to further improve brand value, popularity and influence; second, relying on the existing advantageous resources to establish a number of specialized trading markets for advantageous agricultural and forest products, attending various forms of product exhibitions and trade fairs to enhance brand publicity and product promotion and increase the market share.

2.3.2.5 Speeding up the Development of Clean Energy

1. Actively introducing clean energy policies and regulations

First, further speeding up the making of the *Regulations of Zhejiang Province for the Construction and Protection of Oil and Natural Gas Pipelines*. Convening two meetings of the leading group to solicit opinions on the regulations, with these opinions reported to the provincial people's congress by the provincial people's government, incorporating them into the legislation plan of the year 2014. Second, carrying out the *Regulations of Zhejiang Province for Promoting the Development and Utilization of Renewable Energy*. Conducting law enforcement inspection concerning the *Law of Renewable Energy*, carrying out the survey on the implementation of the *Regulations of Zhejiang Province for Promoting the Development and Utilization of Renewable Energy*. Third, developing and unveiling the *Special Plan of Zhejiang Province for the Network of Natural Gas Pipelines*, the *Plan for the Utilization of Natural Gas*, the *Study of the Natural Gas Utilization Mode*, the *Guiding Rules for the Oil and Natural Gas Safety Assessment Bureau* and the *Feasibility Study of the Regulations of Zhejiang Province for the Protection of Oil and Natural Gas Pipelines*.

2. Adopting multiple measures to promote the development of renewable energy

First, setting up a special fund of 200 million yuan for the development of renewable energy. Second, making all-out efforts to push forward the photovoltaic power generation projects across the province. An installed capacity of 450,000 KW has been completed in the demonstration projects across the province, Zhejiang has ranked among the top two nationwide in the overall progress in such projects as the Golden Sun. Third, adopting every possible means to integrate renewable energy into the grid. Actively coordinating with the Zhejiang Power Grid to take multiple measures for promoting the integration of renewable energy into the grid. Fourth, carrying out technology research and development and project demonstration including the utilization of optothermal energy, ocean tidal current energy and shallow geothermal energy.

3. Energetically pushing forward the development of clean energy projects

Zhejiang has actively sought national policy support in promoting photovoltaic development. The Tonglu Project in Hangzhou, the Binhai Project in Shaoxing and the Hangzhou Bay Project in Ningbo have been included in the first batch of national distributed photovoltaic power generation application demonstration projects, with a gross capacity of 129,000 KW. Moreover, Zhejiang has energetically pushed forward the development of onshore and offshore wind power, approving 9 wind power projects including the GD Power Niujiaoshan Wind Power Project in Fenghua with a gross capacity of 290,000 KW; the total scale of approved wind power across the province has reached 1.15 million KW; 3 wind power projects (100,000 KW), including the GD Power Chashan, have been completed and put into operation, the gross installed capacity of wind power across the province has reached 500,000 KW; the preliminary work on 6 offshore wind power projects, including the GD Power Putuo No. 6 in Zhoushan, the GD Power Xiangshan No. 1 in Ningbo, has been promoted more quickly, with the installed capacity reaching 1.40 million KW, among which the GD Power Putuo No. 6 wind power farm was approved in December, 2013; it is the first offshore wind power project in Zhejiang, its construction began in 2014 and it was partially put into operation in 2015.

2.3.3 *Significant Achievements in the Development of a Green Economy*

2.3.3.1 Green Products Have Developed Rapidly

1. The scale of green agricultural products has increased gradually

During the period 2003–2008, the number of pollution-free agricultural products, green foods and organic foods which were certified increased markedly in Zhejiang. In late 2008, there were 2357 certified pollution-free agricultural products within the effective period in Zhejiang, up 4.16 times compared with 457 at the initial stage of certification in 2003. There were 513 enterprises and 1042 products which effectively used the green food mark, up 3.58 times and 5.51 times compared with 112 green food enterprises and 160 products in late 2003, respectively. There were 41 enterprises and 93 products provided with the organic food mark, up 5.83 times and 10.63 times compared with 6 organic food enterprises and 8 products in late 2003, respectively.[7]

[7]Cai Lei, The Current Development of Pollution-free Agricultural Products, Green Foods and Organic Foods in Zhejiang Province and the Study of Countermeasures, Master's Thesis at Zhejiang University, 2009.

2. The level of industrialization of pollution-free agricultural products, green foods and organic foods has increased continuously

The level of industrialization of pollution-free agricultural products, green foods and organic foods has been on the increase and the agricultural scale has been on the rise in Zhejiang. The base-oriented production, enterprise-based development and brand-based operations and the work on making rural households rich and increasing their income are the leading forms of developing pollution-free agricultural products, green foods and organic foods. As of late 2008, there were 2466 production places for pollution-free agricultural products in Zhejiang; the planting industry covered an area of 8.30 million mu, up 4 times compared with that at the initial stage of certification. Thirty demonstration bases of pollution-free agricultural products and 28 demonstration bases of green (organic) foods have passed the acceptance check. The strength of enterprises involved in certification has been on the increase, there have been 66 leading national and provincial enterprises whose pollution-free agricultural products, green foods and organic foods have been certified and 398 farmers' specialized cooperatives; they have played a very good demonstration effect on promoting agricultural industrialization.[8] As of late 2012, the area of the places of production was 13,437,400 mu in Zhejiang, among which the area of the places of production for pollution-free agricultural products was 12,108,600 mu, the green food monitoring area was 1,217,800 mu and the organic agricultural product area certified by the China Organic Food Certification Center was 111,000 mu. The overall certification work on the places of production for pollution-free agricultural products within grain production function areas and modern agricultural parks has been further strengthened, 48 parks have been newly certified as places of the production of pollution-free agricultural products, the total area involved in the certification of places of production has reached 570,000 mu.

3. The brand influence of green agricultural products has improved constantly

As the enterprises specializing in pollution-free agricultural products, green foods and organic foods have actively participated in agricultural product trade fairs, green food expos and organic food expos, the brand popularity and influence of green agricultural products have improved continuously, more and more certified products have been made available in large supermarkets and on the international market, they have become the widely recognized, safe and high-quality brands of agricultural products. In order to further give play to the guiding role of public brands involving green foods and to increase the level of development of the green food industry in Zhejiang, 12 enterprises were identified as green food demonstration enterprises of Zhejiang Province in 2011, including 10 planting enterprises. These demonstration enterprises have played a benchmarking and guiding role; new philosophies of operations have been adopted, efforts have been made to increase the level of the

[8]Cai Lei, The Current Development of Pollution-free Agricultural Products, Green Foods and Organic Foods in Zhejiang Province and the Study of Countermeasures, Master's Thesis at Zhejiang University, 2009.

standardized production of green foods, making contributions to enhancing the brand influence of green foods in Zhejiang.

2.3.3.2 The Ecological Industry Has Developed Rapidly

1. Green enterprises have been on the increase

In 2002, Zhejiang launched a campaign for turning industrial enterprises into green enterprises. Each year, green enterprises are identified, recognized and examined according to enterprises' economic indicators and environmental protection indicators concerning energy consumption, and relevant preferential policies are introduced for the industrial enterprises granted with the honorary title of green enterprise. In 2011, 90 enterprises were granted with the honorary title of green enterprise. There were 84 and 66 green enterprises (advanced enterprises for clean production) in Zhejiang in 2012 and 2013, respectively.

2. Green technologies have become increasingly mature

The service platforms for energy-saving technology, platforms for industrial scientific and technological innovation, for clean production and other technical resources have been integrated to establish centers for the promotion of green technology in various areas, to improve the system of ecological industrial information and technical services. Green technology research and development bases have been built at the institutions of higher learning; measures have been taken to encourage the establishment of key laboratories and R&D centers for green technology; breakthroughs have been made in the technologies for water pollution prevention and control, comprehensive utilization of water resources, industrial solid waste and sewage sludge recycling technology, high-efficiency energy-saving technology and green chemical technology.

3. The energy conservation and environmental protection industry has begun to take shape

In 2013, Zhejiang ranked No. 2 nationwide in the scale of the energy conservation and the environmental protection industry. In 2011, the added value of the energy conservation and environmental protection industry reached 77.6 billion yuan in Zhejiang, accounting for 31.1% of that of the strategic emerging industries in Zhejiang; its total profit was 23.1 billion yuan, accounting for 26% of the total profit of the strategic emerging industries in Zhejiang, and it has become an important hi-tech industry and strategic emerging industry.

2.3.3.3 Some Gratifying Achievements Have Been Made in Ecotourism

1. A number of ecotourism brands have been built

Zhejiang has actively pushed forward the building of a strong ecotourism province and has made a number of green Zhejiang tourism brand products. As of late 2013, with respect to the tourist resort brands, there was 1 national tourist resort—the Hangzhou Zhejiang National Tourist Resort and 29 provincial tourist resorts. With regard to the scenic area brands, there were 19 national scenic areas and 40 provincial scenic areas. With respect to the forest park brands, there were 37 national forest parks and 72 provincial forest parks. With respect to the natural reserve brands, there were 10 national natural reserves and 9 provincial natural reserves. With respect to the geological park brands, there were 3 national geological parks, 4 provincial geological parks and 1 world-class geological park.

2. The ecological and environmental resources have been effectively protected

The development of ecotourism is conducive to protecting the wetland, biology, water areas and other ecological resources. The resources of tourism in natural reserves have been protected better, there are 10 national natural reserves and 9 provincial natural reserves; the resources of tourism in forest parks have been protected better; there are 37 national forest parks; the resources of tourism in wetland parks have been protected better, there are 7 national wetland parks and 10 provincial wetland parks. In Xixi Wetland in Hangzhou, with a focus on the development of ecotourism, plant landscape gives prominence to nature and rustic charm; more than 300 mu of prune trees have been restored on the Xixi Prune Tree Bank, and waterfowl areas have been built.

3. Ecotourism has delivered significant economic returns

Ecotourism has become an important form of the development of tourism in Zhejiang in the past 10–20 years; the economic returns from ecotourism are very apparent. During the period 2002–2013, the total amount of revenue from tourism kept increasing, the economic aggregate of tourism soared from 71.07 billion yuan in 2002 to 553.6 billion yuan in 2013; the proportion of revenue from tourism in Zhejiang's GDP was on the increase, and that proportion was only 9.3% in 2002, while it reached 14.7% in 2011.

2.3.3.4 The Building of Forest Product Brands Has Been Accelerated

1. The level of the quality of safety has increased significantly

During the period 2005–2011, random inspection was conducted with respect to 7357 batches of edible forest products, including edible bamboo shoots (bamboo shoots in spring and winter, rhizome shoots), dry fruits (hickory nuts, Chinese torreya and Chinese chestnuts); the percent of passes shown by random inspection on the production link of edible forest products in Zhejiang increased from 92.6% in 2005 to 96% in 2011. As of October 25, 2012, local authorities in Zhejiang had put in place 1284 quality safety supervisors for edible forest products in towns (sub-districts), there were 2370 production entities specializing in edible forest products (enterprises and specialized cooperatives) in Zhejiang, 100% of the persons in charge received training, were interviewed, signed a letter of commitment and experienced the establishment of production record files; a rectification plan was developed for solving the identified salient problems concerning pesticide residue in fresh shoots and 100% of the hidden troubles were rectified; 1424 batches of edible forest products underwent random inspection at the municipal and county levels and the percent of pass was 97.3%. In June, 2013, Zhejiang conducted the first monitoring of the quality of the safety of edible forest products, 455 batches of edible forest products underwent random inspection, 449 batches passed that inspection, and the percent of pass was 98.7%.

2. Brand building has been intensified incessantly

Zhejiang has placed great importance on the building of forest product brands and has made great efforts to promote scientific and technological services and the standardization and industrialization of the forest industry; Zhejiang has actively cultivated forest brands and has more rapidly developed forest foods, green foods, organic forest products and the forest products for geographical indication registration and protection; Zhejiang has intensified brand marketing and promotion, brand supervision and management, and it has built a number of brand forest products which are distinctive, enjoy stable quality, have a good reputation and a high market share; Zhejiang has made active progress in the building of forest brands. Zhejiang appraised, chose and identified five batches of famous brand forest products from 2008 to 2013, and 199 forest products were granted the title of Zhejiang Famous Brand Forest Product in various parts of Zhejiang.

2.3.3.5 The Clean Energy Industry Has Developed Extensively

1. The level of use of clean energy has increased continuously

The level of use of clean energy (including hydropower energy from other provinces) in Zhejiang was about 28.57 million tons of standard coal in 2013, up 7.6% compared with 2012; it is estimated that it accounted for 15.1% of the total energy consumption in Zhejiang. Among clean energy, the level of use of natural gas was 5.55 billion cubic meters, up 17.6% compared with 2012, while that of wind power was 1 billion KWH, up 28.2% compared with 2012; solar power generation grew immensely, its level of use reached 290 million KWH, up 314.3% compared with 2012; the level of use of marsh gas and other methods of use was 1.55 million tons of standard coal, up 3.3% compared with 2012. The level of use of non-local hydropower was 20 billion KWH, up 23.3% compared with 2012.

2. The installed capacity of clean electric power has increased steadily, the electric power structure has been further optimized

As of late 2013, the installed capacity of clean energy was 24.12 million KW in Zhejiang, accounting for 37.0% of the total installed capacity of electric power in Zhejiang, up 2.6 percentage points or 2.58 million KW compared with the previous year. The newly-added combined heat and power generation units based on natural gas amounted to 2.05 million KW, accounting for 76.5% of the newly-added units in Zhejiang. Such combined heat and power generation units as Datang Jiangshan and Zheneng Changxing No. 1 unit were completed and put into operation; a number of major clean energy projects, including Fangjiashan Nuclear Power, Sanmen Nuclear Power Phase I and Xianju Pumped Storage, were pushed forward in an orderly fashion.

3. The construction of natural gas storage and transportation facilities has been steadily promoted, the capability for storage and transportation has been further enhanced

A 760-km provincial network of a natural gas pipeline and urban pipeline network were newly added. As of late 2013, there were 834 km of long-distance natural gas transportation pipelines in the provincial pipeline network and 17,955 km of urban pipeline network in Zhejiang, up 38.8 and 3.0% compared with the previous year. The provincial urban natural gas emergency storage capacity exceeded 10 days. All-out efforts were made to push forward the project for the Ningbo-Taizhou-Wenzhou natural gas pipeline; construction fully began with respect to the Jinhua-Lishui-Wenzhou natural gas pipeline project. The Xiaoshan Contact Station in the supporting pipeline project at the Shaoxing-Hangzhou-Jiaxing section of the No. 2 West-East Natural Gas Transmission Line was completed and put into operation; Nanwang Changshan and Yiwu natural gas branches were basically completed and can be put into operation.

2.4 Vigorous Development of the Circular Economy

The development of the circular economy is a systematic project involving the natural, economic and social fields, the production, circulation and consumption links, areas, industries and enterprises. In an effort to make ecological progress and build the pilot province for a circular economy, with a focus on efficient and circular utilization of resources, the driving forces are derived from scientific, technological and institutional innovations to more rapidly shape the industrial structure, growth mode and consumption mode conducive to saving energy and resources and protecting the ecological environment, and efforts are made to explore the path of developing a circular economy with Zhejiang's characteristics, with a view to achieving sustainable economic and social development and harmonious development of the people and nature.

2.4.1 *The Philosophy of the Development of a Circular Economy*

On May 11, 2005, the then Secretary of the Party Committee of Zhejiang Province, Xi Jinping, pointed out, "The development of a circular economy is an important carrier for venturing down the new-type industrialization path and is also essential for fundamentally transforming the economic growth mode; it is necessary to deeply study the technical support and guarantee for developing a circular economy, develop the science and technology for making production clean, the environment pollution-free, and for saving energy, as well as carrying out information consulting, technology promotion and training service in this respect".[9] A circular economy is an important force for the industrial upgrading and transformation of China's economic developmental mode; it is also one of the important means for China to save energy and reduce emissions.

On November 6, 2005, the 9th Plenary Session of the 11th Party Committee of Zhejiang Province adopted the *Recommendations of the Party Committee of Zhejiang Province for Developing the 11th Five-Year Plan of Zhejiang Province for National Economic and Social Development*. The recommendations called for earnestly developing a circular economy, strengthening the improvement in local regulations and in the evaluation system for the development of a circular economy, working out the master plan for conservation and comprehensive utilization of resources and special plans for a circular economy including the recycling of renewable resources and actively promoting waste recovery and recycling, improving the renewable resource recovery, processing and utilization system. A circular economy is different from the conventional manner of economic development; it cannot be developed merely by relying on the market mechanism, and it is necessary to make laws and regulations for specifying the code of conduct governing different players within the society, and to rely on government forces to promote its development. On September 26, 2008, when delivering the report in the 1st Plenum of the 4th Plenary Session of the 12th Party Committee of Zhejiang Province, Zhao Hongzhu vowed to energetically develop a circular economy, actively push forward clean production, improve the work on building the pilot province for a circular economy, become the first to turn Zhejiang into a national demonstration area for developing a circular economy, more quickly build the industrial structure and manner of consumption which are beneficial to saving energy and protecting the ecological environment; he also vowed to promote the building of a pilot province for a circular economy, vigorously develop that kind of economy, ensure that more than 94% of industrial solid wastes would be utilized comprehensively, 72% of the water in the enterprises above the designated size would be recycled and that more than 85% of the straw would be utilized comprehensively in Zhejiang. As an economically large province and a small province regarding resources, Zhejiang is actively exploring a path of developing a circular economy with local characteristics.

---

[9]Xi Jinping, *Zhijiang Xinyu*, Zhejiang People's Publishing House, 2007, p. 140.

On June 30, 2010, the 7th Plenary Session of the 12th Party Committee of Zhejiang Province adopted the *Decision of the Party Committee of Zhejiang Province for Promoting Ecological Development*. The decision called for more rapidly developing ecological circular agriculture and moving faster to promote an ecological mode of stereoscopic agricultural circulation with a combination of cropping and raising, of farming and animal husbandry, of forest and animal husbandry, pushing forward a number of ecological circular agricultural demonstration areas and projects, actively developing an industrial circular economy, strengthening the building of key enterprises, demonstration parks and bases for that economy, gradually promoting its development in the chemical, petrochemical, papermaking and other fields across the board, forming a circular economic industrial chain, thus going down the path of building the province through ecological development, deeply pushing forward the development of an ecological province and energetically increasing the level of ecological development.

On November 19, 2011, Zhao Hongzhu vowed to promote the building of a pilot province for a circular economy, to continue implementing the "991" action plan for a circular economy, push forward the "733" project for an industrial circular economy, the "4121" project for ecological circular agriculture and six projects for intensive land use, and to propel the building of 25 pilot bases for a circular economy, drive the reduction of resource consumption and resource recycling, make enterprise production clean and the regional economy ecological, as well as more rapidly build a circular economic industrial system. The directions for Zhejiang's optimization and adjustment of its industrial structure are economic circulation, industrial ecologicalization, industrial symbiosis, clean production, resource recycling and waste reduction.

On June 6, 2012, Zhao Hongzhu stressed that we should earnestly develop circular ecological agriculture, green manufacturing, ecological service and ecotourism, as well as leisure and health preservation industries. Zhejiang has comprehensively expanded the building of a pilot province for a circular economy, carried out the "991" action plan for a circular economy and it has taken more actions to make new achievements in developing a circular economy. Zhejiang has rationally developed and effectively protected resources and the environment, energetically developed a circular economy, moved faster to build a sustainable national economic system with high returns, low consumption and fewer emissions and a resource-conserving and environmentally-friendly society; this is essential for carrying out the Scientific Outlook on Development and is fundamental to overcoming the contradictions between economic development and population, resource and the environment.

On May 23, 2014, the 5th Plenary Session of the 13th Party Committee of Zhejiang Province adopted the *Decision of the Party Committee of Zhejiang Province on Building a Beautiful Zhejiang and Creating a Good Life*. The decision called for actively promoting the development of a circular economy, actively pushing forward the circular renovation of parks and comprehensively increasing the level of comprehensive utilization of renewable resources. In Zhejiang, significant achievements have been made in developing a circular economy, a circular industry has become a large one, marked improvement has been made in the efficiency of resource utilization and the level of renewable resource utilization, the emission of main pollutants has been effectively controlled, the philosophy of green consumption has been deeply rooted among the people, the environment for development has been further optimized, a circular economic developmental mode with Zhejiang's characteristics has basically taken shape.

2.4.2 Four Major Measures for Promoting the Development of a Circular Economy

2.4.2.1 Vigorously Promoting a Circular Development of the Primary, Secondary and Tertiary Industries

1. Pushing forward the development of a circular industrial economy

Actions are being taken to thoroughly carry out the "733" project for a circular industrial economy, build 30 demonstration parks and 300 demonstration enterprises for the circular industrial economy in seven key fields in Zhejiang, gradually promote the circular industrial economy in the chemical, printing and dyeing, pharmaceutical, papermaking, metallurgical, building material and other fields across the board in Zhejiang, and give shape to the circular industrial chain.[10] Efforts are being made to reduce resource consumption in key industries, including the chemical, iron and steel, electric power, chemical, building material, nonferrous metal, textile printing and dyeing and papermaking industries. Great efforts are being made to build a circular industrial chain in the industrial sector, and establish circular industrial chains with local characteristics in industrial clusters, industrial agglomeration areas, development zones (industrial parks), to promote waste recycling within enterprises, among enterprises and among parks. The existing and planned large oil refining and ethylene projects and fluorine chemical bases are relied upon to vigorously develop the downstream industries; vigorous actions are being taken to build a number of circular economic industrial chains in the petrochemical industry and its downstream fields including the textile, plastics, fine chemical, chemical fiber and pharmaceutical fields. The advantages of Zhejiang's industrial clusters are being fully leveraged to promote further integration between renewable resource recycling and regional characteristics.

2. Promoting the development of circular ecological agriculture

Intensified efforts are being made to promote resource-saving agricultural technologies and agricultural waste recycling technologies, actively boost wide application of organic fertilizers, push forward comprehensive utilization of crop straw, livestock excrement and by-products of agricultural product processing, agricultural and forest

---

[10] Xie Liqun, Developing an Ecological Industry to Promote Industrial Transformation, *Zhejiang Daily*, October 12, 2010.

remainders and abandoned agricultural film. Actions are being taken to energetically push forward agricultural standardization and clean production, vigorously promote highly-efficient ecological agricultural developmental modes, including facility cultivation, ecological breeding, multistoried cropping and raising, integration of cropping, raising and processing, leisure agriculture, and the farming systems, including a combination between cropping and raising, between grain crop and cash crop, between agricultural machinery and agronomy, intercropping, relay intercropping and crop rotation; great efforts are being made to build circular ecological agricultural industrial chains in such fields as ecological planting and breeding with a combination a agriculture (forestry) and animal husbandry, the development and application of biomass resources and fine and deep processing of agricultural products; a circular ecological agricultural production operation system has been established and put into operation. Steps are being taken to promote the development of ecological marine fishery, to build up scallop and algae cultivation and propagation areas, to explore and promote ecological propagation and cultivation techniques and modes at the scallop-algae composite and multi-trophic level and to boost the absorption and utilization of carbon dioxide and terrestrial organic matters in water bodies.

3. Speeding up the development of a circular service industry

Actions are being taken to actively, steadily and in an orderly fashion promote the building of the industrial bases for comprehensive utilization of renewable resources, regional distribution markets and specialized sorting centers; actions are also being taken to develop the recovery networks at communities, villages and towns, online waste recovery, to support the enterprises with the national and provincial qualification of recovery and dismantlement of waste household appliances and scrapped cars at fixed points to promote inter-regional development of the modern chain network. Great efforts are being made to optimize the structure of transportation, actively develop the means of transportation with low energy consumption and less pollution and to push forward the development of the low-carbon transportation service industry. Steps are being taken to encourage the development of the service sectors relating to the circular economy, including training, certification, investment and consulting, actively promote the development of the energy contract management service, ecological logistics, green tourism, environmental service and circular economic science and technology service industries. Actions are also being taken to guide citizens to consciously boycott white pollution and excessive packaging, encourage the use of the products made of recycled materials, green products, the products with an energy efficiency label, the products certified for water saving and the products with an environmental mark.

2.4.2.2 Continuously Increasing the Level of Comprehensive Utilization of Resources

1. Strengthening the recycling of renewable resources

A renewable resource recycling system has been established and improved. Actions are underway to comprehensively promote the recycling of renewable resources, including waste metals, waste plastics, waste household appliances and electronic products, waste cars, waste tyres, waste textiles, waste bamboo and wood products, waste paper, waste oils and fats, and to turn "urban minerals" into important sources of resources in Zhejiang. The urban and rural garbage classification, collection and disposal system has been established and improved; efforts are underway to actively promote the rational layout and healthy development of waste incineration power plants in Zhejiang, comprehensively push forward the recycling and harmless treatment of sludge, urban kitchen wastes, agricultural and rural wastes and to prevent secondary pollution from resource recycling.

2. Encouraging the comprehensive utilization of water resources

Actions are being taken to vigorously promote industrial water saving and encourage industrial enterprises to push forward deep treatment and recycling of wastewater and the recycling of cooling water. Saving water is energetically promoted in agriculture and rural areas; efforts are being made to strengthen the renovation of the saving of water in agricultural sprinkling irrigation and trickle irrigation in rural irrigation areas. Saving water is actively promoted in urban areas, the maintenance of the network of the urban water supply pipeline has been intensified, the leak rate in the network of the urban water supply pipeline has been reduced. The use of recycled water is encouraged; the collection and storage of rainwater, the reuse of reclaimed water and water supply of a different quality are actively promoted in the areas where water is in great demand and there is a shortage of water resources. Such industries as municipal environmental sanitation, ecological landscaping and car washing are guided and encouraged to use recycled water; the building of a regional system of rainwater collection and use and a system for the reuse of recycled water is being encouraged in large urban public buildings and residential quarters. The use of seawater and subsea water is actively guided and encouraged; the development of the seawater desalination industry has been accelerated.

3. Promoting the development and utilization of clean energy

The development of clean energy is planned scientifically; actions are underway to build clean energy demonstration bases, continuously enhance the capacity for a supply of clean energy, and to optimize the structure of energy consumption. The construction of nuclear power bases is being promoted solidly. Measures are being taken to encourage the development and utilization of renewable resources, including wind energy, biomass energy, solar energy, ocean energy and shallow geothermal energy; building a number of coastal and inland renewable resource development and utilization bases is also being promoted. Efforts are being made to actively seek natural gas sources, volume and shares, to expand the utilization of natural gas, and gradually reduce the dependence on coal, oil and other traditional energy sources.

2.4.2.3 Vigorously Promoting the Development of Three Main Carriers

1. Making great efforts to boost the development of the circular economic industrial agglomeration areas

Actions are being taken to speed up the planning and development of the Taizhou Bay circular economic industrial agglomeration area, encourage first-ever implementation of pilot programs, turn it into the provincial even a national circular economic development demonstration area. The existing industrial bases and comparative advantages in different areas are relied upon to build a number of pilot bases for a circular economy. Actions are taken to comprehensively promote the ecological renovation of development zones (parks), and establish a number of industrial circular economic demonstration areas; with a focus on developing the function areas of grain production, modern agricultural parks and leading agricultural industries, actions are being taken to build a number of ecological circular agricultural demonstration areas; with an emphasis placed on ecological logistics and ecotourism, great efforts are being made to build a number of circular economic demonstration areas for the service industry.

2. Actively cultivating the circular economic demonstration enterprises

With a focus on the metallurgical, electrical power, pharmaceutical, papermaking, building material, light textiles and other industries, efforts are being made to cultivate a number of clean production demonstration enterprises; with priority given to the enterprises with heavy energy and electric power consumption, a batch of energy-saving demonstration enterprises are being cultivated; with a focus on the industries with heavy water consumption including electric power, textiles, papermaking and chemical industries, a number of water-saving demonstration enterprises are being developed; a batch of seawater desalination demonstration enterprises are cultivated by focusing on island areas and coastal areas; a number of demonstration enterprises for recovery and comprehensive utilization of resources are being cultivated by centering on the metallurgical, petrochemical, building material, electric power, papermaking, printing and dyeing and leather industries as well as other industries; a number of circular ecological agricultural demonstration enterprises are being cultivated by focusing on agriculture-led industrial development, fine and deep processing of agricultural products and the recycling of agricultural wastes. The building of green enterprises and conservation-oriented enterprises is being promoted earnestly, and the level of circular economic development is taken as an important basis for identifying the outstanding enterprises through appraisal.

3. Carrying out key projects for a circular economy

Continued efforts are underway to implement the "991" action plan for a circular economy in Zhejiang; each year, more than 100 key projects for a circular economy are carried out on a rolling basis in key fields, including the recycling of renewable resources, the utilization of kitchen wastes, sludge recycling, reuse of recycled water, seawater desalination, waste heat utilization, garbage power generation and the recycling of agricultural wastes.

2.4.2.4 More Rapidly Making Breakthroughs in the Key Technologies Relating to Three Main Fields

Actions are underway to more quickly make breakthroughs in the key technologies for the comprehensive utilization of water resources. Priority is given to researching and developing the key technologies for sea-water desalination and its comprehensive utilization, for the reuse of recycled water, the efficient treatment of industrial wastewater and municipal sewage, the control of pollution in the main water systems and ecological restoration in water areas.

Steps are being quickened to make breakthroughs in the key technologies for energy conservation and the utilization of renewable energy. Priority is given to researching and developing the efficient low-cost technologies for electromechanical energy saving, building energy saving, new energy, green lighting, solar optothermal and photovoltaic energy, wind power generation, biomass power generation, the utilization for geothermal energy, and the technologies for efficient and environmentally-friendly power generation based on domestic garbage incineration.

Efforts are underway to more quickly make breakthroughs in the key technologies for the comprehensive utilization of solid wastes. Emphasis is placed on researching and developing the key technologies for harmless treatment and recycling of industrial wastes, sewage sludge, urban domestic garbage and kitchen wastes, the technologies for the harmless treatment of toxic and harmful gases from power generation based on domestic garbage incineration, the technologies for comprehensive utilization, including livestock and poultry breeding excrement, crop straw, rural clean energy and organic fertilizer processing and application.

2.4.3 The Preliminary Effect in the Development of a Circular Economy

2.4.3.1 Building a Number of Circular Economic Industrial Chains

1. Building a number of circular economic industrial chains

A number of circular industrial chains with characteristic industries have been built in industrial clusters, industrial agglomeration areas, development zones (industrial parks) and pilot bases; the vertical extension and horizontal expansion of the circular industrial chains have been promoted; the integration of related industries has been intensified; the circular economic industrial chains have been first built in the petrochemical, iron and steel, pharmaceutical, textile, building material fields as well as in other fields. Based on the development of a number of petrochemical industrial parks, including the Ningbo Petrochemical and Taizhou Large Petrochemical Project, actions have been taken to push forward the refining-chemical integration project, enhance the oil refining and the capacity for the production of ethylene, extend and develop the petrochemical industrial chains covering synthetic resin, synthetic rubber, polyester, polyurethane, specialty fiber and polycarbonate. With a focus on the recycling of waste resources, on the reuse of recycled water and waste heat utilization, the circular industrial chains for the gradient utilization of energy and mass have been built in such industries as textile printing and dyeing, leather making and papermaking; the industrial chains for renewable resource recovery and utilization have been built by focusing on the development and utilization of "urban minerals", including waste metals and waste plastics.

2. Building a number of modes for the development of a circular agricultural economy

With a focus on the leading agricultural industries, actions are being taken to actively promote highly-efficient ecological developmental modes, including ecological cultivation, multistoried cropping and raising, integration of cropping, raising and processing, leisure agriculture, and the farming systems, including combination of cropping and raising, of grain crop and cash crop, of agricultural machinery and agronomy, intercropping, relay intercropping and crop rotation; efforts are being made to build modes for the development of circular ecological agriculture in such fields as ecological planting and breeding with a combination of agriculture (forestry) and animal husbandry, the development and application of biomass resources, fine and deep processing of agricultural products and an operational system for circular ecological agricultural production has been established and put into operation.

2.4.3.2 Effectively Increasing the Level of the Comprehensive Utilization of Resources

1. Increasing the level of renewable resource recovery and utilization

In 2012, more than 200 waste metal processing enterprises in Yongkang City utilized more than 1.50 million tons of renewable resources, the municipal government built three pollution-free, environmentally-friendly sorting processing centers, more than 85% of the waste materials made at the local level and recovered in various areas were sorted in a pollution-free way, 800,000t of iron and steel scrap, more than 100,000t of waste paper and plastics and 500,000t of waste of nonferrous metals were sorted. The enterprises engaged in power generation based on urban domestic garbage incineration increased year by year and their number rose from 25 in 2009 to 38 in 2013. As of late 2013, 38 enterprises engaged in power generation based on urban domestic garbage incineration in Zhejiang had generated 2.98 billion KWH of electricity, up 30.7% compared with the previous year, and they had incinerated and disposed of about 8 million tons of garbage; 5 sludge incineration power generation enterprises had generated 760 million KWH of electricity, up 4.1% compared with the previous year, and they incinerated 1.36 million tons of sludge. In 2013, the daily handling capacity for kitchen wastes reached about 240t in Ningbo, 80% of the kitchen wastes were disposed of in a unified way in six districts in Ningbo and the Ningbo National New and Hi-tech Zone, putting Ningbo in the leading position nationwide in this regard.

2. Increasing the rate of comprehensive utilization of water resources to some extent

According to the communiqué of Zhejiang on water resources, the overall level of water use in Zhejiang was 21,707 million m^3 in 2009, including 8.32 billion m^3 of farmland irrigation water, accounting for 38.3% of the total water supply. In 2010, the level of Zhejiang's total water use was 22,008 million m^3, including 8061 million m^3 of farmland irrigation water, accounting for 36.6% of the total water supply, the level of use of farmland irrigation water decreased gradually, the rate of water resource utilization increased markedly; in 2013, the level of the reuse of industrial water in parks was 5.33 billion tons, up 400 million tons or 8.1% compared with the previous year; the rate of the reuse of industrial water was 71.4%, up 1.2 percentage points compared with the previous year.

3. Continuously increasing the level of the use of clean energy

The level of the use of clean energy (including hydropower from other provinces) in Zhejiang was about 28.57 million tons of standard coal in 2013, up 7.6% compared with the previous year; it is estimated that it accounted for 15.1% of Zhejiang's total energy consumption. Specifically, the level of the use of natural gas was 5.55 billion m^3, up 17.6% compared with the previous year; the level of the use of wind power was 1 billion KWH, up 28.2% compared with the previous year; the generation of solar power grew substantially, the level of its use reached 290 million KWH, up 314.3% compared with the previous year; marsh gas and other utilization modes were equivalent to 1.55 million tons of standard coal, up 3.3% compared with the previous year.

2.4.3.3 Cultivating a Number of Circular Economic Demonstration Points

1. Increasing the scale of circular economic demonstration parks

In 2010, Zhejiang preliminarily identified the first batch of 10 demonstration parks in the "733" project of Zhejiang for the circular industrial economy; in 2011, 5 parks, including the Quzhou New and Hi-tech Industrial Park in Zhejiang, were rated as Zhejiang's second batch of circular industrial economic demonstration parks; in

2012, 10 parks, including the Jinhua Economic and Technological Development Zone, were chosen from among 15 parks for which application had been filed and rated as Zhejiang's third batch of circular industrial economic demonstration parks; in 2013, 8 parks, including the Zhejiang Chun'an Economic Development Zone, were rated as Zhejiang's fourth batch of circular industrial economic demonstration parks.

2. Increasing the number of circular economic demonstration enterprises

In 2010, Zhejiang preliminarily identified the first batch of 100 demonstration enterprises in the "733" project of Zhejiang for the circular industrial economy; in 2011, 76 enterprises, including Zhejiang NHU Company Ltd., were rated as Zhejiang's second batch of circular industrial economic demonstration enterprises; in 2012, 59 enterprises, including Zhejiang Wazam New Materials Co., Ltd., were chosen from among 105 enterprises for which an application had been filed and rated as Zhejiang's third batch of circular industrial economic demonstration enterprises; in 2013, 46 enterprises, including Hangzhou Yokohama Tyre Co., Ltd., were rated as Zhejiang's fourth batch of circular industrial economic demonstration enterprises.

2.5 Vigorous Development of a Low-Carbon Economy

With the goal of becoming the first to build a moderately prosperous society in all respects at a higher level and more rapidly promoting the basic realization of modernization, based on the main driving forces from the government-promoted and market-driven development with the active role of enterprises and public interaction, Zhejiang relies on institutional, scientific and technological innovations, pushes forward the low-carbon adjustment of the industrial structure, enhances the efficiency of energy utilization, optimizes the energy structure, increases carbon sink storage and transforms the outlook on life and consumption, so as to build an economic system with Zhejiang's characteristics in which economic development is decoupled from carbon emissions, and achieve coordinated development among the population, the resources, the environment and the economy and the society.

2.5.1 The Developmental Philosophy of a Low-Carbon Economy

The *Recommendations of the Party Committee of Zhejiang Province for Developing the 11th Five-Year Plan of Zhejiang Province for National Economic and Social Development*, adopted during the 9th Plenary Session of the 11th Party Committee of Zhejiang Province on November 6, 2005, called for controlling the projects with heavy energy consumption, prohibiting the high-pollution projects, phasing out the outdated production processes and equipment which waste resources and pollute the environment, supporting and developing the resource-conserving industries with a smaller amount of pollution, low consumption and high returns. Actions are being taken to promote the environmentally-friendly building materials with high performance and low material consumption, build the energy and land-saving residential and public buildings. The environmentally-friendly means of transportation are developed and utilized, and a green transportation system is being built. Developing a low-carbon economy is essential for Zhejiang to cope with the challenges from a shortage of fossil energy and global warming and to carry out its strategy for sustainable development.

On October 21, 2009, Zhao Hongzhu vowed to improve the energy conservation, emission reduction monitoring, supervision and evaluation system, to strengthen ecological renovation of development zones (parks), more rapidly cultivate a circular economy, a low-carbon economy and an ecological economy, improve the mechanism for ecological environmental compensation, further intensify transfer payment for the areas of ecological functions, earnestly push forward ecological cultural development, speed up the formation of the industrial structure, growth mode and consumption mode conducive to saving resources and energy and protecting the ecological environment.

On June 30, 2010, the 7th Plenary Session of the 12th Party Committee of Zhejiang Province stressed that actions had to be taken to actively develop low-carbon technologies, to comprehensively promote energy conservation in various fields of the national economy and various links of production and life, to focus on phasing out and renovating the equipment with high energy consumption in the electric power, iron and steel and nonferrous metal fields as well as in other fields, to intensify the utilization of industrial waste heat, to enhance the efficiency of energy utilization and to further reduce the energy consumption per unit of GDP as well as to research and develop the technologies for carbon capture and carbon sequestration, and to continuously lower the intensity of carbon dioxide emissions per unit of GDP. The government is in the process of establishing special funds for energy conservation, adopting interest subsidies, allowances, rewards and other means for guiding enterprises and social capital to support the research and development of energy-saving technologies and technical renovation and to improve the efficiency of energy utilization.

On June 6, 2012, Zhao Hongzhu vowed to comprehensively push forward energy conservation and consumption reduction, give prominence to energy conservation in such key fields as industry, transportation, public institutions and residents' life and the key energy-consuming enterprises, actively develop green buildings and promote pilot and demonstration work on low-carbon development.

On May 23, 2014, the 5th Plenary Session of the 13th Party Committee of Zhejiang Province stressed that actions had to be taken to develop a green circular low-carbon economy, more rapidly phase out the outdated production facilities with high energy consumption and high emissions, actively develop new energy, such as solar energy and wind energy, and renewable energy; they also stressed that it was necessary to encourage enterprises to develop green low-carbon products, develop and implement a green purchasing consumption policy, build a low-carbon economic developmental mode based on low energy consumption, low pollution and low emissions and to energetically push forward the building of an integrated low-carbon transportation network, effectively reduce energy consumption in road transportation and greenhouse gas emission.[11] A low-carbon economy is the way to achieve sustainable development of the human society and transformation towards a low-carbon economy has become the general trend of the world economic development. As a coastal economically developed area in China, Zhejiang is able and duty-bound to, and it is also necessary for, Zhejiang to take actions as early as possible to move towards the new economic operational mode—the low-carbon economy, become the first to build a large low-carbon province, and provide a model for the rest of the country.

2.5.2 The Building of a System for Low-Carbon Development

2.5.2.1 The Building of a Low-Carbon Industrial System

1. Promoting low-carbon upgrading of industries

The economic servitization is an important trend in the adjustment of the world's industrial structure and is also an important way to achieve carbon reduction in the industrial structure. With respect to the development of Zhejiang's service industry, more rapidly developing the modern service industry will push forward the coordinated development among industrial transformation, upgrading and the productive service industry, the interactive development among the optimization, upgrading of the resident consumption structure and the consumer service industry. Prominence is given to cultivating and developing innovative modern service sectors, including e-commerce, research, development and design, and digital media, and increasing the speed and enhancing the quality in productive service sectors, including financial insurance, modern logistics, information services, education in science and technology, with efforts made to increase the proportion of the service industry. In the meantime, actions are being taken to actively transform and upgrade the massive economy towards the modern industrial clusters, integrate and improve the industrial chain, value chain and supply chain, and achieve low-carbon upgrading through the scale economy, scope economy and large circular economy in industrial clusters.

2. Renovating the traditional industries with high energy consumption

Zhejiang is carrying out 11 main plans for industrial transformation and upgrading, covering the automobile, iron and steel and petrochemical industries, to promote interactive development between the traditional industries and the new and hi-tech industries. Efforts are being made to earnestly renovate the industries with high

[11] Shen Zhengxi, The Decision of the Party Committee of Zhejiang Province on Building a Beautiful Zhejiang and Creating a Good Life, *Zhejiang Today*, May 29, 2014.

energy consumption, reduce the output of the products with high energy consumption and the embodied energy in products.

3. Cultivating the new energy industry

Developing new energy is an important measure for reducing carbon, so it is necessary to cultivate the new energy industries (consistent with the requirement of low energy consumption), such as wind power and photovoltaic power, by taking them as the emerging industries, so as to promote industrial transformation and upgrading. It is essential to plan and build the new energy industrial bases to cultivate and develop low-carbon industrial clusters by focusing on new energy and new energy equipment manufacturing industries.

4. Developing the carbon sink industry

With respect to natural carbon sink development, emphasis is placed on protecting and developing the ecological system and enhancing the capacity of absorption of natural carbon sink. Actions are being taken to push forward the key forest projects, covering ecological public-welfare forests, coast protection forests, broadleaf improvement and protection of wild animals and plants, to strengthen water quality monitoring, mitigate the pollution of areas of water, prevent the invasion of alien species, reinforce the ecological and environmental protection of wetland and enhance the capacity of the ecological system for carbon sink absorption.

2.5.2.2 Building a Low-Carbon Energy System

1. Actively developing low-carbon energy

Efforts are underway to actively push forward nuclear power development. Zhejiang has a good foundation and conditions for nuclear power development. Developing nuclear power can reduce Zhejiang's dependence on coal, thus mitigating various problems in Zhejiang's coal power development. Nuclear power projects are being carried out in the region around Hangzhou Bay, the Wenzhou-Taizhou Region and western Zhejiang; it takes ten years to greatly increase the proportion of nuclear power in the electric power structure and really replace partial coal power.

Great efforts are being made to earnestly develop natural gas and renewable energy. Natural gas, wind power, photovoltaic power and other types of renewable energy are important measures for making energy clean. During the period of the 12th Five-Year Plan, with an increasing supply of gas, the proportion of natural gas in energy will increase to some extent. Zhejiang has an excellent foundation for the generation of wind power and photovoltaic power. In the next five years, the installed capacity of inland and offshore wind power generation will increase greatly; with such carriers as the Golden Sun Demonstration Project, the one-million rooftop photovoltaic power generation program is being carried out.

2. Carrying out the energy-saving project for technical innovations

With technological advancement as an important means, actions are being taken to organize and carry out the development, demonstration and promotion of major technologies, more rapidly adopt the new and high technologies to transform the traditional industries, improve the efficiency of energy utilization, especially the efficiency of terminal energy utilization. First, carrying out the project regarding innovations to transportation technology, actively researching and developing the technologies for hybrid vehicles, clean gas vehicles, fuel-cell vehicles and urban rail transit; also making breakthroughs in high-efficiency transportation technologies to enhance the efficiency of energy utilization in transportation; second, carrying out the projects for energy production, conversion and transmission technologies, such as clean coal power technology and ultra-high voltage power transmission and transformation technology, improving the efficiency of energy production and conversion and reducing the consumption of energy resources; third, making innovations in the technology for low-cost and large-scale utilization of renewable energy, including the integration technology for solar buildings; promoting building energy conservation and improving the efficiency of energy utilization.

3. Earnestly promoting and applying energy-saving technologies

The advanced and sophisticated technologies which cater to market demand and deliver significant returns are being promoted and applied. They mainly include the technologies for the comprehensive utilization of thermal energy—covering combined heat and power generation, centralized heating, cold accumulation and heat storage, combined cycles, clean coal processing and distribution—and surplus energy and heat recovery technology, conservation technology for building energy, automatic control technology, electricity saving technologies dominated by economical operations of high-voltage motors, adjustable-speed motors and transformers as well as green lighting, advanced combustion technologies dominated by the renovation of industrial boilers and kilns, upsizing of the circulating fluidized bed boiler and petroleum saving and replacement technologies, mainly the well-proven technologies for plasma ignition, low-oil pulverized coal ignition and the replacement of fuel oil with natural gas. The demonstration projects regarding energy-saving technical renovation are being organized and carried out, technical renovation is being conducted to save energy and improve the efficiency in those industries with high energy consumption, such as electric power, metallurgical, building material, chemical industries and to improve the market competitiveness of energy-saving products.

4. Carrying out key demonstration projects for making progress in energy-saving technologies

First, making the enterprises that are energy-saving and clean development-oriented produce a demonstration effect. Choosing a number of enterprises from the industries with high energy consumption, such as the metallurgical, petrochemical and chemical industries, to conduct systematic energy-saving renovation and to produce a demonstration effect. Second, making the small and medium-sized cities with high energy efficiency produce a demonstration effect. Choosing the small and medium-sized cities with better conditions to carry out the plan for the high-efficiency urban challenge for improving the overall energy efficiency of cities, adopting an integrated method for the management of resource planning and demand to push forward technical innovation, to gradually ensure that most enterprises within a region increase their output without increasing energy consumption, create expansion without an increased amount of energy consumption and that they rely on energy saving to increase capacity, and that zero growth of energy consumption and zero emission of wastes are achieved in the entire city (or region). Third, making integrated solar energy saving buildings produce a demonstration effect. Applying the technologies for door and window sealing strips, multilayer thermal insulating windows, the wall made up of external thermal insulation, household-based heat metering and temperature control, the regulation and control of the network of heat supply pipelines, heat reflection heat preservation and insulation, solar buildings, efficient lighting systems and computer simulation to develop the real estate projects that are integrated with solar energy saving aspects, to build energy-saving and environmentally-friendly real estate brands, improve the quality of the dwellings in cities and the life of the people. Fourth, carrying out wind power localization and large-scale marsh gas demonstration projects and the high-quality clean coal processing and distribution project.

2.5.2.3 Advocating a Low-Carbon Lifestyle

1. Advocating green travel

First, reducing the demand for oil in transportation by improving the fuel economy in vehicles through compulsory standard or voluntary agreements, vigorously developing the alternative fuels, rationally guiding and planning transportation demand, increasing the costs for private car purchases and use, and developing the technologies for the utilization of renewable energy. Second, energetically developing the non-motor vehicle transportation system and public transportation, focusing on developing rail transportation, promoting electric vehicles and bicycles; furthermore, strictly executing the low emission standard for motor vehicles and more rapidly pushing forward intelligent traffic management.

2. Advocating green buildings

The design standard of 65% energy saving is strictly executed in new civil buildings across the province, the proportion of renewable energy in buildings has been increased and the energy-saving standard of buildings has been improved gradually. The system for the monitoring of energy utilization has been comprehensively adopted in the office buildings of government agencies and large public buildings; renewable energy is used universally in buildings. The standard for energy-saving building renovation has been improved; energy saving renovation is being carried out in more than 50% of the public buildings with high energy consumption. A green rating of energy-saving buildings is being promoted. In coordination with the construction of a new countryside, actions are being taken to actively guide the adoption of new technologies for energy saving in the new rural houses.

3. Advocating green consumption

The energy-saving standards for household appliances with high energy consumption, such as air conditioners and refrigerators, are being continuously raised; the minimum standard of energy efficiency has been adopted in most household energy facilities. Residents are encouraged to choose energy-saving household appliances and to replace the old household appliances with high energy consumption.

2.5.3 *Preliminary Achievements in Low-Carbon Economic Development*

In Zhejiang, preliminary achievements have been made in low-carbon economic development, the top-level design has become better, the basic work has been solidly pushed forward, and the low carbon philosophy has been deeply rooted among the people. As shown by an initial estimation, the carbon emissions per unit of GDP across the province decreased by 3.54% in 2013, down 13.37% compared with 2010, which was higher than the average national decrease of 10.68; 70.4% of the goal set by the State under the 12th Five-Year Plan was achieved within three years.

2.5.3.1 Top-Level Design Has Become Better

First, the *Plan of Zhejiang Province for Addressing Climate Change (2013–2020)* has been developed, and it has passed the acceptance check conducted by the National Development and Reform Commission. Second, the implementation plan has been unveiled. The People's Government of Zhejiang Province has released the *Implementation Plan of Zhejiang Province for Greenhouse Gas Emissions* (Zhe Zheng Ban Fa (2013) No. 144), specifying the overall requirement for and main goal of controlling greenhouse gas emissions in Zhejiang by 2015 and breaking down the goal and main tasks for reducing carbon emissions per unit of GDP. Third, relevant systems have been improved. Zhejiang has issued the *Opinions on Strengthening the Statistical Work on Addressing Climate Change*, and has more rapidly established the sectoral statistical system (trial) covering the data in various fields of climate change, and has identified the evaluation of the responsibility of cities and counties towards the goal of carbon intensity reduction and the system of reporting the carbon emissions in key enterprises and public institutions as the priorities of the reform involving the system of an ecological civilization.

2.5.3.2 The Pilot Work on Low-Carbon Development Has Been Promoted Actively

First, the national pilot cities for low-carbon development have respective characteristics. Hangzhou City strives to become a six-in-one low-carbon city by building a comprehensive demonstration city for fiscal policies relating to energy conservation and emission reduction; Ningbo City leverages its advantages of a pilot city for low-carbon transportation to become the first nationwide to put forward a carbon emission peak value; Wenzhou City relies on the comprehensive financial reform experimental area to make breakthroughs in low-carbon financing. Second, the pilot work on low-carbon industrial parks has been actively carried out. Since 2014, the National Development and Reform Commission, the Ministry of Industry and Information Technology carried out the pilot work on low-carbon industrial parks; the Hangzhou Economic and Technological Development Zone, the Ningbo Economic and Technological Development Zone, the Wenzhou Economic and Technological Development Zone and the Jiaxing Xiuzhou Industrial Park were included in the first batch of national low-carbon industrial parks. Third, Zhejiang has become the first nationwide to carry out carbon sink forestry demonstration and pilot work, Zhejiang has established the first provincial fund—the China Green Carbon Foundation Zhejiang Carbon Sink Fund, the first prefecture-level city-level fund—the Wenzhou Carbon Sink Fund and the first county-level special fund—the Zhejiang Carbon Sink Fund Yinzhou Special Fund; Zhejiang has built carbon sink forests covering an area of 137,200 mu, accounting for 10% of the national area; Zhejiang has established the first forestry carbon sink trading platform in China on the East China Forestry Exchange.

2.5.3.3 The Work Foundation Has Been Consolidated Continuously

First, efforts have been made to promote the development of the greenhouse gas lists at the provincial and municipal levels and the evaluation against the goals of the reduction of carbon intensity at the municipal and county levels. The *Greenhouse Gas Emission List of Zhejiang Province (2006–2011)* was developed; Zhejiang released the *Work Plan of Zhejiang Province for Developing the Greenhouse Gas List for Cities and Counties of Zhejiang Province* (Zhe Fa Gai Zi Huan (2014) No. 103) and the *Guide for Developing the Greenhouse Gas List for Cities and Counties of Zhejiang Province (Trial)*, and it has initiated the pilot work on developing the list involving 11 cities divided into districts and the county-level list. Zhejiang has developed a preliminary plan for evaluation against the goal of reduction of carbon intensity at the municipal and county levels. Second, the work on reporting the greenhouse gas emissions in enterprises has been promoted gradually. According to the national line of thought of accelerating the establishment of a national carbon trading market and the work requirements for pushing forward the reporting of the greenhouse gas emissions in key enterprises and public institutions, based on the line of thought of first-ever implementation of pilot programs and step-by-step promotion, Zhejiang has initiated the work on the reporting of the greenhouse gas emissions in key enterprises in Zhejiang. Third, a research and communication platform for climate change has been built. Zhejiang's climate change research and communication platform has been built; Zhejiang has integrated the greenhouse gas list data at the provincial, municipal and county levels, enterprise carbon emission data, product carbon footprint data and other basic data on climate change.

2.5.3.4 The Low Carbon Philosophy Has Been Promoted in a Thorough and Complete Way

In order to further enhance the awareness of the whole society about coping with climate change and to foster a good social atmosphere for promoting green low-carbon development, Zhejiang has actively carried out various activities, including National Low Carbon Day and the Low-carbon China Academician Tour. During the low carbon day activities in 2014, Zhejiang announced the establishment of the first provincial expert committee on climate change in China, released Zhejiang's first low-carbon development report, and invited the experts from the National Climate Center to give lectures on the 5th report of the Intergovernmental Panel on Climate Change (IPCC); Zhejiang intensively presented Zhejiang's work on addressing climate change and its pilot work on three national low-carbon cities including Hangzhou, Ningbo and Wenzhou, and publicized Zhejiang's typical cases in low-carbon energy, low-carbon industry and low-carbon technology.

Chapter 3
The Protection of the Ecological Environment with Beautiful Mountains and Rivers

Zhibin Li, Qingkai Su and Hongyan Tian

The ecological environment is the foundation for economic development and the guarantee for the quality of life. Thus, the building of an ecological environment which features marvelous scenery south of the Yangtze River with lush mountains, clean, clear waters, a blue sky and clean land has become the basic project for building a beautiful Zhejiang; the "811" action plan, "governing five waters in tandem", "three renovations and one demolition" and "forested Zhejiang" have become the famous brands for Zhejiang's ecological and environmental development.

3.1 Mountains, Waters, Forests, Farmland and Lakes Make up a Community of Life

3.1.1 The Origin of the Philosophy that "Mountains, Waters, Forests, Farmland and Lakes Make up a Community of Life"

When delivering his report during the 3rd Plenary Session of the 18th Central Committee of the Communist Party of China, Xi Jinping pointed out, "We should realize that mountains, waters, forests, farmland and lakes make up a community of life, the people's lifeline lies in the farmland, the lifeline of the farmland lies in the water, that of the water lies in the mountain, that of the mountain lies in the soil, that of the soil lies in the trees; the control of usage and ecological restoration must follow the law of nature; if those who plant trees only pay attention to planting trees, those who govern water only pay attention to the governance of water, those who protect

Z. Li (✉) · Q. Su · H. Tian
The Zhejiang Research Center for Ecological Development, Zhejiang Sci-Tech University, Hangzhou, China

J. Pan and M. Shen (eds.), *Chinese Dream and Practice in Zhejiang – Ecology*, Research Series on the Chinese Dream and China's Development Path,
https://doi.org/10.1007/978-981-13-7209-4_3

the farmland only pay attention to protecting the farmland, it is very easy to only care for particular things and lose sight of the importance of the other things, finally the ecological environment is damaged systematically; it is absolutely necessary to determine one department that is responsible for regulating all usage of national land spaces within the territorial limits and protecting and restoring mountains, waters, forests, farmland and lakes in a unified way."[1]

"Mountains, waters, forests, farmland and lakes make up a community of life" means that the several aspects—mountains, waters, forests, farmland and lakes—the movement of those aspects, and the transfer of energy are interdependent (mutually dependent), correlated and stimulate each other, they organically make up a community of life. The farmland yields cereals to sustain human life; water wets the farmland to make it sustainably useful; the mountains gather moisture to nourish the soil; mountains, waters and land (covering climate and landform) make up an environment in the ecological system, while trees rely on sunshine and rain and dew, they are the most basic producers in the ecological system. The philosophy that "mountains, waters, forests, farmland and lakes make up a community of life" requires us to further value the rational allocation of these natural resources, scientifically set the intensity of resource utilization, manage and protect the resources according to local conditions, maximize the functions of ecological services of the community of life, promote harmony among the sustainable use of natural resources, the people and the land.[2]

The sustainable use of natural resources is the foundation for sustaining the community of life. There is a certain total quantity in the matter and energy exchange among the mountains, waters, forests, farmland and lakes. If the exploitation or consumption by mankind exceeds that limit, the operation of the community will change greatly and may even become intermittent. Mountains, waters, forests, farmland and lakes are tangible and substantial entities, while the community of life consisting of these entities certainly varies with time and local conditions. To better manage and utilize natural resources, it is essential to follow the basic principle of adapting measures to local conditions. As ecological elements, mountains, waters, forests, farmland and lakes have an extremely close symbiotic relationship with the people, and reinforce one another, they make up an organic and orderly community of life. The rational allocation of ecological elements directly determines the prosperity, health and sustainability of that community of life.

[1]Xi Jinping, Explanatory Notes for the Decision of the Central Committee of the Communist Party of China on Some Major Issues Concerning Comprehensively Intensifying the Reform, *People's Daily*, November 15, 2013.

[2]Yu Xindong, "Focusing on Life", Seizing the "Common Ground" – Zhejiang's Explorations in the Philosophy That "Mountains, Waters, Forests, Farmland and Lakes Make up a Community of Life", *Zhejiang Daily*, September 3, 2014.

3.1.2 *Zhejiang's Experiments in the Philosophy that "Mountains, Waters, Forests, Farmland and Lakes Make up a Community of Life"*

The significant achievements in Zhejiang's ecological and environmental development cannot be made without the guidance of the philosophy that "mountains, waters, forests, farmland and lakes make up a community of life". In Zhejiang, there are about 70% mountains, 10% waters and 20% farmland; the area of mountains and forests accounts for 65.6%, but economic development has always been slow in mountainous areas, mountainous areas appear to be weaknesses restricting Zhejiang's economic and social development. However, if this problem is observed with the philosophy that "mountains, waters, forests, farmland and lakes make up a community of life", it can be found that mountainous areas are not a burden on Zhejiang; on the contrary, they are a valuable wealth. With coordinated operations and rational development, it is completely possible to ensure that mountainous areas contribute more to Zhejiang's overall development.

To build an ecological Zhejiang and protect an ecological environment with beautiful mountains and rivers, the key lies in profoundly understanding and carrying out the important philosophy that "mountains, waters, forests, farmland and lakes make up a community of life", thoroughly grasping the words "a community of life" and taking the path of symbiosis, coexistence, joint promotion and development, co-management, co-building and joint governance.

First, it is necessary to profoundly understand and foster the basic philosophy of symbiosis and coexistence of mountains, waters, forests, farmland and lakes. Symbiosis and coexistence means interdependence, absolute necessity, interconnection and the sharing of a common destiny. For instance, as Zhejiang strengthens the development of forests and the ecological environment, Zhejiang also pays attention to the conservation of water and soil, gradually incorporates the upstream section of large rivers, important water sources, natural reserves and other important areas into the scope of ecological public-welfare forests, comprehensively adopts various engineering and biological measures, including slope modification, bank protection, grass planting, return of farmland to forests and the development of water conservation forests, to greatly promote the governance of water and soil erosion in small basins.

Second, the overall line of thought regarding joint promotion and development of mountains, waters, forests, farmland and lakes is developed and improved. Joint promotion and development means that various forces should play their roles in a coordinated way—work neck and neck at the source—and they should be balanced and orderly as much as possible to deliver unified and coordinated development outcomes. As Zhejiang has diverse topographical features and a complicated landform, this provides excellent natural space for developing mountains, waters, forests, farmland and lakes in a coordinated way. For instance, the highly-efficient ecological agriculture is developed, priority can be given to developing ecological agriculture which is based on grain, livestock and fishery and integrates cash crops including vegetables, melons and fruit in the network of water within the plains areas; ecological agriculture which integrates multistoried cropping, agriculture, animal husbandry and fishery can be developed in hilly basins; on mountains and hills, emphasis can be placed on developing ecological agriculture that is dominated by multistoried cropping of crops including well-known tea, well-known fruits, shoot bamboo, medicinal materials and high-mountain vegetables; in the coastal harbor areas and on the plains by the bay, priority can be given to developing ecological agriculture dominated by coastal planting and mudflat aquaculture; in the island areas, ecological agriculture dominated by ecological fishery and water-saving planting can be developed earnestly.

Third, the measures for the co-management and co-building of mountains, waters, forests, farmland and lakes are researched and introduced. Co-management and co-building means management should be combined with building and promoted through building with respect to natural resources so as to ensure sustainable development and utilization of resources on the one hand; the control of usage and ecological restoration should take into full account the integrality and systematicness of nature, the practice of valuing some aspects while losing sight of other aspects is inadvisable on the other hand. It is necessary to enhance the capability for providing ecological products—such as the conservation of water sources, water and soil conservation and the maintenance of biodiversity—and co-management and co-building are conducted to ensure that the ecological system remains intact. Actions are being taken to promote natural forest protection, to govern soil erosion, to maintain and rebuild the ecological system including wetlands and forests, to strengthen the conservation of water sources, to energetically push forward water-saving irrigation and to develop dry-farming and water-saving agriculture. Steep slope reclamation and cultivation and overgrazing are restricted, efforts are underway to maintain biodiversity, achieve benign circulation and the sustainable use of wild animal and plant resources.

Fourth, the system and mechanism of joint governance of mountains, waters, forests, farmland and lakes are established and improved. Joint governance means that the system of supervision is improved, the responsibility for regulating all usages of land spaces is assumed in a unified way so that the owner of the natural resources and the national natural resource administrator are mutually independent, supervise, coordinate and promote each other on the one hand, a unified authoritative department is established for regulating all usages of land spaces within the areas involved, protecting and restoring mountains, waters, forests, farmland and lakes in a unified way on the other hand. Zhejiang is institutionally establishing a long-term mechanism which combines incentive with restraint so that the means for environmental protection are no longer dominated by administrative means and the economic, technical, administrative and other means are comprehensively adopted for environmental protection. With regard to the prevention and control of water and soil erosion, an effective system for comprehensive protection against water and soil erosion is being established by integrating prevention, protection, supervision, governance and restoration, acting according to local conditions, adopting preventative measures in light of harm and optimizing the engineering, biological and farming measures. The management mode characterized by joint defense and control in regions, river basins and offshore areas is being built; the mechanism for regional coordinated and cross-regional comprehensive governance regarding environmental protection has been improved to regulate the resource environment in a more unified and systematic way.[3]

3.2 Promoting the Building of an Ecological Province Through the "811" Action

The "811" action is a brand action taken by Zhejiang for environmental protection, and it is the basic and landmark project for making ecological progress in Zhejiang. In January, 2003, with the approval of the State Environmental Protection Administration, Zhejiang became the 5th pilot province for building an ecological province in China. To achieve the goal of building an ecological province, in 2004, the People's Government of Zhejiang Province creatively developed the "811" three-year action plan for environmental pollution control, so a sweeping environmental improvement action kicked off, there was a firmer means and more vigorous guarantee for the building of an ecological province; preliminary achievements were made in three-year environmental improvement. In 2007, the People's Government of Zhejiang Province continuously pushed forward the "811" Action, and put forward the new "811" three-year action for environmental protection; Zhejiang unswervingly took the path of civilization development for more advanced production, a better life and excellent ecology. In 2011, the Party Committee and the People's Government of Zhejiang Province comprehensively arranged the "811" action for promoting ecological development, vowed to take five years to basically make sure that economic and social development adapts to the bearing capacity of resources and the environment, and that the improvement in the environmental quality is commensurate with the needs for improving the people's well-being. Zhejiang has persisted in the "811" action for a decade and has stayed ahead nationwide in the building of an ecological province and ecological development.

3.2.1 The Line of Thought and the Philosophy of the Building of an Ecological Province

3.2.1.1 The Origin of the Building of an Ecological Province

Situated in China's southeastern coastal areas and the south wing of the Yangtze River Delta, Zhejiang is a small province with fewer resources, it is a province

---

[3]Yu Xindong, Focusing on "Life", Seizing the "Common Ground" – Zhejiang's Explorations in the Philosophy That "Mountains, Waters, Forests, Farmland and Lakes Make up a Community of Life", *Zhejiang Daily*, September 3, 2014.

with a large population and an economically large province. In Zhejiang, the environmental capacity is small, the population is distributed unevenly, and industrial distribution is very unbalanced. In particular, since the beginning of the new century, Zhejiang has entered a new stage of more rapidly promoting industrialization and urbanization, building a moderately prosperous society in all respects and basically achieving modernization; economic and social development has brought more pressure on resources and on the environment. If there is a failure to fundamentally change the traditional extensive economic growth mode, this will certainly cause severer resource wasting and environmental problems and will greatly hinder sustained economic and social development. Therefore, it is essential to develop new developmental philosophies according to the Scientific Outlook on Development, attach greater importance to ecological development and environmental protection, the coordinated and sustainable development among the economy, the society and the population, resources and the environment. Given such basic conditions in Zhejiang and philosophies of development, the Party Committee and the People's Government of Zhejiang Province have closely followed the objective law, sized up the situation, and made great strategic decisions about further leveraging Zhejiang's ecological advantages, and building an ecological province and a green Zhejiang.

3.2.1.2 The Overall Line of Thought for the Building of an Ecological Province

Building an ecological province is a strategic decision made by the Party Committee and the People's Government of Zhejiang Province to carry out the principles adopted during the 16th National Congress of the Communist Party of China, to build a moderately prosperous society in all respects and to basically achieve modernization in advance; it is a great measure for venturing down the new-type path towards industrialization and implementing the strategy of sustainable development by proceeding from Zhejiang's conditions and following the objective law. The building of an ecological province calls for persisting in the strategy of sustainable development, adopting the principle of ecology, the systems engineering approach and the philosophy of a circular economy, proceeding from promoting a change in the economic growth mode and improving the environmental quality to fully leverage the regional advantages regarding ecology, resources, industries and mechanisms, vigorously develop an ecological economy, improve the ecological environment, cultivate the ecological culture, basically achieve regional coordinated development among the economy, the society and the population, resources and the environment. The governments at various levels in Zhejiang should increase their input, and adopt a reform-oriented line of thought to broaden the investment and financing channels, push forward ecological development and environmental protection programs in a market-oriented and industrialized way.

Zhejiang's overall line of thought regarding its work of building an ecological province is as follows: follow the guidance of the Deng Xiaoping Theory and the Important Thought of Three Represents, earnestly implement the principles adopted during the 4th Plenary Session of the 16th Central Committee of the Communist Party of China and the 7th Plenary Session of the 11th Party Committee of Zhejiang Province, firmly keep to and carry out the Scientific Outlook on Development, follow the requirements of the Central Committee of the Communist Party of Chin for building a harmonious socialist society, the "Eight-Eight Strategies" and the strategic arrangement of building a safe Zhejiang, focus on carrying out the *Plan of Zhejiang Province for Building an Ecological Province*, give priority to the "811" improvement project and the work on the "1000-village demonstrations and 10,000-village improvements", provide classified guidance, bring innovations to the working mechanism, work harder, stress implementation, strengthen policy guarantees, promote project construction, develop models for the demonstration effect, make breakthroughs in key fields, and vigorously move forward with the building of an ecological province.

3.2.1.3 The Philosophy of the Building of an Ecological Province

Based on the work of the previous Party Committees of Zhejiang Province, Xi Jinping earnestly carried out the principles adopted by the Central Committee of the Communist Party of China, gave due considerations to Zhejiang's reality, fully respected the creative practice of Zhejiang's cadres and people, drew upon foreign and domestic beneficial practices and experience in environmental protection and ecological development and gradually developed the philosophy of building an ecological province with Zhejiang's characteristics.[4]

"If ecological development is achieved, progress will be made; if ecological degradation occurs, retrogression will come about." This is Xi Jinping's dialectical view on the relations between the ecology and progress, and retrogression. According to Xi Jinping, there is an eternal contradiction between human needs for pursuing development and the limited supply of earth's resources, so we must resolve the contradiction between the limited resources and the unlimited desire of the people to achieve great harmony between the people and nature. He has advocated the philosophy of ecological development which stresses respecting, conforming to and protecting nature. In his view, attaching great importance to sustainable development, the issues of resources and ecology, developing a circular economy and building an ecological province have a vital bearing on the overall situation, the future and the people's well-being in Zhejiang.

"Equal emphasis should be placed on the 'gold and silver mountains' and clean, clear waters and lush mountains." In a meeting of the leading group of Zhejiang Province for ecological development, Xi Jinping stressed, "We should seek harmony between the people and nature and between the economy and the society; in other words, equal emphasis should be placed on the 'gold and silver mountains' and

[4]The Party History Research Office under the Party Committee of Zhejiang Province: For Zhejiang's Brighter Future – A Review of Zhejiang's Efforts at Promoting Ecological Development, *Zhejiang Today*, 2010(12).

clean, clear waters and lush mountains; these two mountains are contradictory, but they can become a dialectical unity; clean, clear waters and lush mountains can endlessly bring a 'gold and silver mountains', evergreen plants are the source of money; ecological advantages can be turned into economic advantages, the environment can bring wealth, this represents a higher realm and reflects the requirements of scientific development."

"If early actions are taken with regard to environmental protection and ecological progress, more achievements will be made with less effort; otherwise, fewer achievements will be made with greater efforts; if actions are taken too late, the situation will become more passive." According to Xi Jinping, we cannot allow the developmental mode which only takes into account short-term interests in disregard of long-term interests, applying treatment after pollution and taking restoration actions after destruction.

"Environmental protection is equivalent to the protection of wealth." In the opinion of Xi Jinping, destroying the ecological environment disrupts the productive forces; protecting the ecological environment protects the productive forces; improving the ecological environment develops the productive forces. The development of Anji County is a good example. With the Anji Model, which is the national pilot, Zhejiang has stayed ahead nationwide in ecological protection. In Anji County, a mountainous county, the per capita net income of farmers has exceeded 13,000yuan, which has reached the average level of the province, sustained economic development has been achieved and the people's well-being has improved amidst the improvement in the ecological environment.

"As the economy further develops, greater importance should be attached to environmental protection and ecological development." According to Xi Jinping, a government which does not value ecology is a government which lacks vision; a leader who does not value ecology is an incompetent leader; an enterprise which does not value ecology is a futureless enterprise; a citizen who does not value ecology is a citizen who does not have the awareness about modern civilization.

3.2.2 Continuously Pushing Forward the "811" Action for Environmental Protection and Ecological Developments

Since the beginning of the new century, the Party Committee and the People's Government of Zhejiang Province have actively explored the new path for ecological and environmental protection, incessantly promoted three rounds of the "811" action, resolutely overcome the difficulties in environmental protection and made great efforts to improve the ecological and environmental quality. Three rounds of the "811" action have shown the process of the gradual development of Zhejiang's ecological and environmental protection programs; this is also Zhejiang's trajectory for making progress in ecological development.

3.2.2.1 The "811" Action for Environmental Pollution Control

The first stage: the "811" action for environmental pollution control (2004–2007).

In 2004, the People's Government of Zhejiang Province unveiled the *Action Plan of Zhejiang Province for Environmental Pollution Control*, specifying the main tasks of the prefecture-level cities and the provincial departments and bureaus in pollution control; the People's Government of Zhejiang Province signed, with the prefecture-level cities, a letter of responsibility of provincial governor and mayors for the control of environmental pollution. In the meantime, the tasks for pollution control were broken down in the annual evaluation against the letter of responsibility for the building of an ecological province, evaluation was conducted at the end of the year, a one-vote veto was cast against key tasks of pollution control, such as sewage treatment projects and the improvement projects in key supervision areas. Therefore, the "811" action for environmental pollution control kicked off.

In the "811" action for environmental pollution control, "8" refers to 8 major water systems, canals and lowland river networks in Zhejiang, "11" means 11 cities divided into districts in Zhejiang, which are the 11 provincial-level key areas for environmental supervision. These areas of supervision include the Waisha, Yantou chemical and pharmaceutical bases in Jiaojiang, the Huangyan Chemical and Pharmaceutical Base, the Shuiyang Chemical and Pharmaceutical Base in Linhai, the Shangyu Fine Chemical Park, chemical enterprises in the Nanjiang River Basin in Dongyang, the Xinchang-Shengzhou Section of the Xinchang River Basin, chemical enterprises in the Shenjia Industrial Park in Quzhou, printing and dyeing and dyeing industries in the eastern area of Xiaoshan, the Shuitou Leather Making Base in Pingyang, the electroplating industry in Wenzhou, and the storage battery industry in Changxing. This control action focused on 8 industries with heavy pollution—including chemical, pharmaceutical, leather making, printing and dyeing, gourmet powder, cement, smelting and papermaking industries—and 573 provincial-level key enterprises for environmental supervision and 27 key enterprises as ammonia nitrogen emission sources in the Qiantang River Basin. In 2006, 5 additional areas, including Fuyang (the papermaking industry), were identified as the provincial-level key areas for environmental supervision subject to supervision based on the requirements for key areas of supervision. In the meantime, each year different key polluting industries and polluting enterprises were identified for special control.

After the "811" action for environmental pollution control began, each year the environmental protection departments at various levels in Zhejiang worked with functional government departments for development and reform, economic and trade affairs and construction to exercise special supervision, inspect the completion of the main tasks, including the construction of environmental protection infrastructures and the improvement in key supervision areas; where the control work was obviously left undone, the provincial office for pollution control promptly gave feedback to local governments and raised the requirements of supervision. In early 2005, the leading group of Zhejiang Province for environmental pollution control signed, along with 11 cities, the *Statement of Tasks of Zhejiang Province for Key Environmental Improvement Projects*, intensifying the investigation on responsibility and identifying the improvement of the Qiantang River as the priority of improvement in eight major water systems in Zhejiang during the current year. In a meeting of the leading group for environmental pollution control, the then Governor of Zhejiang Province, Lu Zushan, pointed out that Zhejiang would incorporate Zhejiang's work on environmental pollution control into the evaluation of responsibility against the annual goal of government departments at various levels, and include the intensity of the main pollutant emissions per 10,000 yuan GDP and the level of environmental quality into the system for evaluating the performance of Zhejiang's Party and government cadres. To promote pollution control in key supervision areas, in 2006, the former provincial environmental protection bureau adopted the on-the-spot supervision system under which the bureau's leaders lead teams to carry out the specified tasks in different areas; in 2007, Zhejiang introduced an on-the-spot supervision system under which relevant member units of the leading provincial group for environmental pollution control conduct the specified work in different areas, the persons in charge of the units lead teams to supervise pollution control on the spot each month. Dynamic management was performed in key polluted areas, polluting industries and polluting enterprises.

In three years, Zhejiang actively took a number of measures, including technical renovation, shutdown, transfer and relocation and introduction of new pollution control facilities, to intensify pollution control in nine key polluting industries, including the chemical, pharmaceutical, leather-making, printing and dyeing, gourmet powder, cement, smelting, papermaking and solid waste dismantling industries, producing noticeable effects. As of 2007, pollution control work had been completely finished in the gourmet powder industry across the province, Zhejiang had become the first nationwide to ensure that the whole gourmet powder industry within the province met the COD and ammonia nitrogen emission standards; the structural adjustment in the cement industry had basically been completed and the task of shutting down mechanical shaft kilns had been completed a year in advance; all of the straw pulp production lines in the papermaking industry had been closed; the electric power industry had completed the desulfurization project for 2502 MW units and had reduced about 90,000t of sulfur dioxide emissions within two years; the imported solid waste dismantling industry had basically achieved park-based development.

In three years, Zhejiang further reduced the total emissions of the main pollutants. In the first half of 2007, the General Office of the State Council, the State Environmental Protection Administration and the East China Inspection Center investigated, inspected, checked and examined the work on reducing the emissions of the main pollutants in Zhejiang many times; as shown by the results, in the first half of 2007, the national COD emission increased by 0.24%, sulfur dioxide emissions decreased by only 0.88% compared with the same period of the previous year, while Zhejiang's COD emissions dropped by 2.48%, ranking Zhejiang no.1 nationwide; sulfur dioxide emissions fell by 4.05%, ranking Zhejiang no.4 nationwide.

Amidst rapid economic development, Zhejiang's environmental quality further improved. Zhejiang stayed ahead nationwide in the state of its ecological environment; in the *National Ecological Environmental Assessment Report* released by the China National Environmental Monitoring Center in November, 2006, Zhejiang ranked no.1 nationwide with 87.1 points. As of late 2007, the overall goal of the "811" action for environmental pollution control—basically finishing the work on two fronts and becoming the first to make progress in two respects—had been achieved on schedule. As of late October, 2007, all of 11 provincial-level key environmental supervision areas and 5 quasi-key supervision areas in Zhejiang had passed the field acceptance check.

3.2.2.2 The New Three-Year "811" Action for Environmental Protection

The second stage: the new three-year "811" action for environmental protection (2008–2010).

Given that great achievements had been made in the first round of the "811" action for environmental protection and the ecological environment had been further protected, developed and improved, the People's Government of Zhejiang Province unveiled the *Implementation Plan for the New Three-year "811" Action for Environmental Protection* early in 2008, vowing to take three years to basically solve the salient environmental pollution problems in different areas, continue staying ahead nationwide in the building up of capacity in environmental protection and the ecological and environmental quality. At that time, "8" meant the work goals for the following eight aspects: pollution and emission reduction, prevention and control of industrial pollution, treatment of urban and rural sewage, disposal of garbage and other solid wastes, prevention and control of agricultural non-point and soil pollution, building up the capacity for environmental monitoring, ecological protection and restoration, environmental quality and the comprehensive index of ecological environmental quality. The following goals were also included: a decrease in the total COD emissions by more than 15.1% compared with 2005, a drop in the total sulfur dioxide emissions by more than 15% compared with 2005, about 1/3 of the counties (county-level cities) should reach the provincial-level ecological county (county-level city) standard, Zhejiang continues staying ahead nationwide regarding the ecological environmental index.

"8" also means the work tasks regarding 8 aspects: finishing the task for reducing the emission of the main pollutants, continuing to promote the prevention and control of water pollution, continuing to intensify the prevention and control of industrial pollution, continuing to make profound overall improvements in the urban environment, to comprehensively push forward the prevention and control of agricultural and rural environmental pollution, to actively propel the prevention and control of pollution in offshore areas, to more rapidly press ahead with ecological restoration in soil, mines and rivers, persistently and profoundly carrying out the activity of ecological development.

"11" means 11 key environmental problems to be solved under the supervision of the provincial departments. The first batch of 11 key environmental problems to be solved under the supervision of the provincial departments include: small-scale smelting pollution along the Hangzhou-Xin'an'jiang-Jingdezhen Highway, industrial waste gas pollution in the Ningbo port area, environmental pollution in the Wenruitang River in Wenzhou, organic glass pollution in Jiuguan Town in Nanxun, Huzhou, livestock and poultry breeding pollution in Jiaxing, printing and dyeing and chemical pollution in the Xiaoshao area, water environmental pollution in the Dongyang River Basin, solid waste dismantling pollution in Taizhou, environmental pollution in the Changshan Chemical Park in Quzhou, environmental pollution in the leather-based cloth and synthetic leather industries in the Lishui Economic Development Zone, and environmental pollution in some development zones and industrial parks across the province. "11" also means 11 safeguard measures: speeding up the transformation of the economic developmental mode, solving environmental problems at the source, comprehensively carrying out the plan for ecological environmental function areas by strictly executing the area-specific environmental access policy, continuing to strengthen environmental law enforcement by strictly exercising environmental law enforcement supervision, continuing to improve the environmental protection infrastructure by really strengthening operational supervision, more rapidly building a modern environmental monitoring and supervision system by increasing the level of environmental supervision, enhancing the capacity for environmental emergency response thus practically guaranteeing environmental safety, reinforcing the building of a platform for the science and technology in environmental protection and conscientiously increasing the scientific and technological support for environmental protection, accelerating the development of the environmental protection industry, increasing the professional level of environmental protection; improving the economic policy for environmental protection by guiding and encouraging social forces to actively participate in environmental protection and ecological development, energetically cultivating and developing the ecological culture by continuously enhancing the awareness about ecological development of the whole society, and bringing innovations to the working system and mechanism for ecological and environmental protection and strengthening the responsibility evaluation against goals.

Zhejiang will shift its focus from prevention and control of industrial pollution to comprehensive prevention and control of industrial, agricultural and domestic pollution, and has further put forward the goals of finishing the tasks specified in the 11th Five-Year Plan for environmental protection, basically solving the salient environmental pollution problems in different areas, thus continuing to remain ahead nationwide in the capacity for environmental protection and ecological and environmental quality. This is a more complete and extensive action of control and protection, from the treatment of rural domestic sewage to the regulation of livestock and poultry breeding, from the disposal of urban garbage to the protection of the quality of surface water, from the reduction of the emission of the main pollutants to deepening the prevention and control of industrial pollution, from the execution of ecological restoration and protection projects to accelerated prevention and control of offshore areas.[5]

---

[5]Xu Zuomin, "811", Zhejiang's Brand for Ecological Development, *Zhejiang Today*, 2011(13).

In three years, 85.6 billion yuan was allocated to complete 579 water pollution prevention and control projects; each year, nearly 8 billion yuan was earmarked to carry out more than 100 key circular economic projects in industrial enterprises; the urban environment was improved in a comprehensive way, sewage treatment plants were built in counties and even in central towns and a solid waste disposal system was improved; the project of "1000-village demonstrations and 10,000-village improvements" was carried out perseveringly, agricultural and rural environmental protection was strengthened comprehensively; the development of ecological public-welfare forests was promoted vigorously; great efforts were made to push forward the control of water and soil erosion; and the building of ecological models was conducted extensively.

In late 2010, the new three-year "811" action for environmental protection was concluded successfully. Except for a low rate of reaching the water quality standard at cross sections between cities and counties in the Qiantang River Basin, 20 work goals in 8 aspects specified in the action plan were fully achieved. In the meantime, as confirmed by the Ministry of Environmental Protection of China, as of late 2010, COD and sulfur dioxide emissions decreased by 18.15% and 21.15% in Zhejiang compared with 2005, which were 120.2% and 141% of the tasks specified in the 11th[h] Five-Year Plan and thus overfulfilling the emission reduction goals specified in the 11th Five-Year Plan.

3.2.2.3 The "811" Action for Promoting Ecological Development

The Third Stage: "811" Action for Promoting Ecological Development (2010–now).

After the 7th Plenary Session of the 12th Party Committee of Zhejiang Province made the *Decision of the Party Committee of Zhejiang Province on Promoting Ecological Development* in June, 2010, the Party Committee of Zhejiang Province conducted intensive surveys on the implementation of the principles adopted in the plenary session and the progress in ecological development, and stressed that a concrete action plan, feasible means and carriers were necessary for promoting ecological development. Thus, the Party Committee of Zhejiang Province vowed to continue the "811" action and take it as the main means and important carrier for promoting ecological development. The action, called the "811" action for promoting ecological development, lasts for five years. The term "811" was given new content; "8" now means ecological economy, energy conservation and emission reduction, environmental quality, pollution prevention and control, ecological protection and restoration, the building up of the capacity for an environmental safety guarantee, ecological development and institutional improvement in the ecological field; "11" now means 11 special actions for energy conservation and emission reduction, a circular economy, green towns, a beautiful countryside, clean water sources, clean air, clean soil, forested Zhejiang, blue barrier, fire protection and disaster reduction, green development; "11" also means safeguard measures in 11 aspects. Eleven special actions are the mainstay. The progress made through the "811" action for promoting ecological development was ultimately embodied by the effects produced by 11 special actions.

Since the "811" action for promoting ecological development was taken, the people across the province have made concerted efforts to tackle great difficulties, and staged achievements have been made.

The third round of the "811" action is being expanded under the banner of "vigorously promoting ecological development" put forward by the 18th National Congress of the Communist Party of China, and it is being promoted steadily through concerted efforts across the province. Three rounds of the "811" action have inherited Zhejiang's philosophy of respecting, conforming to and protecting nature, and have also demonstrated Zhejiang's determination to promote green, circular and low-carbon development, as well as laying a foundation for building an ecological Zhejiang and a beautiful Zhejiang. Three rounds of the "811" action have led efforts in a consistent and progressive way to intensify Zhejiang's ecological development from environmental pollution control to ecological development, from "becoming the first in two respects" to "staying ahead on two fronts", from attention to the environmental quality to more attention to the improvement of the people's well-being.

3.2.3 The Achievements and Experience in the Protection of the Ecological Environment

Three rounds of the "811" action have vigorously promoted the leapfrog development of Zhejiang's environmental protection and ecological improvement programs. With more than 10 years of efforts, the environmental quality, such as the quality of water and air, has remained generally stable across the province, it has improved to some extent in some areas, and Zhejiang has always stayed ahead nationwide in the ecological environmental index. The construction of the environmental protection infrastructure has been further strengthened, Zhejiang has become the first nationwide to achieve full coverage by municipal sewage treatment plants and garbage disposal facilities above the county level, and to promote the construction of the town-level sewage treatment facilities. The capacity for environmental supervision has improved significantly, Zhejiang has become the first to build a system for automatic monitoring of the environmental quality and key pollution sources, and basically establish an environmental monitoring network covering full coverage. The mechanism for ecological and environmental protection has improved more and more, Zhejiang has become the first nationwide to have the opinions developed by the government on improving the ecological compensation mechanism, push forward the practice of ecological compensation in a systematic and comprehensive way, initiate a management system of evaluating the quality of the water at cross sections between administrative divisions, and link the results of the evaluation with the assessment of the environmental impact of regional construction projects, a water resource feasibility study, examination and approval, economic reward and punishment, put in place government responsibilities.

3.2.3.1 Main Achievements in Ecological and Environmental Protection

1. The ecological and environmental quality has improved significantly

(1) Zhejiang has successfully reached the standards for environmental pollution and emission reduction. The ecological development focuses on achieving coordination between economic and social development and the environment, and changing the traditional developmental mode characterized by high input, high consumption and heavy pollution. The local authorities and departments in Zhejiang have closely followed the decisions and arrangements made by the Party Committee and the People's Government of Zhejiang Province and have addressed environmental protection according to the overall situation of economic and social development, seized upon the situation to promote ecological development, pollution and emission reduction, and to fully finish the tasks for environmental protection.

(2) Marked effects have been produced in improving the environment. The quality of the water has improved significantly. In Zhejiang, the cross sections where the quality of the water reached or surpassed the standard environmental quality for surface water accounted for 62.9% of the 221 cross sections controlled at the provincial level, the sources of centralized drinking water above the county level where the quality of the water reached the relevant standard accounted for 86.4%; the cities above the county level where the quality of the air quality the Grade II standard accounted for 92.8%; the cities divided into districts where the quality of the air reached the Grade II standard accounted for 90.9% of the 11 cities divided into districts; the cities above the county level where regional environmental noise exceeded 55 dB accounted for more than 75%. With continuous improvement in environmental infrastructures, Zhejiang has fully built the network for automatic monitoring of environmental quality and key sources of pollution, and the capacity for disposing of urban pollutants has further improved. In late 2011, Zhejiang's municipal sewage treatment rate reached 82%, the towns where sewage treatment facilities were built accounted for 72.4%; the rate of harmless treatment of domestic garbage in the cities divided into districts reached 97%; in Zhejiang, the handling capacity of the municipal sewage treatment plants above the county level was 8.42 million t/d and the municipal sewage treatment rate above the county level reached 94%.

2. Ecological and environmental development has been earnestly promoted

(1) Active progress has been made in ecological development, such as forest, wetland, plains and natural reserves. The *Decision of the Party Committee of Zhejiang Province on Promoting Ecological Development* specified the following

ecological developmental goals: forest coverage, forest stock volume and the afforestation area on the plains will be increased steadily, the system for the guarantee of ecological safety will basically take shape, the urban and rural environment will be optimized continuously, and it will become more livable. During the period of the 12th Five-Year Plan, Zhejiang will undertake a 9% increase in the national forest stock with 1% of the national land area; this will become a great contribution made by Zhejiang to the State in addressing climate change.

Regarding wetland development, the People's Government of Zhejiang Province issued the *Circular Concerning Strengthening the Management of the Protection of the Wetland* in 2005; the *Regulations of Zhejiang Province for Wetland Protection* was adopted in 2012; emphasis was placed on protecting the diversity of the wetland resources and ecosystem through the construction of wetland nature reserves and wetland parks as well as the wetland restoration project and the project for the building up of capacity. According to the results of the second national wetland resource survey in 2013, Zhejiang's wetland area (except the area of paddy fields) reached 1,110,100 hectares, accounting for more than 1/10 of Zhejiang's total area.

Regarding the plains afforestation, Zhejiang launched the "1818" plains afforestation campaign in 2011. According to the relevant plan, by 2015, more than 1.80 million mu forest area in plains will be newly added and the rate of forest coverage on the plains will be higher than 18% across the province. During the period 2010–2014, the newly-added plains afforestation area was 1,818,000 mu, the per capita newly-added afforestation area was 138,000 m^2, the rate of forest coverage on the plains reached 18.1%; 8 national forest cities, 46 provincial forest cities, 177 provincial forest towns and 1094 forest villages were successfully built across the province.

Regarding the development of natural reserves, the People's Government of Zhejiang Province issued the *Administrative Measures of Zhejiang Province for Natural Reserves* in 2006; in 2014, the People's Government of Zhejiang Province issued Decree No.321 of the People's Government of Zhejiang Province—the *Decision of the People's Government of Zhejiang Province on Amending Nine Regulations Including the Administrative Measures of Zhejiang Province for Forest Land*, amending the *Administrative Measures of Zhejiang Province for Natural Reserves*. Such departments as the provincial environmental protection bureau issued the *Circular of Zhejiang Province on the Indicators for the Evaluation Concerning the Standardized Construction of Natural Reserves (Trial)* in 2008, calling for annually evaluating the compliance of the standardized construction of different types of reserves with relevant standards. As of late 2013, there were 10 national natural reserves and 9 provincial natural reserves across the province.

(2) A number of ecological cities, counties and model cities in environmental protection have been built. Zhejiang has energetically carried out a campaign for building three major systems including ecological cities and counties, model

cities in environmental protection and "green cells", delivering a lot of outcomes. In late 2014, there were 16 national ecological counties, 8 national model cities in environmental protection, 581 national ecological towns, 57 provincial ecological counties, 10 provincial model cities in environmental protection and 1038 provincial ecological towns in Zhejiang.

3. The environmental protection policy has constantly improved

(1) Zhejiang has developed and improved a number of environmental protection regulations. The Standing Committee of the Party Committee of Zhejiang Province has issued the *Regulations of Zhejiang Province for Environmental Protection against Radiation from Nuclear Power Plants*, the *Regulations of Zhejiang Province for Prevention and Control of Air Pollution* and other relevant regulations. Such cities as Hangzhou City and Ningbo City have also speeded up local legislation. These regulations have further strengthened the legal guarantee for ecological development and environmental protection.
(2) The guarantee system for ecological development and environmental protection has improved gradually. Zhejiang has issued the *Several Opinions of the Party Committee and the People's Government of Zhejiang Province on Carrying out the Scientific Outlook on Development and Strengthening Environmental Protection*, taking strengthening ecological development and environmental protection as an important measure for thoroughly and extensively carrying out the "Eight-Eight Strategies" and as an important part of the efforts in comprehensively building an ecological Zhejiang, and basically giving shape to the working mechanism in which the Party committee plays the leading role, the government assumes responsibilities, there is overall coordination among departments and extensive social participation. In recent years, Zhejiang has preliminarily built and put in place the leadership responsibility system for environmental protection, and has signed, on a level-by-level basis, a letter of responsibility against the goals of building an ecological province and achieving environmental protection; Zhejiang has broken down and put into practice the goals, indicators and annual plan, rigorously carried out evaluation, reward and punishment, incorporated environmental protection into the indicator system for evaluating the performance of the leading groups and leading cadres of local Party committees and governments.
(3) Market-oriented means have been adopted to bring innovations to the environmental management mechanism, and significant progress has been made. Zhejiang has issued the *Several Opinions of the People's Government of Zhejiang Province on Further Improving the Mechanism for Ecological Compensation*, calling for following the principle of first-ever implementation of pilot programs and gradual promotion to progressively establish and improve a well-regulated and orderly mechanism for ecological compensation operations in which powers and interests are commensurate with responsibilities. Zhejiang has explored pilot compensated distribution of COD and sulfur dioxide emission rights and

pilot trading of emission rights in several counties (county-level cities) in Zhejiang. The centers for the trading of emission rights established in such areas as Jiaxing City have been under sound operations, offering an effective early platform for further expanding the scope of the trading of emission rights and establishing a market for the trading of emission rights in the future. Moreover, Zhejiang has given full play to the role of the market mechanism in resource allocation, it has adopted the market-oriented means to establish and improve diverse financing channels, encourage and support social capital to participate in ecological development and environmental protection by means of sole proprietorship, joint ventures, contracting, leasing, auctions, the join-stock system, the joint stock cooperative system and BOT.

4. The public awareness about environmental protection has been constantly enhanced

Ecological and environmental development calls for active public participation and enhanced public awareness about environmental protection. Zhejiang has attached great importance to carrying out publicity and education among the general public, and has enhanced the behaviors of environmental protection and green consumption through publicity at ecology festivals. On the one hand, Zhejiang has fully utilized ecology festivals to conduct publicity and provide guidance. With the important occasions, such as the World Environment Day, the China Water Week, the China Land Day and the China Tree-Planting Festival, Zhejiang has promoted a low-carbon life and has encouraged green consumption. On the other hand, Zhejiang has fostered an ecology day atmosphere with Zhejiang's characteristics. Anji County established the first county-level ecology day in China in 2003, and it was a successful experience; in 2009, Zhejiang established the first provincial-level ecology day in China, and identified each June 30 as Zhejiang's ecology day.

Zhejiang has attached great importance to internalizing the philosophy of green consumption through green organizations, and has actively improved the "green series" work. Zhejiang has energetically built green schools, green communities, green homes, green hospitals, green enterprises and green hotels, so that the players at different levels in the society can participate in environmental protection.

Moreover, Zhejiang has adopted the restraint policy for non-green products and the incentive policy for green products to guide residents to consciously advocate green consumption. The restrictive measures for the vehicles with larger engines, the subsidy policy for electric vehicles, the total quantity control of fossil energy, encouragement of the development of renewable energy, strict supervision over polluted foods, high quality and the high price of green foods guide residents' consumption behaviors, and the ecological and environmental protection philosophy.

3.2.3.2 The Basic Experience in Ecological and Environmental Protection

1. Close coordination among end-of-pipe treatment, process management and source control

The end-of-pipe treatment addresses how to effectively control pollutants, how to reduce pollutants through process renovation; process control deals with how to reuse and recycle resources, improve the efficiency of resource utilization, pursue harmony between economic growth and environmental protection by considering the problems concerning resources in production, use and consumption; source control proceeds from industrial orientation and product design and stresses the entrepreneurial spirit of being responsible for the environment, seeks the optimal strategy for sustainable development and the strategy for comprehensive management for the economic, social and environmental fields, and incorporates environmental management into various levels of enterprise organization.[6]

Regarding end-of-pipe treatment, Zhejiang has comprehensively arranged for the "811" environmental pollution control action focusing on 8 major water systems and provincial-level key areas of supervision, and has intensified the work on improving the key industries—including chemical, pharmaceutical, leather-making, printing and dyeing, gourmet powder, cement, smelting, papermaking, solid waste dismantling industries—and 573 key pollution sources controlled at the provincial level. During the period of the 10th Five-Year Plan, more than 6000 enterprises which violated environmental laws and caused heavy pollution were shut down, more than 4000 construction projects which did not comply with national industrial policy and ecological and environmental requirements were denied or their sites were required to be re-selected. Zhejiang adopted rigid means of constraint for pollution and emission reduction, shut down, transferred and renovated a batch of outdated facilities to promote total quantity reduction, quality improvement and development optimization. Zhejiang also started with resource conservation and environmental protection to make structural adjustments, and carried out ten major projects for energy conservation and consumption reduction. During the period of the 11th Five-Year Plan, Zhejiang shut down 5.31 million KW of small thermal power units, 2.31-million-ton outdated steelmaking facilities, 23.97-million-ton outdated cement facilities, 578,000t of outdated papermaking facilities, 2100 inefficient industrial boilers and 2547 clay brick and tile kilns. Zhejiang continuously carried out compulsory clean production in key industries and key enterprises, and the enterprises' awareness of assuming social responsibilities for environmental protection improved significantly.[7] In the meantime, Zhejiang speeded up the construction of environmental protection facilities, the capacity for urban sewage treatment and harmless treatment of domestic

[6]Shen Manhong et al., *Annual Report on Zhejiang's Ecological Economic Development in 2012*, China Financial & Economic Publishing House, 2012, p. 259.

[7]Shen Manhong, From a Green Zhejiang to an Ecological Zhejiang – Five Years of Brilliant Ecological Development in Zhejiang, *Zhejiang Daily*, May 25, 2012.

garbage improved greatly; the construction of facilities for the centralized disposal of hazardous wastes was also initiated in an all-round way; the facilities for harmless centralized disposal of medical wastes in such cities as Hangzhou, Ningbo, Huzhou and Taizhou were completed and put into operation.

With regard to process management, Zhejiang has constantly improved the efficiency of resource utilization. On the one hand, with a focus on the industries with heavy energy consumption and enterprise energy conservation, Zhejiang has strengthened the management of raw material consumption in the key material-consuming industries, pushed forward economical and intensive use of urban construction land and industrial land, and carried out the strictest water resource management system. On the other hand, Zhejiang has moved faster to promote and apply new technologies and new products for energy conservation, push forward pilot promotion of methanol gasoline, reinforce the construction of smart power grids and the development of technology for the generation of clean coal power, it has become a national clean energy base demonstration province.

Regarding source control, Zhejiang has focused on policies of planning and industrial access to exercise strict source control management of the newly added production projects. With the promulgation and implementation of the *Work Program of Zhejiang Province for the Planning of Main Function Areas* (2007), the *Technical Guide of Zhejiang Province for the Planning of Ecological and Environmental Function Areas in Counties (County-level Cities, Districts)* (2006) and the *Plan of Zhejiang Province for the Division of Water Function Areas and Water Environmental Function Areas* (2006), the functional orientation among regions within the province was determined. With regard to industrial distribution and development, the *Guiding Catalogue of Zhejiang Province for the Development of the Manufacturing Industry* (2008), the *Plan of Zhejiang Province for Shutdown of the Outdated Production Facilities (2013–2017)* (2013) and other industrial policies and access rules further identified encouraging development and restricting development as industrial directions, and determined the indicators for the constraint for access, including the output per unit of land, energy consumption per unit of output value, water consumption per unit of output value, thus eliminating the newly-added production facilities with heavy pollution and low efficiency at the source, exercising effective control and management of the newly-added pollution.

Zhejiang is at the stage of transformation from the middle period of industrialization to the middle and later periods of industrialization. The accumulative effect of environmental pollution has been quite obvious; with respect to the polluted ecological environment, prompt and effective end-of-pipe treatment must be made to repay debts; in the meantime, in response to the problems concerning inefficient resource utilization and other respects during the existing production, Zhejiang is committed to the management of the efficiency of resource utilization, the application of energy-saving technologies, the reuse and recycling of waste resources, and has actively pushed forward clean and circular production. Moreover, Zhejiang has, by taking the long-term perspective, incorporated source control of pollutants into its routine work, carried out the regional main functional planning, industrial development plans and industrial access system to render, at the source, the newly-added projects clean, prevent new polluting projects and guide new capital to support green industrial development. As shown by the course of Zhejiang's ecological and environmental development, these three segments basically feature respective prominent roles, close connection and coordination.

2. Cooperation and coordination among three main players—the government, enterprises and the public

Ecological and environmental protection involves diverse fields, so the participation of the whole society is required. Another successful experience from Zhejiang's ecological and environmental protection lies in building a working pattern of ecological and environmental protection in which the Party and the government assume responsibilities, there is inter-departmental coordination and the participation of the whole society, and protection also involves establishing a mechanism of cooperation and coordination among the three main players, including the government, enterprises and the public.

At the governmental level, the People's Government of Zhejiang Province has built a number of mechanisms for regulation and guidance to promote ecological and environmental protection. First, ideologically, Zhejiang has become the first to make the great decision to build an ecological Zhejiang which is affluent, beautiful, harmonious and peaceful, and has comprehensively taken the "811" action for promoting ecological development, established the developmental line of thought of building an ecological Zhejiang and building the province through ecological development. Regarding organizational leadership, Zhejiang has established a leading group for building an ecological province in which the Secretary of the Party Committee of Zhejiang Province serves as the head, the Governor of Zhejiang Province serves as the executive deputy head and the people in charge of 40 departments are members; leading groups have also been established at the municipal and county levels. Second, Zhejiang has established a number of incentive and restraint mechanisms conducive to ecological and environmental protection, including the planning of the main functional areas, the plan for the building of an ecological province, the plan for the development of a circular economy, policies for industrial orientation and ecological compensation; these mechanisms have vigorously stimulated and guided the production and consumption behaviors of enterprises and residents. Third, the People's Government of Zhejiang Province has also built a system of scientific comprehensive evaluation, under which each year the leading group for building an ecological province holds a meeting to draw up some summarizing conclusions, evaluate the performance of cities and departments in ecological and environmental protection and determine their ranking, while the results of the evaluation will be taken as the important basis for evaluating the performance of the Party and the government's leading groups, appointing, rewarding and punishing the leading cadres.

At the enterprise level, with the combined action of the market mechanism and the government's mechanism of incentives and restraints, enterprises have seen an effective allocation of resources. In Zhejiang, the market mechanism developed early, the foundation for economic development is good, the government adopts a mechanism for incentives and restraints at the macro level to guide enterprise development characterized by low input, low consumption, low emissions and high returns. In particular, as some innovative mechanisms—including the ecological compensation system and the system for the trading of emission rights—have been implemented and the reforms involving the prices of resource products—including water, electricity and gas—have further expanded in Zhejiang, resources and ecological environment have been given value, enterprises have started to consciously adjust the input-output structure and the R&D direction, thus more economically allocate market factors and resources, dispose of pollutants and combine their production behaviors with the government's macro regulation and control. Moreover, a circular economy has developed robustly within enterprises, among enterprises and in industrial parks; with the execution of the "911" action plan and the "1421" demonstration project in Zhejiang, a large number of demonstration projects have been promoted and applied, clean production has extended from key industrial enterprises to the whole society and great progress has been made in developing highly-efficient green ecological agriculture.

At the level of the public, they have more enthusiasm about participation in ecological and environmental protection, and have actively supervised the government and enterprises in environmental improvement, thus the effective supervisors and supervisory atmosphere have emerged. With an increasing level of economic income, the public have more urgent needs for improving the ecological environment; with extensive environmental publicity and education by government departments, Zhejiang's ordinary people have a deeper understanding of ecological and environmental protection and a greater awareness about participating in ecological and environmental protection, a good environmental protection atmosphere has been fostered in the whole society, the public have effectively supervised and participated in the government's environmental management and enterprises' production behaviors through the media, the system for the disclosure of environmental information and through participation in environmental affairs. In the meantime, with the government's fiscal and tax policies for green consumption and under the strong publicity atmosphere, the public in Zhejiang have preliminarily built a green manner of consuming, internalized environmental protection into their behavior and promoted environmental protection in their daily lives.

Under the principle that environmental protection is regulated mainly by the government, pollution is prevented and controlled mainly by enterprises and environmental protection is supervised mainly by the public, Zhejiang has built a three-in-one system of an institutional framework for environmental protection; as the government plays a restraining and guiding role, the government has promoted environmental protection and ecological development from sectors to the society and from the governmental level to the non-governmental level; with a number of measures taken by the government, Zhejiang's people have increasingly enhanced their awareness of environmental protection.

3. An institutional structure which integrates the compulsory means, the selective means and the guiding means

As the province where the market-oriented reform was first carried out and the degree of marketization is the highest, Zhejiang has made a lot of institutional innovations and contributions in ecological and environmental protection, and has comprehensively adopted a number of policy tools, including the compulsory means, the selective means based on weighing advantages and disadvantages, and the guiding means for moral education.

As rigid institutions, the compulsory means are the foundation and baseline for other institutional arrangements. Zhejiang has issued decisions and opinions on a green Zhejiang and an ecological Zhejiang on many occasions, which shows the government's determination and attitude towards ecological and environmental protection at the institutional level; Zhejiang has subsequently established relevant policies, regulations, standards and plans to build its complete framework at the institutional level. Furthermore, Zhejiang has also linked the making and implementation of compulsory policies in different areas with the mechanism for the evaluation of performance. The *Implementation Measures of Zhejiang Province for the Evaluation of the Total Reduction of the Emission of the Main Pollutants*, issued in 2007, required, for the first time, a quantitative evaluation of the work carried out by governments at various levels involving the reduction of environmental emissions. According to these measures, the results of the evaluation are taken as the important basis for a comprehensive evaluation of the leading groups and leading cadres of local governments, a strict accountability system and a one-vote veto system are carried out.

Regarding the means of selection, Zhejiang is the first province in the country to introduce the system of ecological protection compensation. In 2005, the People's Government of Zhejiang Province issued the *Several Opinions on Further Improving the Mechanism for Ecological Compensation*, making a historic change from uncompensated ecological protection to compensated ecological protection. During the period 2007–2011, the cumulative ecological compensation fund from transfer payments made by the provincial finance reached 5.1 billion yuan. Based on many years' practice, Zhejiang has also continuously expanded the mechanism for ecological compensation: First, expanding the mechanism for the single ecological compensation into a scientific system which combines ecological compensation with indemnity for damages, and determining whether compensation or indemnity is made according to the results of the monitoring concerning the quality of the water in trans-boundary rivers; second, turning regional ecological compensation into inter-regional ecological compensation, developing and implementing the *Pilot Implementation Plan for Water Environmental Compensation for the Xin'an River Basin*. During the period 2007–2012, a cumulative amount of 6.6 billion yuan in transfer payments was made from the provincial finances. The mechanism for ecological compensation has enhanced the enthusiasm for ecological protection in the ecological barrier area, and has safeguarded ecological safety in the whole area, thus bringing about all-round regional coordinated and sustainable development in the economic, social and ecological fields. Therefore, Zhejiang has remained ahead nationwide in ecological development. Moreover, as the national pilot province for compensated use and trading of emission rights, the People's Government of Zhejiang Province has issued the *Interim Measures of Zhejiang Province for Pilot Work on Compensated Use and the Trading of Emission Rights* and other relevant regulations and policy documents; relevant departments have introduced a slew of supporting policies. As of late June, 2014, compensated use and the trading of emission rights was piloted in 68 counties (county-level cities, districts) in 11 prefecture-level cities.

Regarding the guiding means, Zhejiang has attached great importance to carrying out publicity and education concerning green development and ecological and environmental protection, so as to guide, encourage and stimulate participants to purchase green products, choose a green manner of consumption, and strengthen their awareness and behavior in environmental protection. The guiding means can effectively reduce the costs for environmental protection, cultivate the public's preference for green consumption, and gradually guide the society to foster an atmosphere of green consumption and of ecological and environmental protection. For instance, local authorities in Zhejiang have built education bases for ecological development, they have compiled and published a series of textbooks concerning ecological development for different social groups, including state organs, schools, enterprises, public institutions and rural areas (communities), they have piloted the compilation of local teaching materials concerning ecological development, so that the knowledge about ecological development is promoted at state organs, enterprises, communities (rural areas), schools and households.

Zhejiang has paid a great amount of attention to institutional adaptability, compatibility and effectiveness, and has built an institutional framework in which the compulsory means play the leading role, the selective means represent the direction and the guiding means play the supporting role. With continuous improvement in the market mechanism, the roles of the selective means and the guiding means will be strengthened.

3.3 Promoting the Building of a Beautiful Zhejiang by Governing "Five Waters" in Tandem

Zhejiang becomes famous, prosperous and beautiful thanks to water. At present, Zhejiang is facing several problems concerning water at the same time: it is overwhelmed by water—flood disasters frequently occur; water becomes dirty—water is severely polluted; water becomes less abundant—there is a shortage of water resources, thus the region of rivers and lakes south of the Yangtze River becomes "thirsty", mountains remain lush, but the water is no longer clean and clear, and this poses severe threats to the safety of the lives of the people of Zhejiang, their production and the ecology around them. Therefore, the 4th Plenary Session of the 13th Party Committee of Zhejiang Province vowed to start with treating sewage, controlling floods, discharging the accumulated water, guaranteeing a supply of water and saving water—governing "five waters" in tandem—to push forward transformation and upgrading. The decision to govern "five waters" in tandem has initiated Zhejiang's large-scale campaign of governing water. Governing "five waters" in tandem is the concrete action for extensively carrying out the "Eight-Eight Strategies" in line with the conditions of Zhejiang as a region of rivers and lakes and public appeals for water; it is the strategic priority and important means for building a beautiful Zhejiang; it is of important significance for promoting development in the economic, political, cultural, social and ecological fields.[8]

3.3.1 Promoting the Building of a Beautiful Zhejiang by Governing "Five Waters" in Tandem

The strategic decision of governing "five waters" in tandem, made by the Party Committee and the People's Government of Zhejiang Province, has drawn a high degree of attention from and enthusiastic participation by the people; it has become a great strategic decision having a vital bearing on economic and social development. This fully indicates that the ultimate goal of the strategy of governing "five waters" in tandem does not lie in governing water, but that it is aimed at transforming the lifestyle, production and developmental modes by governing "five waters" in tandem.

3.3.1.1 Promoting Industrial Transformation

The protection of Zhejiang's water environment urgently calls for industrial upgrading. There are many causes for the pollution of the water environment, but the largest source of pollution is industrial pollution caused by low-level industries. The water environment mirrors the industries in an area—if the water is clean and clear, the industrial level is high; if water is black, the industrial level is low. According to statistical data, in 2012, the traditional sectors—including textiles, chemicals, rubber, iron and steel, nonferrous metals and papermaking—still made up a large proportion of the manufacturing industry in which the main business income of the industries above the designated size exceeds 100 billion yuan in Zhejiang, while these sectors mostly feature high water consumption and heavy pollution. In particular, some sectors are generally small, scattered, disorderly and at low levels, their production processes are backward, and pollution prevention and control is less vigorous; it is very easy to cause environmental problems in these sectors.

Driving forward transformation and upgrading and promoting the building of a beautiful Zhejiang by governing waters is an important objective of governing "five waters" in tandem. The real breakthrough in governing "five waters" in tandem consists in not only governing waters, but also unswervingly shutting down the outdated production technologies, processes and products with heavy pollution, high consumption and high emissions, making room for the "birds" which eat less, lay more eggs and fly high, making available environmental space, land space and transformational space.

[8]Shen Manhong, The Strategic Significance and Realistic Path of Governing "Five Waters" in Tandem, *Zhejiang Daily*, February 10, 2014.

Under the principle of shutting down a batch of facilities, standardizing and upgrading a batch of facilities and transferring a batch of facilities to parks, Zhejiang is improving six major industries with heavy pollution, including the electroplating, printing and dyeing, papermaking and chemical industries, in a differentiated way. In 2013, more than 14,000 crystal processing enterprises (workshops) were shut down in Pujiang; Yinzhou District allocated 1.6 billion yuan to comprehensively improve the largest electroplating park (Yinzhou Electroplating Industrial Park) in the province. In an all-out effort to govern waters, more energy has been dedicated to studying structural transformation, and the ecologically economic, spatial pattern and industrial layout have been carefully planned according to the bearing capacity of the regional water environment and resource endowment. More efforts have been made to promote industrial upgrading; only in this way can the achievements in governing "five waters" in tandem be fundamentally consolidated, can the mountains become evergreen and can the water become clear forever.

3.3.1.2 Improving the Water Environment

Water is the basic natural resource, strategic economic resource and common social resource; it is the controlling element in ecology and the environment; it is the most important material foundation for the survival and development of mankind and all organisms. "Mountains, waters, forests, farmland and lakes make up a community of life." Whether in a mountain forest, a wetland, farmland, rivers or lakes, the most basic element is water. As long as there is moderate, ecological and dynamic water, the community of life made up of mountains, waters, forests, farmland and lakes is healthy. The objectives of governing "five waters" in tandem are to turn sewage into clear water, by restoring the water ecology, turn disadvantages into advantages by safeguarding the supply of water, protect the running water at the source thus making sure that the people drink ecological water and turn water wasting into water saving, thus guaranteeing ecological water utilization. Therefore, governing "five waters" in tandem is aimed at protecting the water ecology and improving the water environment.

Since the strategy of governing "five waters" in tandem was carried out, local authorities in Zhejiang have initiated an extensive campaign of governing waters, developed the roadmap for doing that and carried out solid work to govern waters smoothly. Based on the goal of eliminating black and odorous water within one year, improving the quality of the water within two years and making swimming possible within three years, Yongkang City started from rivers and rivulets, and in coordination with the actions of clearing riverways and cleaning the countryside and "cleaning, greening and beautifying the areas beside highways, railways, rivers and mountains", Yongkang City took the action of removing black, stinking and garbage-ridden rivers throughout the city, promoted the work by assigning tasks to the cadres at town and village levels and allocating tasks by area and block. Jiaxing City earmarked about 67.5 billion yuan for governing waters and carried out 509 projects involving water governance, with a planned allocation of 13.9 billion yuan in funds in 2014, thoroughly implemented the river chief system and made special improvement efforts in key areas. With the implementation of the strategy of governing "five waters" in tandem in different areas, the water environment will improve in an all-round way, Zhejiang will really return to a region of lush mountains and clean, clear waters.

3.3.1.3 Promoting the Development of the Water Governance Industry

Zhejiang has become the first nationwide to govern "five waters" in tandem; this is based on the foresight that water resources are strategic resources; this is an important sign of social progress. Governing "five waters" in tandem will bring a huge market demand and broad developmental space to the water governance industry. Zhejiang has closely seized the good opportunity for governing "five waters" in tandem, has acted ahead of others to govern waters and make arrangements in the water governance industry, thus going beyond Zhejiang by exporting equipment and technologies, resulting in a good pattern of improving the environment in other areas and stimulating the development of a batch of industries in other areas.

First, the prospects for the manufacturing of water governance equipment, key technologies and operation and maintenance services are broad. According to incomplete statistics, Zhejiang has 270 suitable technologies for preventing and controlling sewage, saving and supplying water, controlling floods and draining flooded areas, and has ascertained more than 30 provincial-level outstanding new industrial products (new technologies) in the field of water governance, it has identified and recognized more than 10 major first-ever (complete) equipment products for sewage treatment and the reuse of recycled water; after widespread integration and all-round intelligent connection between this equipment, materials, technologies and IT applications, unlimited business opportunities will emerge in the research and development of new equipment, new materials and new processes. The technology for the Internet of Things has been adopted to save water and reduce emissions, sensor products are being used to strengthen the monitoring of water resources and sewage discharge, early warning of floods, clean production and water environmental governance; there are good realistic needs and application prospects for them. Yuyao has introduced national sensor products to strengthen the monitoring of water resources and of sewage discharge, the early warning of floods, clean production and water environmental governance, which suggest great realistic demand the costs can be saved by and good application prospects. The experts under the Recruitment Program of Global Experts have adopted the advanced technology for ecological restoration to carry out the ecological restoration of riverways, effectively reduce the eutrophic matters in bodies of water, such as nitrogen and phosphorus, improve the quality of the water environment, conduct ecological desilting to enhance the capacity for self-cleaning and reduce the desilting work, while the cost of pollution control per square meter is about 70% lower than that of the method of general pollution control. One drinking straw enterprise in Yiwu has invested 2 million yuan in building "watertight" ecological recycling facilities; as a result, each year more than

700,000 yuan can be saved, the expenditure on water and electricity consumption can be reduced from 10% to 8%.

Second, the polluting enterprises have made a gorgeous turn and have become the leading enterprises for pollution control and environmentally-friendly enterprises. Subject to pressure from governing "five waters" in tandem, the large pollution emitters have no longer been in a passive position and taken a beating, they have examined themselves and worked with scientific research institutions to research special equipment and technologies for pollution control. As a Chinese saying goes, "prolonged illness makes a doctor of a patient". They may rise in a short time to become the water governance leaders in certain industries, and the consumers of a large amount of water will become water saving experts. The Juhua Group is a large pollution emitter on the upstream section of the Qiantang River, it has undergone ecological renovation at the industrial and spatial levels, it has also become the first nationwide to carry out circular ecological economic renovation in the national manufacturing industry, it has built a "resources—products—renewable resources—products" circular economic cycle, developed the "venous economy" by "eating and squeezing" raw the raw materials, products, intermediate goods and by-products from companies, thus greatly reducing energy consumption and pollution emission; in the meantime, it has engaged in major cooperation with institutions that do research on environmental protection; it has also forged an alliance with other giants, and it has made great efforts to build a platform for the development of the emerging environmental protection industry for proceeding from Juhua to cover Quzhou and the rest of the country. Shaoxing Binhai can handle 900,000t of sewage from printing and dyeing and other sectors each day, which is the largest scale of a single facility in China. Shaoxing Water Treatment Development Co., Ltd. is dedicated to scientific and technological innovations, and has cooperated closely with scientific research institutions and the institutions of higher learning, it has built R&D centers and pilot scale test basses to conduct a number of research tests, including microbial augmentation, deep treatment of sewage, heterogeneous catalysis and oxidation and the biotoxicity test, which have resulted in staged achievements.

3.3.1.4 Building a Beautiful Zhejiang

The report delivered at the 18th National Congress of the Communist Party of China vowed to build a beautiful China inherited the Chinese dream of the great rejuvenation of the Chinese nation and presented a bright future for ecological development. In order to thoroughly carry out the principles from the 18th National Congress of the Communist Party of China, the 3rd Plenary Session of the 18th Central Committee of the Communist Party of China and General Secretary Xi Jinping's major addresses, to actively push forward Zhejiang's practice in building a beautiful China, speed up ecological institutional development and march towards a new era of socialist ecological civilization, the Party Committee of Zhejiang Province made, in good time, the decision to build a beautiful Zhejiang and create a good life. Building a beautiful Zhejiang which has a blue sky, clean, clear waters, lush mountains and clean land and that is affluent, beautiful, harmonious and peaceful, culturally advanced, business-friendly and livable is the vision of building a beautiful Zhejiang; it accurately encapsulates the look of ecological development; it is an ultimate and ideal pursuit. To realize this vision, it is necessary to better utilize the important means—governing "five waters" in tandem.

When it comes to governing "five waters" in tandem, with the spirit of painstakingly taming the floods, the ambition of "reshaping mountains and rivers" and great resolve, under the principle of scientifically and resolutely governing waters with efforts from the entire population according to laws, various forces across the province have pooled together to vigorously treat sewage, control floods, discharge the accumulated water, guarantee a supply of water and save water. Water governance is being carried out to promote transformation and upgrading, to restore the lush mountains and clean, clear waters left over by our ancestors, tame the floods, remove waterlogging, make sure that the people live and work in peace and contentment, remain in harmony with nature, embark on the new developmental path in which clean, clear waters and lush mountains are the "gold and silver mountains", build a beautiful Zhejiang which has a blue sky, clean, clear waters, lush mountains and clean land and that is affluent, beautiful, harmonious and peaceful, culturally advanced, business-friendly and livable.

3.3.2 The Realistic Path of Governing "Five Waters" in Tandem

On December 23, 2013, the Standing Committee of the Party Committee of Zhejiang Province made the great decision regarding governing "five waters" in tandem, and identified the roadmap and timetable for governing "five waters" in tandem: solving salient problems within three years (2014–2016), producing marked effects, basically solving problems within five years (2014–2018), making improvements in an all-round way and basically guaranteeing no occurrence of problems within seven years (2014–2020), thus making a qualitative change.

3.3.2.1 First Controlling Pollution, Making Breakthroughs in Key Fields

The salient problems relating to pollution are an insufficient capacity for sewage treatment, underdevelopment of the pipeline network, a low rate of reaching the emission standard, severe water pollution on the river surface and the riverbed. Efforts focus on "improving three rivers, providing coverage in two respects, achieving two transformations". "Improving three rivers" means improving the black, stinking and garbage-ridden rivers. Zhejiang planned to eliminate garbage-ridden rivers and effectively improve the quality of the water in the stinking and black rivers by late 2014.

"Providing coverage in two respects" means basic coverage by urban sewage pipes, rural sewage treatment and centralized disposal of domestic garbage. "Two transformations" mean industrial transformation and agricultural transformation. Under the principle of shutting down a batch of facilities, standardizing and upgrading a batch of facilities and transferring a batch of facilities to parks, Zhejiang has taken a differentiated approach to improve six industries with heavy pollution, including the electroplating, printing and dyeing, papermaking and chemical industries. With "one policy for one factory", actions have been taken to transform and upgrade sewage treatment facilities and standardize clean production in enterprises. According to regulations, all enterprises which discharge wastewater beyond the discharge standard of sewage pipes to sewage treatment plants have been ordered to make improvements within the specified time; all of the enterprises which directly discharge wastewater that is inconsistent with relevant standards into the environment have been ordered to halt production and make improvements; all the enterprises without sewage treatment facilities and connection to a sewage pipe network have been shut down; all the enterprises which discharge pollution in violation of laws and severely in excess of relevant standards are subject to a penalty based on the maximum limit; all suspected criminals in the environmental field are prosecuted for criminal liability by a judicial organ according to laws. With a focus on ecological and intensive agricultural development, Zhejiang has vigorously promoted aggregated and large-scale operations of the planting and breeding industry and the animal husbandry industry, the centralized and pollution-free treatment of pollutant emissions, and the control of agricultural non-point source pollution.

3.3.2.2 Strictly Preventing Floods and Reinforcing Stagnant Water Drainage

Flood disaster is being overcome to ensure the people's well-being; water conservation is harnessed to develop farming and sericulture. Zhejiang has always attached importance to water conservation. Given that some cities are less able to address waterlogging and drain flooded areas, some dykes are based on low standards and are weak and there are other salient problems, local authorities in Zhejiang have taken various measures to reinforce dykes, dredge channels and intensify drainage. For reinforcing dykes, Zhejiang reinforced 100 dangerously weak reservoirs, comprehensively improved 1200 key hilly ponds with a capacity of 10,000–100,000 m^3, and reinforced 100 km of the dykes in trunk streams at key river sections separately leading to the sea in 2014. Dredging channels centered on unchoking the blocked channels, opening new channels, dredging riverways and smoothing out the pipelines. Zhejiang has continuously promoted the building of 5000 km of clear water riverways, it has initiated pilot comprehensive building of 21 national key counties for improving small and medium-sized rivers; it has planned to complete the comprehensive improvement of 2000 km of riverways and 350 km of river bank greening. Hangzhou has unveiled the three-year action plan for controlling floods, it has arranged 109 water conservation projects, and has planned to invest 19.6 billion yuan; moreover, it has carried out the project for comprehensive improvement of the Qiantang River, built some new reinforcing dykes, and raised the flood control standard; it has also completed hilly pond reinforcement above 3.50 million m^3 in 60 reservoirs within three years, built a number of new drainage and irrigation stations in such areas as Binjiang and Zhijiang and has carried out the project for expanding the Qibao Drainage Pump Station and the flood control project in Tangqi, Yuhang.

3.3.2.3 Guaranteeing a Supply of Water, Actively Promoting Water Saving

To guarantee a supply of water, it is essential to stabilize the source of water. With accelerated promotion of the 10-billion-yuan project for guaranteeing water resources, Zhejiang has completed a number of projects regarding sources of water, water diversion and supply, including the Laohutan Reservoir, the Muchen Reservoir, the mainland water diversion project in Zhoushan and the project of water diversion from the Nanxi River; Zhejiang's overall pattern of water resource allocation has basically taken shape. In 2014, Zhejiang quickened the pace of promoting the project on the allocation of water resources; breakthroughs will be made in some regional water diversion projects, such as more rapidly pushing forward Phase III of the mainland water diversion project in Zhoushan, the Shuangjiang water conservation hub project in Yiwu and the Xin'an River water diversion project.

The strictest system for the management of water resources has been carried out. In March, 2013, the People's Government of Zhejiang Province issued the *Opinions of the People's Government of Zhejiang Province on Carrying out the Strictest System for the Management of Water Resources and Comprehensively Promoting the Building of a Water-saving Society*, calling for establishing a relatively complete system for the management of water resources and a system for the management of supervision, specifying the goal, tasks and path of basically finishing the building of a water-saving society. It outlined the working measures for the quantitative indicators—the red line of control for the development and utilization of water resources, the red line of control for the efficiency of water utilization and the restrictive red line for pollution control in water function areas.

The red line of control for the development and utilization of water resources has been established. Such means as suspension or restrictive examination and approval have been adopted to exercise strict management involving total control of the quantity of water intake and utilization; the plans which fail to pass the feasibility study of water resources cannot be approved for implementation. The red line of control for the efficiency of water utilization has been established. The construction in the projects for water diversion, water intake, supply and utilization must first take into account the requirements of water saving; water saving measures and plans must be developed for the new projects, expansion projects and reconstruction projects; the industrial projects and service sectors with high water consumption are restricted in the areas with a shortage of water resources and weak ecology. The restrictive red line for pollution control in water function areas has been established. The system for the management of supervision for water function areas has been improved; for the areas where the quality of the water at the cross sections of the rivers across administrative divisions fails to pass evaluation, the examination and approval authority for the water intake permit and pollution discharge outlets of rivers is transferred to the department at the next higher level, the examination and approval of the new water intake and pollution discharge outlets of rivers is restricted.

Zhejiang has also incorporated the main indicators concerning the management of water resources into the evaluation conducted by local governments at various levels, while the results of the evaluation are taken as an important basis for the comprehensive evaluation of relevant leading cadres of local governments, fiscal transfer payments for ecological protection and the arrangement of water conservation funds.

Water must be saved while opening up the sources of water. Zhejiang has comprehensively promoted the building of a water-saving society by taking it as the main carrier. In May, 2013, Zhejiang initiated the work on building a water-saving society in the first batch of counties (county-level cities, districts). Based on pilot work and the summary of experiences, Zhejiang initiated, in batches, the building of standardized cities and counties (county-level cities, districts) for building a water-saving society. The goal is that 2/3 of the cities and counties (county-level cities, districts) across the province will reach the standard of building a water-saving society by 2020. By that time, a water-saving society will basically have taken shape in Zhejiang.

3.4 Promoting the Development of Forestry for Building a Forested Zhejiang

The forest is hailed as the main regulator of nature; it sustains the global ecological balance. The forest plays a basic and crucial role in regulating the dynamic balance of the biosphere, the atmosphere, the hydrosphere and the pedosphere; it plays the main role in the exchanging of energy and matter in the biological world and the non-biological world; it serves as the hub and lever in maintaining the overall functions of the global ecosystem. Forestry performs the ecological functions of preventing wind and fixing sand, nourishing the source of water, conserving water and soil, regulating the climate, cleaning the air, reducing noise and beautifying the environment. Developing an ecological forestry is a strategic choice for addressing climate change and is an important means for developing countries to promote sustainable development.

3.4.1 The Strategy of the Development of Forestry

The forest is the main part of the terrestrial ecosystem; it is the indispensable natural resource for human development. The forest is the main player in and the main undertaker of ecological development; it is a basic industry and an undertaking for the public benefit for sustainable economic and social development. With rapid economic development and an increasingly high standard of living for the people, forestry has played an increasingly important role and has increasingly drawn wide attention from the society.

On June 12, 2002, the 11th Party Congress of Zhejiang Province officially put forward the strategic goal of building a green Zhejiang, and stressed that the building of a green Zhejiang was a great event for Zhejiang's sustainable development. It is necessary to better develop ecological public-welfare forests, to strengthen the comprehensive improvement of river basins, to establish a mechanism for the compensation of ecological protection, and to bring about beautiful mountains and rivers. It is essential to rationally develop, utilize and protect the land, water, minerals, forests and other natural resources, and build a resource-conserving society.[9]

On December 22, 2003, when delivering his report during the 5th Plenary (Enlarged) Session of the 11th Party Committee of Zhejiang Province, Xi Jinping vowed to comprehensively plan and initiate the building of an ecological province, foster a new atmosphere of building a green Zhejiang, to adopt the new philosophy of development to actively promote a circular economy, to carry out clean production, strictly control and vigorously curb environmental pollution, attach great importance to the comprehensive improvement of the rural environment, modernize forestry across the board, pay more attention to the rational development, utilization and protection of natural resources, and to further intensify the management and construction of ecological demonstration areas, natural reserves and important ecological function areas.[10]

On June 30, 2010, the 7th Plenary Session of the 12th Party Committee of Zhejiang Province adopted the *Decision of the Party Committee of Zhejiang Province on Promoting Ecological Development*. The decision called for increasing forest coverage and improving the quality of forests, vigorously developing ecological public-welfare forests, economic forests with valuable species of trees, focusing on strengthening the tending of young and middle-aged forests, improving the low-efficiency forest and the forest form, promoting afforestation in the plains areas and more rapidly building a forested Zhejiang.

On June 6, 2012, when delivering a report during the 13th Party Congress of Zhejiang Province, Zhao Hongzhu vowed to speed up the building of green barriers in western Zhejiang and blue barriers in eastern Zhejiang, to practically strengthen the development and management of natural reserves and important ecological function areas as well as to extensively carry out the planting of forests, energetically develop

[9]Zhang Dejiang, Practicing the Important Thought of Three Represents, Comprehensively Promoting Socialist Modernization in Zhejiang Province The Report Delivered during the 11th Party Congress of Zhejiang Province, *Zhejiang Today*, 2002(12).

[10]Xi Jinping, Giving Full Scope to Eight Strategies, Thoroughly Implementing Eight Measures, Solidly Promoting All-round, Coordinated and Sustainable Development in Zhejiang Province The Report Delivered during the 5th Plenary (Enlarged) Session of the 11th Party Committee of Zhejiang Province, *Zhejiang Today*, 2004(1).

modern forestry, protect the forests, the wetland and wild animal and plant resources and to intensify the protection of biodiversity.

On May 23, 2014, the 5th Plenary Session of the 13th Party Committee of Zhejiang Province adopted the *Decision of the Party Committee of Zhejiang Province on Building a Beautiful Zhejiang and Creating a Good Life*, calling for strengthening the development of green ecological barriers, extensively promoting the project of stimulating the farmers on high and remote mountains to descend from the mountains and migrate to other areas, intensifying the protection of forest resources, comprehensively promoting the afforestation of the plains, forest coverage and quality enhancement, improving the quality of forest stands and increasing the volume of forest stock, thus enhancing the functions of the forest ecosystem; it also called for improving the reform of the system of collective forest rights, establishing a mechanism for adjusting the provincial fiscal compensation standard for ecological public-welfare forests and studying the compensation measures for ecological benefits in the wetlands.[11]

On October 10, 2014, the Party Committee and the People's Government of Zhejiang Province issued the *Opinions of the Party Committee and the People's Government of Zhejiang Province on Accelerating the Forestry Reform and Development, Comprehensively Putting into Practice Afforestation in the Plains and the Region of Rivers and Lakes in Five Years and Finishing the Building of a Forested Zhejiang in Ten Years*, making major strategic arrangements for the forestry work at present and in a certain period of time to come by proceeding from the overall situation from A strategic perspective.

3.4.2 *The Development of the Modern Forestry Industry for Building a Forested Zhejiang*

3.4.2.1 Key Ecological Projects Related to the Forestry Industry

1. The building of the compensation mechanism for ecological public-welfare forests

Zhejiang's minimum compensation standard[12] for public-welfare forests above the provincial level increased year by year, the compensation standard for public-welfare forests was raised from 8 yuan/mu in 2004 to 19 yuan/mu in 2011; the minimum compensation standard for public-welfare forests above the provincial level was further raised to 25 yuan/mu in 2013 after it increased from 19 yuan/mu to 21 yuan/mu in early 2013; it increased from 25 yuan/mu/year to 27 yuan/mu/year in

[11] Shen Zhengxi, The Decision of the Party Committee of Zhejiang Province on Building a Beautiful Zhejiang and Creating a Good Life, *Zhejiang Today*, May 29, 2014.

[12] Shen Zhengxi, The Decision of the Party Committee of Zhejiang Province on Building a Beautiful Zhejiang and Creating a Good Life, *Zhejiang Today*, May 29, 2014.

2014; in 2015, it increased to 30 yuan/mu/year, while the rental price of collective forests in natural reserves above the provincial level rose by 2 yuan and 3 yuan, and the incremental funds came from provincial finances. Afterwards, Zhejiang readjusted the mechanism for the growth of forest-related ecological compensation in which adjustment is made once every two years and the extent of that adjustment has been 2 yuan at a time. The area of public-welfare forests covered by the fiscal compensation above the provincial level across the province exceeded 40 million mu, and the amount of compensation was more than 1.1 billion yuan. In Zhejiang, more than 1.09 million rural households, more than 18,000 village-level collective economic organizations and more than 300 state-owned units received benefits.

2. The development of the system for the protection of forests

In 2010, 1,043,300 mu of afforestation was completed, including 207,800 mu of artificial afforestation; 205,000 mu of reforestation was completed, 382,400 mu of low-production and low-efficiency forests were improved. In 2011, 633,000 mu of afforestation renewal was completed; 200,000 mu of reforestation was completed, 404,000 mu of low-production and low-efficiency forests were improved. In 2012, 39,695 hectares of afforestation was completed in Zhejiang, including 12,561 hectares of afforestation in a project for the protection of key forests; 16,587 hectares of reforestation was completed, 16,878 hectares of low-production and low-efficiency forests were improved. In 2013, 39,455 hectares of afforestation was completed in Zhejiang, including 17,369 hectares of afforestation in a project for the protection of key forests; 14,380 hectares of artificial reforestation was completed, 8574 hectares of low-production and low-efficiency forests were improved.

3. Vigorously pushing forward the "1818" action for the afforestation of the plains areas

Given that forests were lacking in the areas where they were most needed, the Secretary of the Party Committee of Zhejiang Province, Xia Baolong, worked on the afforestation of the plains, and held a provincial working conference on it once each year for five consecutive years since 2010, thus fostering and consolidating the fine tradition that the Secretary of the Party Committee works on afforestation. The "1818" action for the afforestation of the plains was initiated during the period of the 12th Five-Year Plan. At present, the cities and counties across the province have earnestly carried out the principles adopted at the working conference of the Party Committee of Zhejiang Province on the afforestation of the plains, the "1818" action for the afforestation of the plains has been initiated in an all-round way, it was promoted more quickly by taking a series of feasible measures, including training classes, on-the-spot meetings, organized learning and visits as well as special supervision. Many cities and counties have annually allocated on average more than 100 million yuan from their public finances for the afforestation of the plains, and the increased afforestation area exceeded 10,000 mu.

3.4.2.2 Strengthening the Protection of Forest Resources in an All-Round Way

1. The management of the quota of cutting down forests

In order to reinforce the management of cutting down forests, effectively protect, cultivate and rationally utilize forest resources, Zhejiang has drawn upon a number of national and provincial reform policies and feasible practices in pilot work regarding the management of cutting down forests, amended the *Administrative Measures of Zhejiang Province for the Cutting Down of Forests*, and has made great breakthroughs in the cutting down of Moso bamboo, tree cutting and excavation, the disclosure of cutting information, the control of cutting areas, and streamlined administration and delegation of powers to the lower levels.

2. Organizational safeguard

Regarding the management of forest resources, the Department of Forestry of Zhejiang Province has established the division of forest resources and a provincial resource management station, and the resource management station assists the division of forest resources in work; cities, counties and towns have set up the department of forest administration, the section of forest administration and the post of forest administrator, respectively, which are responsible for resource survey and forest administration. Relevant provincial departments have also approved the establishment of timber inspection stations and patrol brigades, including timber transportation patrol brigades, national highway inspection stations, provincial highway inspection stations, county and town road inspection stations, non-road inspection stations and water inspection stations. There has been an effective mechanism for management at the upper and lower levels for effectively protecting forest resources.

3. The protection and management of wild animals and plants

Zhejiang has actively carried out salvage-oriented protection of endangered species, and has built 19 wildlife resource and disease monitoring stations above the provincial level. In the meantime, Zhejiang has carried out large-scale publicity activities across the province, covering wildlife protection laws, regulations and policies, bird knowledge and the popularization of science. Zhejiang has conducted various forms of distinctive activities, including lectures by experts, photo displays, distribution of materials, award-winning quizes, which have played an active role in promoting bird protection, resource recovery and growth in Zhejiang and in enhancing public awareness about loving and protecting birds.

4. Wetland protection

Zhejiang has made efforts to strengthen supervision, curb illegal occupation, reclamation, landfilling and wetland pollution according to laws, and hold to account the units and individuals severely damaging wetland ecology according to laws; moreover, Zhejiang has made efforts to increase financial support for the construction of necessary infrastructures, ecological restoration and improvement, biodiversity protection, scientific research and monitoring in important wetland areas, including wetland reserves (small reserves), wetland parks, and demonstration areas for the rational utilization of wetland, guaranteeing the funds for wetland protection and management; it also attempts to extensively mobilize various types of media to publicize relevant laws and regulations, including the *Convention on Wetlands*, the *Wildlife Protection Law*, the *Regulations of Zhejiang Province on Wetland Protection* and the *Regulations on the Administration of Wetlands*, popularize the knowledge about wetland protection, enhance the awareness of the entire society about wetland protection, consciously fulfilling their obligation for wetland protection and jointly promote it.

3.4.2.3 Actively Promoting Industrialized Operations of the Forestry Industry

1. Accelerating the development of modern forestry parks

Zhejiang has concentrated its efforts on building a number of modern forestry parks where planning and layout are rational, production factors join together, science, technology, facilities and equipment are advanced, the operational mechanism is well-functioning and where economic returns and demonstration effects are apparent. Zhejiang has moved faster to promote the transformation and upgrading of six leading industries—including forestry, bamboo, flower and nursery stock, forest food, wild animal and plant taming and breeding, forest tourism—and actively develop a forest economy and further enhance the capacity for making the people rich. Zhejiang has more rapidly pushed forward road construction in forest areas, developed the rational standards for road grades and the intensity of road networks in light of the reality in the forest areas, it has strengthened its policy support and put in place construction funds.

2. Further developing the industry of deep processing of forest products

Zhejiang has supported the enterprises for the processing of forest products in introducing advanced technologies and equipment, in speeding up technical transformation, enhancing their capability for independent innovation, and in increasing the level of deep processing of forest products. Zhejiang has actively promoted the circular utilization of bamboo and timber and the comprehensive utilization of resources, strengthened the reuse of bamboo and timber processing wastes, bamboo and timber waste products, and pushed forward the development of a circular economy in the forestry industry. Zhejiang has guided industrial agglomeration, cultivated the industrial clusters of bamboo and timber processing with regional characteristics, and reinforced the function of industrial clusters in bringing together enterprises, capital and technologies, it has transformed and upgraded the massive economy in the forestry industry towards modern industrial clusters.

3. Vigorously developing the forest tourism industry

Zhejiang has tapped various functions of the forest, including ecology, leisure and health preservation, and has energetically developed the forest-related ecological leisure tourism industry; Zhejiang has built a system of forest ecotourism which is dominated by forest parks and supplemented by natural reserves wetland parks and forestry sightseeing parks; and it features a rational layout, rich content, scientific management, complete functions and good returns. Forest tourism festivals are held to promote the forest ecotourism routes and choicest forest tourist attractions with Zhejiang's characteristics. The forest resources and excellent endowment of tourism and landscape resources in the mountain forest development areas in central-southern Zhejiang and the middle and lower mountain forest development areas in western Zhejiang are utilized to deeply tap the unique forest-related ecological culture connotation, build a series of famous brand forest ecotourism products which are well-known within the province, famous in the country and regionally influential, cultivate the agglomeration areas of forest tourism industries with contiguous development, a high service level and a high capability for stimulating the development of others, and to establish a system of forest eco-tourism industries which centers on forest eco-tourism products.

3.4.2.4 Strengthening the Building up the Capacity for the Support of Forestry

1. Promoting the rule of law in the forestry industry

Smooth progress has been made in legislation projects, including the *Administrative Measures of Zhejiang Province for the Cutting Down of Forests*, the *Regulations of Zhejiang Province on Wetland Protection*, the *Measures of Zhejiang Province for the Protection of Wild Plants*, the *Regulations of Zhejiang Province on the Prevention and Control of Pine Wilt Disease*, the *Regulations of Zhejiang Province on Forest Fire Protection* and the *Administrative Measures of Zhejiang Province for Public-welfare Forests.*

2. Promoting the science and technology-based development of the forestry industry

Zhejiang has actively pushed forward scientific and technological innovations, and has energetically organized efforts to make scientific and technological breakthroughs in line with the reality of the modern forestry industry. To further spread science and technology, Zhejiang has promoted a number of projects regarding seedlings and flowers, forest protection, cultivation of an economic forest, management of science and technology, forest product processing, bamboo and forest cultivation, forestry ecology and forest operations.

3. Boosting the development of forestry public security personnel

With regard to the public security of forestry, based on the line of thought of building highly capable forces, purifying the public security circles, fostering good conduct in the field of public security and increasing the level of law enforcement, with a focus on strengthening regularization and with a view to improving the overall quality, the public security organs at various levels have been strict with police governance, police administration, working style, police style and conduct, law enforcement and discipline compliance, so that the cohesive force of the police and its capability to combat are further enhanced. Regarding the basic primary-level work, police go to the primary-level areas, work on the front line, provide services in advance, get familiar with the conditions of the mountains, forests and communities and play is given to the roles of the people, mass defense and mass governance. Laws are enforced strictly in handling cases, the basic services are improved, the primary-level service system is reformed, the first-line safeguard measures are taken, and the regularization of the police force is strengthened, and the basic primary-level work has shown a good momentum of development.

4. Attaching importance to forest fire protection

The Forest Fire Protection Headquarters of Zhejiang Province has carried out the provincial activity to extensively identify potential forest fire hazards, and has organized relevant personnel to go deep into key forest areas to identify those potential hazards and help the primary-level units really remove them. In the meantime, Zhejiang promulgated the *Regulations of Zhejiang Province on Forest Fire Protection* in July, 2007, further improving the forest fire protection organizations, the prevention and extinguishment of forest fires, safeguard measures, supervision and management and legal liability. Moreover, Zhejiang has actively carried out the publicity of forest fire protection.

3.4.3 The Achievements of and Experience in the Development of the Forestry Industry

3.4.3.1 The Main Achievements in the Development of the Forestry Industry

Based on the overall line of thought of intensifying the reform of the forest rights, developing the modern forestry industry, making ecological progress and promoting scientific development, Zhejiang has strengthened ecological development, developed the industries conducive to making the people rich, reinforced resource management and improved policy measures, and it has made significant progress in the development of the forestry industry.

1. The forest resources have grown continuously, significant improvement has been made in the quality and the ecological conditions of the forests

The forest area, the volume of the forest stock and the forest coverage have increased steadily. The area of forestland in Zhejiang was 6,612,700 hectares in 2012, up 1500 hectares compared with 2011. Zhejiang's forest area was 6,040,600 hectares, down 12,200 hectares compared with 2011; the forest coverage was higher than 60%. The volume of standing forest stock reached 282 million m^3, up 13,287,000 m^3 compared with 2011; the quantity of Moso bamboo plants increased by 20,672 compared with 2011. The volume of arbor forest stock per unit area increased by 2.67 m^3/hectare. The volume and quality of forest resource stock have improved to some extent; the structure of age groups and the structure of the tree species have gradually tended towards the rational; the ecological condition of the forests has further improved; further scope has been given to the ecological service function of the forests; Zhejiang's forests are developing from a sustained quantitative increase to a concurrent quantitative increase and there is quality and structural improvement.

2. Key ecological projects have been promoted smoothly, significant achievements have been made in the ecological development of forests

Zhejiang has dedicated more resources to and has promoted the development of key ecological public-welfare forests. The area of public-welfare forests above the provincial level has exceeded 40 million mu, the minimum compensation standard has increased from 25 yuan/mu/year to 27 yuan/mu/year, more than 1.09 million rural households, more than 18,000 village-level collective economic organizations and more than 300 state-owned units have directly enjoyed the compensation relating to ecological public-welfare forests in Zhejiang. Key ecological projects in the forestry industry have been completely carried out, 17,748 hectare of afforestation were completed in Zhejiang in 2013, up 4948 hectares compared with 2012, accounting for 41.9% of the total area of afforestation—the area of the project on forest protection in the Yangtze River Basin was 6535 hectares, up 2760 hectares compared with the previous year; that of the project for the protection of the coastal forests was 11,213 hectares, up 2188 hectares compared with the previous year; the protective forests in the Yangtze River Basin and the coastal protective forests accounted for 15.4% and 26.5% of all the afforestation, respectively.

3. The upgrading of the forestry industry has been pushed forward, new progress has been made in developing the forestry industry to make the people rich

The output value of Zhejiang's forestry industry totaled 396.49 billion yuan in 2013, up 38.89 billion yuan or 10.9% compared with 2012, and the industrial scale increased continuously. The output value of the primary industry was 76.65 billion yuan, accounting for 19.3% of the total output value of the forestry industry, up 4.6% compared with 2012; that of the secondary industry was 229 billion yuan, accounting for 57.8% of the total output value of the forestry industry, up 7.0% compared with 2012; that of the tertiary industry was 90.83 billion yuan, accounting for 22.9% of the total output value of the forestry industry, up 29.0% compared with 2012. The structure of the output value of the primary, secondary and tertiary industries in the forestry industry was adjusted from 28:50:22 in 2010 to 19:58:23 in 2013, and the proportion of the secondary and tertiary industries increased gradually.

4. Resource and forest administration and management has been strengthened, biodiversity has been protected effectively

Zhejiang has reinforced the protection and administration of its forestland, forest trees, wild animals and plants as well as its wetland resources according to laws. Zhejiang has promulgated relevant regulations, including the *Administrative Measures of Zhejiang Province for the Cutting Down of Forests* and the *Regulations of Zhejiang Province for Wetland Protection*. Zhejiang has strictly carried out the system for the management of quotas for land expropriation and occupation, for the dynamic monitoring of forest resources and it has established the early warning mechanism for forest resource growth and decline; Zhejiang has built 19 wildlife resource and disease monitoring stations above the provincial level, and has actively carried out salvage-oriented protection of endangered species; Zhejiang has incessantly promoted the standardization of forest public security and forest fire protection forces, intensified infrastructure input in forest pest quarantine, prevention and control and forest fire protection. The natural reserves of the forest type have been further developed. The *Regulations of Zhejiang Province for Wetland Protection* have been made. As of late 2013, there were 30 wetland reserves, and 17 wetland parks above the provincial level (7 national-level wetland parks); more than 50,000 hectares of wetland was under protection in Zhejiang.

5. The activity "Care for the Forests" has been carried out, new progress has been made in the development of the forest culture

Zhejiang has fully initiated a series activities called "Care for the Forests" to arouse the whole society's enthusiasm about planting trees, protecting and loving green landscapes, thus making improvements to the forests. Zhejiang has actively carried out the activity of building forest cities. As of late 2013, there were 28 forested cities, more than 89 forested towns and more than 487 forested villages above the provincial level. Zhejiang has conducted the activity of appraising and choosing the provincial tree and provincial flower, popular voluntary tree planting, and the voluntary cultivation, management and preservation of trees; a good social conduct of popular tree planting and green landscape protection and ecological protection has gradually taken shape. Zhejiang has successfully held the Yiwu International Forest Product Expo 6 times, the China (Wenzhou) Forest Tourism Festival 5 times, the National Forest City Forum, the Yangtze River Delta International Forum on Flower Industrial Development and the International Forum on Wetland Protection.

6. Institutional reform has been deepened, injecting new vitality into developing the forestry industry

According to statistics, the area for which forest right certificates had been renewed (issued) was 86,545,000 mu in Zhejiang, accounting for 99.7% of the area for which forest right certificates should be renewed (issued), among which the collectively owned area and state-owned area were 82,558,000 mu and 3,987,000 mu, accounting for 95.4% and 4.6%. A total of 5,055,000 forest right certificates in the nationally unified form were renewed (issued), accounting for 99.8% of the forest right certificates which should be renewed (issued); 4,539,000 contracts for the contractual mountains were signed, accounting for 99.5% of the contracts which should be signed. The reform of the management of the cutting down of forests has been carried out, and the approval procedure and the scope of quota management have been further simplified. Great progress has been made in the work on policy-oriented forest tree insurance across the province. According to statistics, as of late December, 2013, more than 60 million mu of forests were covered by insurance across the province except Ningbo, accounting for more than 70% of the total forestland area, up 15 percentage points compared with the same period of the previous year; the total insurance amount was 25.5 billion yuan, the insurance premium subsidy from the finances at various levels exceeded 40 million yuan, accounting for 88% of the total insurance premium—nearly 40 million mu in public-welfare forests above the county level were covered by insurance. In 2013, various types of disasters frequently occurred in Zhejiang, the role of policy-oriented forest tree insurance in disaster relief and in the reduction of losses became more apparent; 39,094 rural households received claim settlement relating to forest disasters throughout the year, the claim settlement was more than 20 million yuan, the simple loss ratio was 44%, up 13 percentage points compared with the previous year.

7. The scientific, technological and infrastructure input in the forestry industry has increased, the scientific and technological support and production conditions have improved

Zhejiang has given priority to the work on science and technology in the forestry industry during the reform and development of the forestry industry by taking it as its basic task. Besides the research, development of and efforts in making breakthroughs in key technologies for supporting the ecological forestry industry and the forestry industry relating to the people's well-being, Zhejiang has established the responsible promotion system and the system of scientific special commissioner, it has built a large number of forestry-related scientific and technological innovation and service platforms, provincial-level key scientific and technological innovation teams, promoted more than 200 fine varieties including the tea-oil tree series, and practical advanced technologies in cloning and more than 100 other areas, including "one bamboo and three shoots", with a focus on the promotion of new varieties, the application of new technologies and the demonstration of new products; the rate of scientific and technological achievement conversion and their rate of contribution have reached 60%.

3.4.3.2 The Basic Experience in the Development of the Forestry Industry

1. Attaching great importance to the plan for development

Thinking determines a path of development. One prominent feature in Zhejiang's development of the forestry industry is that Zhejiang has attached great importance to the plan for development for the ecological forestry industry, it has promptly put forward the strategy and goals for development in light of different stages of development, thus the development of the ecological forestry industry has become a conscious action. After 2002, Zhejiang proceeded from the reality to carefully plan the development of the forestry industry and work out the overall line of thought, goals, work priorities and policy measures. In 2004, the Party Committee and the People's Government of Zhejiang Province set the following goals: by 2010, 2 million hectares of key ecological public-welfare forests will be built, Zhejiang's forest coverage will reach 62%, and the volume of standing forest stock will reach 165 million m^3. The density of the canopy of the forest stand will reach 0.6, the broad-leaved forest and mingled forest will account for 30% of the forest stand, the rate of urban greening will reach 32%, the rate of forestry-related scientific and technological achievement conversion and the rate of contribution from forestry-related scientific and technological advance will reach 60%, and Zhejiang will stay ahead nationwide in the overall level of forestry science and technology. By 2020, 3.33 million hectares of ecological public-welfare forests will be built, the forest coverage rate will be higher than 62%, the volume of standing forest stock will be more than 180 million m^3, and the rate of urban greening will reach 35%; Zhejiang will have a modern forestry system which features abundant resources, a rational layout, complete functions, high quality, high efficiency, advanced management and rich connotation, and will basically satisfy the requirements for building an ecological province with beautiful mountains and rivers, harmony between the people and nature and sustainable economic and social development.

2. Actively pushing forward institutional innovations

Institutional innovation is an important guarantee for boosting industrial development. The Party Committee and the People's Government of Zhejiang Province and the Party committees and governments at various levels in Zhejiang have always taken the development of the forestry industry as a priority in sustainable economic and social development, they have practically strengthened leadership, increased input, attached importance to the development of formal institutions including regulations and rules and improvements in the mechanism, so that there are regulations and rules for governing the development of the forestry industry. First, in recent years, Zhejiang has promulgated relevant regulations, including the *Administrative Measures of Zhejiang Province for Public-welfare Forests* and the *Regulations of Zhejiang Province on Forest Fire Protection,* creating a good legal environment for the reform and development of the forestry industry. Second, Zhejiang has been carrying out the revised *Administrative Regulations of Zhejiang Province for Forests* as from July 1,

2004 and the *Administrative Measures of Zhejiang Province for the Cutting Down of Forests* as from April 1, 2014 according to the *Forest Law* and its implementation regulations and in light of the province's reality. Both regulation documents draw upon the principles adopted in the normative documents concerning the administration of forest resources and the practical work experience in recent years, they address some legal issues that were difficult to address in the past, they have made comprehensive affirmation in the form of local and government regulations, they have improved the administration system for the ecological forestry industry and have provided vigorous support for the administrative enforcement of laws. Third, in order to regulate the reform of the collective forest right system and guarantee that such a reform is smoothly carried out, the Party Committee and the People's Government of Zhejiang Province have introduced relevant supporting systems, including the *Several Opinions on Further Intensifying the Reform of the Collective Forest Rights System*, improving the supporting systems for the reform of the collective forest rights system.

3. Constantly intensifying the reform of the forestry industry

Reform and innovation is the impetus for boosting the development of the forestry industry. With reform and innovation, the development and management of the modern forestry industry has been improved. In particular, since the reform of the collective forest rights system was fully carried out across the province, the ownership of mountain forests has become clearer, the players in the contractual operations of forestland have become more explicit, the system and mechanism of the forestry industry have been gradually rationalized, the productive forces in the forestry industry have been emancipated continuously, so that forest workers have unprecedented enthusiasm for afforestation, forest protection and cultivation. Various elements in the society have gathered in the forestry industry, the non-public forestry industry has developed rapidly, giving shape to a new pattern with the diversification of investments and operational forms, province-wide mobilization and with the participation of the whole province and the whole society.

Zhejiang's experience in intensifying the reform of the forestry industry is mainly as follows: First, with a focus on capital access, innovating the work on mortgage loans based on forest rights, adopting three modes—small mortgage loans based on forest rights, mortgage guaranteed loans based on forest rights and direct mortgage loans based on forest rights—to effectively overcome the financial difficulties for forest workers in developing forestry production; second, with a focus on risk prevention and control, innovating forest insurance—the People's Government of Zhejiang Province has carried out policy-oriented compensation through forestry insurance, introducing the packing joint guaranty, allocating funds from public finances, dealing with risks and making claims settlements in a unified way, overcoming the high insurance costs and claims settlement difficulties for the households of forest workers which arise due to the small area of a single land block and a large number of land blocks, thus greatly enhancing the capacity of forest workers for preventing and controlling risks; third, with a focus on digital management, innovating the forestry information platform, building a system for the management of the information regarding forest rights, introducing the forest rights IC cards in 9 cities and countries, including Lishui, dealing with forest resources and credit insurance of households through information means and chips, so as to make the management of forest rights digital and convenient and really build a digital forestry industry on the mountains and land blocks; fourth, with a focus on serving forest workers, innovating the forestry management body, building 65 service centers for the management of forest rights, 49 centers for the trading of forest rights, resource evaluation centers as well as collection and storage centers, thus providing very convenient service platforms for forest workers.[13]

[13]Shen Manhong, Wei Chu, Cheng Hua, et al. 2012 Zhejiang Eco-economic Development Report. China Financial & Economic Publishing House, 2012, p. 99.

Chapter 4
Inheritance of the Ecological Culture with Harmony Between People and Nature

Hongyu Li

The manners of production and life in the society of a certain area are guided by unique regional cultural forms; culture is the starting point of the philosophy relating to human activities in production and life; a civilization dominated by industrial culture constitutes an industrial civilization, while production and life modes dominated by an ecological culture give rise to an ecological civilization. The thoughts that "harmony between people and nature should be achieved", "the objective law follows nature", "care for all the people and animals should be shown" advocated in traditional Chinese culture reflect the spirits of an ecological culture. In traditional Chinese culture, the philosophical line of thought of the systematic relationship between the people and nature is mainly reflected in the "harmony between people and nature" which is a long-standing issue. According to the *Commentaries on the Book of Changes*, the people and all creatures arise out of heaven and earth, so nature and people are essentially identical. Only when people act in line with the virtues in heaven and earth, brightness in the sun and the moon and the time sequence of the four seasons can the people master the law of nature to achieve freedom. An ecological civilization is a new form of civilization following a primitive civilization, a farming civilization and an industrial civilization; it carries forward and inherits the achievements of the human civilization, it also embodies a profound reflection on and criticism of the issues concerning a traditional civilization, especially an industrial civilization, it discards the value outlook, in an industrial civilization, that people rule nature, and stresses that human development should conform to the law of ecology and that people and nature should coexist harmoniously; it changes the developmental mode of "treatment after pollution", emphasizes the principle of giving priority to ecology, and calls for promoting economic and social development and ecological and environmental development at the same time; it changes the materialistic lifestyle, stresses an abstemious accumulation of material wealth, upholds a moderate rate of consumption and cultural enjoyment, and the lifestyle of meeting human

H. Li (✉)
Chinese Academy of Social Sciences, Beijing, China

J. Pan and M. Shen (eds.), *Chinese Dream and Practice in Zhejiang – Ecology*, Research Series on the Chinese Dream and China's Development Path,
https://doi.org/10.1007/978-981-13-7209-4_4

needs and causing no damage to the natural environment. It is generally held that an ecological culture is a new cultural form which negates traditional culture and stresses coordinated development of people and nature.

Zhejiang has a long history; it is ecologically developed and culturally prosperous. After the national economic center shifted southward to the region south of the Yangtze River during the Eastern Jin Dynasty (317–420) and the Southern Dynasties (420–589), Zhejiang caught up from behind and was far ahead for thousands of years, and it has been prosperous till now. Its economic and social development has always been promoted under the guidance of an ecological cultural spirit. Zhejiang's basic spirit in ecological culture highly tallies with the inherent requirements of an ecological civilization, so Zhejiang has become the first to respond to and take practical actions with respect to ecological civilization.

4.1 The History of an Ecological Culture with Harmony Between People and Nature

Zhejiang has a long-standing ecological culture. Its philosophical thought of harmony between people and nature has been consistent throughout ancient culture, the development of traditional industries, social life and political law.

4.1.1 The Birthplace of the Chinese Ancient Ecological Culture

Zhejiang enjoys an excellent natural environment of mountains and waters and is livable for human beings. As early as the Palaeolithic age 50,000 years ago, the primitive people "Jiande People" were active in Zhejiang (more than 100 ruins in the Neolithic Age have been found within Zhejiang); this was the prelude to a harmonious coexistence between the people and nature. Afterwards, the Kuahuqiao Culture, which has a history of 8,000 years, emerged. The Kuahuqiao Culture originated at the intersection of the Qiantang River, the Fuchun River and the Puyang River. The ruins of the Kuahuqiao Culture are named after a cross-lake bridge between Lake Shangxiang and Lake Xiaxiang which are parts of the ancient Lake Xianghu. The Hemudu Culture, which has a history of 7,000 years, originated at the foot of the scenic Siming Mountain in Hemudu Town of Yuyao City and the north bank of the Yaojiang River; a stilt-style architecture is the typical representative of the Hemudu Culture, its architectural structure facilitates the flow of air and brings a pleasantly cool experience, it can prevent miasma, moisture, water logging, insects and snakes; its construction is simple and it is a typical green architecture. The ancient people in Zhejiang attempted to cultivate crops, including paddies, and to domesticate wild livestock and poultry, make pottery items and lacquerware, utilize nature to improve production tools, while these attempts reflect the ecological cultural wisdom and philosophy of the ancient people in Zhejiang. Zhejiang's Majiabang Culture, which has a history of 6,000 years, was available along the southeastern coast of Lake Taihu; The Majiabang Culture gave birth to the early fishing, hunting and gathering economy in China. As shown by Caixieshan and Chaodun ruins, the paddy had become an economic method of production, and it entered the period of paddy agriculture in the strict sense; the Taihu River basin during the Majiabang period can be considered one of the birthplaces of paddy agriculture in the world. In the meantime, animal domestication reached a stage of high maturity during the period of the Majiabang Culture; pigs, water buffalo and dogs were the domesticated livestock, even the sika deer and elk were reared in pens; the remains of these animals have been massively found on the sites of the Majiabang Culture. Livestock needed fodder, this suggested that agriculture had been able to provide surplus foods, livestock provided the people with meat, their excrement was the best manure for agricultural production and animal bones and horns were the raw materials for making tools. The Majiabang Culture can be considered to reflect a relatively systematic ecological economic thought and it initiated the history in which Zhejiang was part of the country with an ancient civilization, thus it is hailed as the birthplace of the culture south of the Yangtze River. Zhejiang's Liangzhu Culture, which has a history of 5,000 years, was mainly available in the Taihu River basin. The Liangzhu Culture is mainly characterized by the making and application of jadeware; there was jadeware with holes in the center, hollow jadeware with rectangular sides, tomahawk-shaped jadeware, semi-annular jadeware and crown-shaped jadeware, tripod jadeware, jade bracelets, jade tubes, jade beads, jade pendants, cylindrical jadeware, conical jadeware and jade belts and rings; the jade culture arose accordingly, indicating that the people revered nature. Moreover, pottery craftsmanship was more exquisite, it gradually developed into an important industry at that time; it can be considered as the early ecological industry in Zhejiang. Zhejiang was divided into Wu State and Yue State during the Spring and Autumn Period (770 B.C.–476 B.C.). Kuaiji Prefecture was set up in Zhejiang during the Qin Dynasty (221 B.C.–207 B.C.). Sun Quan, a native of Fuyang, founded Wu State during the Three Kingdoms Period (220–280); during these three historical periods, Zhejiang was an area with the most developed a farming culture and a mulberry silkworm culture. During the Spring and Autumn Period, developing agriculture and mulberry was one of the national policies adopted by Goujian, the King of the Yue; silk fabrics made in Shaoxing were well-known in many areas during the Tang Dynasty and the Song Dynasty. According to an ancient book entitled *Yueyou Bianlan*, the ornate areas in northwestern Shaoxing featured a large number of silk fabric machines and a market for silk fabrics where products were soft plain-weave silk fabrics and silk fabrics with a geometric design; Xiafangqiao mostly provided high-quality slender and tenacious tribute silks and they were sold in large quantities nationwide and to southeastern Asia. The silk fabrics unearthed in Liangzhu, which have a history of 4,700 years, suggest a long history of silk in Zhejiang. Zhejiang's mulberry silkworm culture can be regarded as an ecological culture with the most Chinese characteristics, so it is hailed as the starting point of Chinese civilization.

4.1.2 A Variety of Distinctive Philosophical Thoughts Regarding the Ecological Culture

Throughout history, Zhejiang has had a lot of philosophical thoughts on ecological culture. Wang Chong, a thinker in the Han Dynasty, native of Shangyu, Kuaiji (now Shangyu City, Zhejiang), inherited the naive materialistic thought and explained the nature of the heaven and the earth from the perspective of ecological philosophy. He said, "The heaven and the earth are nature which contains vitality; the heaven is a body, it is equivalent to the earth; the heaven, the body and vitality-containing nature fundamentally affirm the natural material attribute of the heaven and the earth." Wang Chong proceeded from the material attribute of the heaven and the earth to elucidate the theoretical view that the heaven and the earth naturally exist and develop, which means that the movement of celestial bodies does not generate things on earth, and things arise by themselves, this is called nature; the application of vitality by celestial bodies does not create things on earth, and things come into being subject to vitality, this is called inaction. In his view, the movement of the heaven and the earth is a process of natural change rather than the effect of any spirit, the people and things naturally emerge amidst the movement of the heaven and the earth; the heaven covers the earth, the earth lies underneath, the vitality from the earth ascends to the heaven, while the vitality from the heaven descends to the earth, things arise between the heaven and the earth; the people do not arise because of the heaven and the earth, things do not come into being due to the heaven. He used the doctrine of vitality and vitality change to explain the generation of and change in things, which was the early systematic theoretical expression of philosophical thoughts of an ecological culture.

Wang Yangming, a famous thinker in the Ming Dynasty, native of Yuyao County, Shaoxing (now Yuyao, Ningbo), once built a house in Yangming Cave on Kuaiji Mountain, Zhejiang; he gave a great deal of attention to national knowledge, the law of nature, and advocated the unity of knowledge and action and conformance to nature. In his opinion, to govern a country, the monarch and his subjects should study and observe the laws of nature to prevent natural disasters. He expressed his views on the laws of nature. He said, "The objective law in heaven and on earth is eternally unchanged; the sun and the moon can bring both day and night repeatedly and immortally because they are governed by objective law; the four seasons change forever for the same reason." This remark shows Wang Yangming's outlook on the laws of nature in which the heaven and earth endlessly and eternally move in cycles, and his view that the things in nature are governed by their own laws.

The book *Dream Creek Essays* written by Shen Kuo, a scientist in the Song Dynasty, native of Qiantang, Zhejiang (now Hangzhou), is a masterpiece in the Chinese history of ancient science and technology and also a valuable piece of heritage in the history of the world's science and technology. This book contains the records and studies of astronomy, geography, physics, mathematics, chemistry, weather, engineering technology, biology and medicine. Moreover, he proposed to revise the calendar, determine the months in 12 solar terms, with alternation between

31-day months and 30-day months; this calendar ensures that the arrangements of agricultural production better tally with the laws of nature.

Li Shutong, also called Master Hong Yi, in the late Qing Dynasty, native of Pinghu, revered nature during his lifetime; his outlook on nature is reflected in not only his poems, books, calligraphy and painting, but also in his thinking regarding the relationship between the people and nature; this is also a line of thought of harmony between the people and nature and a forward-looking type of ecological awareness. Such ancient philosophers in China as Laozi and Zhuangzi put forward the line of thought of harmony between the people and nature early on; their line of thought stresses the unity of people and nature, coordination between the people's behaviors and nature, consistency between moral reason and nature. Li Shutong's understanding and development of the relationship between people and nature, his dialectical line of thought of the relationship between subject and object, between subjective initiative and objective law exerted an important impact on Zhejiang's, and even on Chinese, history of ecological culture.

The *New Population Theory* written by Ma Yinchu, an economist, educator and demographer, native of Shengxian County, Shaoxing, Zhejiang, systematically put forward the line of thought of controlling the total population and improving the quality of the population and called for controlling the total population to make it coordinate with the bearing capacity of nature, paying attention to the quality of the population, human resources and technological updating, coordinated development between the population and employment, between the population and the resource environment; this is the concrete embodiment of ecological culture in the population and in social fields.

The line of thought of protecting the ecological balance has been prominent in China's traditional agricultural production; for instance, leaving land uncultivated, fallowing and crop rotation in agriculture are important measures for combining use with maintenance and safeguarding the agricultural ecological balance. The successive reigns and dynasties in China had relevant laws and decrees for ecological protection and paid attention to preventing heavy fishing, deforestation denudation and indiscriminate killing.

4.1.3 Three Main Distinctive Branches of the Ecological Culture

Since modern times, three main distinctive branches have gradually taken shape in Zhejiang's ecological culture amidst the integrative development of the Wuyue Culture and the Eastern Zhejiang School.

1. Marine culture

It is mainly the marine (maritime) ecological culture in coastal areas represented by Shipu. Zhejiang enjoys unique natural marine conditions, abundant marine resources and a long history of regional coastal culture; they give rise to Zhejiang's marine cultural characteristics. Zhejiang's marine culture was at its initial stage before the Warring States Period (the 5th century B.C.–221 B.C.), which laid a foundation for the formation of Zhejiang's marine culture; the period from the Qin and Han Dynasty (206 B.C.–202 B.C.) to the period of the Sui and Tang Dynasties (581–907) was the period of its initial prosperity; it became greatly prosperous and was established and complete during the Song and Yuan Dynasties (960–1368); it gradually declined during the Ming and Qing Dynasties (1368–1912). Zhejiang's marine culture is characterized by dynamism and progressiveness; this is embodied in an exquisite material marine culture, teamwork-oriented marine behavioral culture, a clear spirit of marine commerce and trade and marine aesthetic culture which combines roughness with tenderness. Shipu Town, characterized by a marine culture, has become the important commercial, trade and cultural center on the eastern Maritime Silk Road from China.

2. River culture

Zhejiang was originally the name of a river and is also called Zhehe, Jianjiang, Qujiang and Zhijiang, it means a wandering river, so the river is the outstanding representative of Zhejiang's image. Zhejiang is divided by the Qiantang River into eastern Zhejiang and western Zhejiang. In western Zhejiang, the water network is dense, river culture is particularly prominent, and it is the river culture represented by the Yuyao River, the Fenghua River and the Yongjiang River. In river culture, commerce is upheld, attention is paid to developing the rice market, mulberry planting and sericulture, and the people have a pioneering spirit.

3. Mountain culture

In western Zhejiang, there are many mountains, the ecological environment is excellent, the transportation is inconvenient, the population is sparse, and the people and nature coexist in a harmonious way. The mountain culture in western Zhejiang is mainly represented by the Tianmu Mountain. In the mountain culture, emphasis is placed on farming and livestock, and such animals as cattle, birds and snakes, the people are modest and conservative, they are attached to their native land and unwilling to leave it. Zhejiang's ecological culture of mountains and rivers is guided by a forest ecological culture.

4.2 The Resource-Conserving and Environmentally-Friendly Manners of Production and Life

Zhejiang has more mountains and less land, it lacks terrestrial resources; it is a traditional small province with regard to resources; its area accounts for only 1.06% of the national area. With 70% mountains, 10% water and 20% farmland in Zhejiang, land is dominated by the mountainous area, the plains available for farming and living account for only 19%. As time passes, the people of Zhejiang have developed the distinctive manners of production and lifestyle that are resource-conserving and environmentally-friendly; they have given importance to economical, intensive and efficient utilization of resources in the development of the traditional industries. In the transformation and upgrading of the industrial structure, Zhejiang has relied on marine resources to build a marine-type Zhejiang; in the meantime, Zhejiang has vigorously developed the non-resource-based industries and an ecological cultural industry; the endowment conditions in a small province of resources do not restrict the building of Zhejiang into an ecologically large province. Zhejiang's experience and philosophies deserve to be summarized and carried forward.

4.2.1 Harmonious Integration Between the Traditional Industries and the Ecological Culture

Historically, Zhejiang's traditional industries and commercial culture had the distinctive feature of integrative development with the ecological culture.

1. The ecological cultural connotation of the traditional industries

The silk, tea, salt, porcelain, shipping and financial industries have been six major traditional industries in Zhejiang throughout its history. These six industries have a rich ecological cultural connotation. Zhejiang's silk and tea industries derived from its ecological agriculture. Zhejiang is hailed as the Capital of Silk and Tea. Historically, the silkworm and tea were the main agricultural products, which gave birth to silk making and tea processing industries and then the silk and tea trade, relevant productive service industries for them—financial and shipping industries. In other words, the traditional historical industries in Zhejiang had the remarkable features of an ecological industrial chain and ecological industrial cluster, and reflected a high degree of integration between industrial development and ecological culture. Since the Tang Dynasty (618–907), Zhejiang has been a large province of salt production and a paramount area for the salt industry; Zhoushan and Ningbo have been the most important salt-producing areas in Zhejiang. Both Zhoushan and Ningbo are surrounded by the sea, they abound in natural resources, their geographical environment is excellent, there are a large number of islands, most of them can withstand strong winds and big waves, the duration of sunshine is long, there are favorable weather and geographical conditions for making salt by distilling sea water; fishing and salt have been the important pillar industries in both areas since ancient times. Thanks to its excellent soil conditions, Zhejiang is suitable for making porcelain. Historically, Zhejiang developed many kiln systems; the Yue kiln (the Yue kiln system) was situated in today's Yuyao, Cixi and Shangyu in Zhejiang, and belonged to the State of Yue in the Tang Dynasty, which was one of the six largest areas for the production of celadon ware in the Tang Dynasty. During the Eastern Han Dynasty (25–220), mature porcelain could be made in this area. In more than 1,000 years from the Eastern Han Dynasty (25–220) to the Northern Song Dynasty (960–1127), porcelain production was never interrupted in this area; the porcelain made during the Tang Dynasty and the period of Five Dynasties and Ten Kingdoms (907–979) was the most exquisite. The Longquan kiln (the Longquan kiln system) was situated within today's Longquan City, Zhejiang; there were kiln sites in many areas including Dayao, Jincun, Xikou and Songxi; there were more than 20 kiln sites during the Northern Song Dynasty and more than 40 kiln sites during the Southern Song Dynasty (1127–1279); there was a large number of kiln sites in Dayao and Jincun. It started during the period of Five Dynasties and Ten Kingdoms and prospered during the Southern Song Dynasty and the Yuan Dynasty, declined during the Ming Dynasty (1368–1644) and ended during the period of Emperor Kangxi (1662–1722) in the Qing Dynasty. The history of porcelain production has covered nearly 800 years. China's financial industry during its early stage developed in Zhejiang, while the financing of old-style Chinese private banks, operated by the people of Zhejiang and linked through kinship and geographical complexity, emerged in the region of Jiangsu and Zhejiang with Shanghai and Hangzhou as the center. Before Zhang Shicheng dug the canal from Beixinguan in Hangzhou to Tangqi in Yuhang at the end of the Yuan Dynasty, the riverway of the Shangtang River from Genshanmen in Hangzhou, via Linping in Yuhang, Xucun and Chang'an in Haining, to Chongfu in Tongxiang served as the main channel of the Jiangnan Canal. Chang'an Town in Haining was the hub of the land and waterways. Chang'an channel at Chang'an Town in Haining City was the main channel of the Jiangnan Canal during the Tang and Song Dynasties (618–1279); there was a dam and a sluice along the riverway within the town. During the Northern Song Dynasty, the Grand Canal was the main artery between Hangzhou and the capital Kaifeng, while Chang'an was an area which it had to pass through. After the southward migration of the Song Dynasty, the capital was founded in Lin'an, while Chang'an became the strategically important area for the capital. Water transportation of grain, trade and official travel were frequent between western Zhejiang and the capital; Chang'an became increasingly prosperous. Moreover, since the Tang and Song Dynasties, with the development of shipbuilding and navigation technologies, the sea routes from China to Southeastern Asia, the Strait of Malacca, the Indian Ocean, the Red Sea and the African continent were opened and extended, the Maritime Silk Road finally replaced the Land Silk Road and became the main channel for China to engage with the outside world. Zhejiang's Ningbo is situated at the center of the Chinese Mainland's coastline. The Port of Ningbo is one of the three main ports and nodes along the South China Sea route of the ancient Maritime Silk Road; it has a long history of engaging with Southeastern Asia and the Indian Ocean Rim. Ningbo also kept cross-sea exchanges, silk and tea trade with Japan and South Korea for a long time. Canal shipping and the Maritime Silk Road are the typical means of green transportation with an ecological cultural philosophy and have left a rich ecological cultural heritage.

2. The ecological cultural philosophy in the history of business history

There were three large groups of businessmen in Zhejiang's history, including the Ningbo group of businessmen in modern times, the Longyou group and the Nanxun silk merchants. These three groups of businessmen made huge contributions to

Zhejiang's economic development in modern times and were also the origin of Zhejiang's ecological economic civilization. The operational thoughts of these three groups of businessmen were closely associated with Zhejiang's characteristics—resource shortage, complicated geographical environment and a large population, and reflected Zhejiang's ecological culture in different ways. The Ningbo group of businessmen advocated the thought of "breaking away from ancient doctrines and refraining from rigidly keeping to customs" and "practicing diligence and frugality, offering benevolence and boon to others".[1] In the Ningbo group of businessmen in modern times, small and medium-sized merchants, barons and magnates mostly started off as apprentices, worked hard and endured great hardships in their pioneer work; many Ningbo merchants have been able to maintain an entrepreneurial spirit of practicing economy and diligence after getting rich by starting businesses, they have lived a thrifty and simple life and strictly avoid extravagance although they are wealthy, so it has long been believed that the people of Ningbo are shrewd. The Longyou group of businessmen originated in the Longyou area (called Gumie in ancient times) during the Southern Song Dynasty; it is situated in the Jinqu Basin in western Zhejiang and is part of Quzhou City; it is the important channel connecting eastern and central Zhejiang with Jiangxi, Anhui and Fujian, and is known as "the channel in the southeast which radiates in various directions". During the Qing Dynasty, the Longyou group of businessmen was active in the economically underdeveloped areas, while the natural environment and living environment with more mountains, less farmland and a large population in small areas forced the people to do business everywhere in order to make a living; they leveraged "the channel in the southeast" and land and water transportation advantages as well as the produce peculiar to mountainous areas to develop a commercial and trade circulation industry at home and abroad. The Longyou group of businessmen advocated the traditional Confucian ecological culture of creating wealth by upholding a good manner of doing business, morality, justice and integrity. The Nanxun silk merchants rose in Nanxun, south of Lake Taihu during the late Qing Dynasty; they were a very unique group of silk merchants in modern Chinese history; they became extremely rich in the region south of the Yangtze River thanks to the operation of silk, so Nanxun became a rare area with a large number of very wealthy people in modern Chinese history. Nanxun merchants had the spirit of blazing new trails, they fully leveraged their advantage of a good location, seized the opportunity from opening Shanghai to the outside world, exported raw silk at low prices to quickly occupy the international market, set up silk stores and storehouses at Nanxun and Shanghai, and they became the forerunners in China's silk industry. After the Nanxun silk merchants built up a fortune through the silk trade, most of them guarded against possible dangers in times of peace, gradually adjusted the direction of their operations, shifted their capital from single silk trade to such traditional industries as salt, pawning, money shops, real estate, medicinal materials, sauce and pickle shops, groceries, modern industries such as silk reeling, cotton spinning, flour, papermaking, electric power, minerals, new financial sectors

[1]Han Yongxue, The Main Groups of Businessmen in Zhejiang's History and a Study of Their Operational Thoughts, *Journal of Suihua University*, 2009(1).

including banking, trust, securities and insurance, new transportation industry such as railways and liner transportation and emerging service sectors such as entertainment and tourism. This not only defused the commercial risks caused by the shrinking exportation of Chinese raw silk, but also cultivated a diversified traditional industrial system. The Nanxun silk merchants actively received Western education, and combined advanced foreign scientific and technological market management means with Chinese traditional ways of business; they also upheld the Chinese traditional ecological cultural philosophy of favoring Confucianism in trade and stressing the unity of scholars and merchants, and combined economic development with cultural quality.

4.2.2 Harmonious Integration Between the Manners of Modern Production and Lifestyles and the Ecological Culture

Zhejiang is a province with a rapid development of the ecological cultural industry in China. In an effort to build an ecological province, the People's Government of Zhejiang Province has supported the ecological cultural industry by taking it as the golden industry which is a developmental priority in Zhejiang, and has attempted to adopt various ways to tapping and developing ecological cultural resources in Zhejiang, continuously explored the relations between production and culture under the particular natural ecological environment in different areas, developed a range of ecological cultural products with local characteristics, such as Zhejiang's cultural products relating to bamboo and wood, silk, tea, Chinese torreya, wetland and sandbeach which are famous at home and abroad. These unique ecological cultural products have, in conjunction with the characteristic ecological cultural tourism projects, laid a solid foundation for the development of Zhejiang's ecological cultural industry, thus Zhejiang has become the province with the fastest development of the creative cultural industry in China.

Xi Jinping led the people of Zhejiang to strictly carry out the basic principle, adopted during the 16th National Congress of the Communist Party of China, that it is essential to give priority to sustainable development and uphold the basic national policy of family planning, environmental and resource protection, and put forward the strategic goal of ecological development for building a green Zhejiang, elevating the ecological value outlook to an unprecedentedly high level. First, in 2002, the 11th Party Congress of Zhejiang Province further put forward the goal and tasks of building a green Zhejiang; in late 2002, the building of an ecological province was officially identified as the main carrier for the new ecological value outlook—a green Zhejiang, with efforts to keep coordinated development between population, resources, the environment and the economy and the society. Second, in late 2002, the People's Government of Zhejiang Province officially applied to the State Environmental Protection Administration for the national pilot province for building an ecological province. In early 2003, Zhejiang was identified as the 5th national pilot province for building an ecological province. In the same year, the People's Government of Zhejiang Province developed the *Plan of Zhejiang Province for the Building of an Ecological Province*, calling for vigorously developing an ecological economy and cultivating the ecological culture. Third, in 2004, Zhejiang initiated the "811" action for environmental improvement to push forward the construction of infrastructures in environmental protection. In 2005, Zhejiang further initiated the "991" action plan for developing a circular economy, with actions to develop a circular economy in 9 fields and carry out ten projects for environmental protection and ecological development. The above measures for ecological development show that Zhejiang has changed from "passive" to "active" in ecological development, and ecological development has become an important means for Zhejiang to protect the ecological environment and rationally utilize natural resources; the awareness of a green Zhejiang and an ecological province has been deeply rooted among the people. In 2007, the 17th National Congress of the Communist Party of China identified ecological progress as a new part of China's efforts at building a moderately prosperous society in all respects and incorporated ecological progress as one of Zhejiang's specific goals for finishing the building of a moderately prosperous society in all respects in advance. In 2012, the 13th Party Congress of Zhejiang Province was officially held, which put forward the goal of building a materially affluent and culturally advanced province and marked the official formation of Zhejiang's ecological value outlook and elevation from ideology to value consciousness. First, in 2007, the 12th Party Congress of Zhejiang Province officially identified the ecological developmental goal of a moderately prosperous society: the environment will become more beautiful, the ecological quality will improve significantly, people and nature will coexist harmoniously, and people will enjoy a good living environment. In 2010, the 7th Plenary Session of the 12th Party Committee of Zhejiang Province made the *Decision on Promoting Ecological Development*, putting forward the connotation of an ecological Zhejiang: affluent, beautiful, harmonious and peaceful. Second, in 2010, at the 7th Plenary (Enlarged) Session of the 12th Party Committee of Zhejiang Province, Hu Jian, the Executive Deputy Director of the Department of Publicity under the Party Committee of Zhejiang Province, specially studied and arranged for the issue of ecological culture. Third, in 2012, the 13th Party Congress of Zhejiang Province explicitly put up the banner of ecological development for building a beautiful Zhejiang, and vowed to leave a beautiful homeland with a blue sky, green land and clean, clear water to the future generations in the process of implementing the developmental strategy of making the people rich by starting businesses and of building a strong province through innovation. Xi Jinping pointed out, "In order to create glory again, Zhejiang must better raise 'two birds': 'achieving the rebirth of the phoenix amidst fire' and 'vacating the cage to change birds'; it is essential to combine 'going global' with 'bringing in' to introduce high-quality foreign capital and domestic capital, promote the adjustment of the industrial structure, make up for the deficiency in the industrial chain, cater to the international market, thus cultivate and introduce 'good birds' which eat less, lay more eggs and fly high; "two mountains" means that equal emphasis must be placed on the 'gold and silver mountains' and clean, clear waters and lush mountains and that clean, clear waters and lush mountains are the 'gold and silver mountains'; Zhejiang has realized in practice that clean, clear waters and lush mountains can endlessly bring a 'gold and silver mountains', and clean, clear waters and lush mountains themselves are the 'gold and silver mountains'; evergreen plants are the source of money, ecological advantages become economic advantages, and a harmonious integral whole is formed; this stage represents a higher realm and reflects the requirements of scientific development, the philosophy of developing a circular economy and building a resource-conserving and environmentally friendly society.[2]"

Since then, Zhejiang has, under the theme of consolidating the cultural foundation for ecological development, increasingly improved the method of incorporating ecological culture into the modern ecological cultural system.

1. The development of marine ecological culture

The Party Committee and the People's Government of Zhejiang Province have attached great importance to the development of the marine ecological culture and have made arrangements for developing the marine economy and making ecological progress in Zhejiang. Zhejiang has always paid a great deal of attention to the development of marine ecology; it has consciously protected the natural marine ecological environment, it has become the first to build an ecological province, actively built the national experimental area for comprehensive marine development and the marine ecological development demonstration area, and explored the path of harmonious marine ecological development. Based on the unique marine ecological system and its advantages of marine and port resources, Zhejiang has optimized the layout of "one core, two wings, three circles, nine areas and multiple islands", rationally developed and protected its marine resources, more rapidly developed the marine ecological economy, and taken the new path of marine economic development with land and sea coordination. Actions have been taken to correctly deal with the dynamic relations between sustainable marine industrial development, the improvement in the people's living standard and the protection of the marine ecological resource environment, develop and implement the plan for ecological environmental function areas, identify the functional orientation and industrial priorities in different areas, promote coordination between the layout of productive forces and the bearing capacity of the ecological environment, so comprehensive marine development has increasingly tended to be rational. Zhejiang has fully realized that the ecological environment on islands is associated with industrial development and urban construction, and has practically ensured the comprehensive utilization, development and protection of sea areas, land, water sources, uninhabited islands and other resources on the basis of scientific feasibility studies and law-based examination and approval. Zhejiang has weighed the plan and options for the marine industry according to the environmental conditions and resource characteristics of islands at sea; Zhejiang has more quickly adjusted its economic structure, so a number of advantageous marine indus-

[2]Xi Jinping, *Carrying out Concrete Work to Stay Ahead – Line of Thought and Practice in Promoting New Development in Zhejiang*, The Party School of the CPC Central Committee Press, 2006.

tries with low resource consumption, slight environmental pollution, great impetus for stimulating the development of others and high comprehensive returns, such as port logistics, marine tourism and modern fishery, have developed rapidly. Zhejiang has vigorously carried out the "991" action plan for a circular economy, endeavored to cultivate the emerging marine industries, promoted the ecological development of coastal industrial parks, and fully initiated a batch of energy-saving projects, including the generation of wind power and seawater desalination. Zhejiang has been dedicated to developing a modern type of ecological fishery, strictly controlling the intensity of fishing, working on a marine summer fishing moratorium, enhancement and releasing, ecological restoration and reducing predatory exploitation of natural marine ecological resources. In the meantime, Zhejiang has energetically publicized and carried forward the marine ecological culture so that the awareness of a marine ecology has been deeply rooted among the people and marine ecological ethics have been widely recognized.

Zhejiang carried out the *Regulations of Zhejiang Province on the Administration of the Use of Sea Areas* as from March 1, 2013; on World Oceans Day—June 8, relevant personnel from the marine surveillance and law enforcement brigades and ocean sections of the ocean and fisheries bureaus of coastal cities and counties took ships and embarked on islands to conduct publicity activities.

Xiangshan County, Zhejiang is the national experimental area for the ecological protection of marine fishing culture and has held the China Fishing Festival for the 13 consecutive times. The festival arises out of the fishing moratorium, and its fundamental purpose is to embody the people's ecological philosophy of protecting the ocean, revering nature and achieving common prosperity of people and the ocean. The ocean worshipping ceremony, the sailing ceremony and the tour commemorating the blessing of the Chinese sea goddess Matsu give expression to the rich connotation of the age-old fishery culture. "Clean, clear waters and lush mountains are the 'gold and silver mountains'; ecological advantages are late-mover advantages". Under the guidance of the ecological culture, Anji County, the first national ecological county in Zhejiang, has taken the construction of a beautiful countryside as the carrier to ensure that ecology and the economy promote each other and develop together; it has become Zhejiang's model for the construction of a new countryside.

Zhejiang has also attached importance to theoretically studying and summarizing the development of the marine ecological culture. Being included in the first batch of key research bases for social science in Zhejiang, the Zhejiang Marine Culture & Economic Research Center was officially established at Ningbo University. The center focuses on such research fields as Zhejiang's marine economy and management, Zhejiang's overseas economic and cultural exchanges, Zhejiang's marine cultural and regional social change, and is aimed at generating a number of high-quality and influential academic research achievements and countermeasure-oriented application achievements through interdisciplinary and joint cross-department research.

2. The development of a forest ecological culture

In Zhejiang, the area of mountains accounts for 70% of the province's total area, the forest coverage reaches 60.2%, there are "70% mountains, 10% water and 20%

farmland", thus a forest ecological culture is an important part of Zhejiang's ecological culture. Zhejiang's forest cultures are rich and colorful; they are mainly such ecological cultures as bamboo, tea and flower cultures. The Party Committee and the People's Government of Zhejiang Province have vigorously carried forward the core value outlook that there should be harmonious coexistence between people and nature. A good atmosphere of respecting, loving and being kind to nature has been fostered all across the province.

The bamboo culture in Anji. Anji has successfully held the first China Bamboo Cultural Festival, and has been home to Anji China Bamboo Expo Park, the largest bamboo culture carrier with the most complete varieties and functions and the best returns at the highest level in China, and even throughout the world. It has the Grand National Bamboo Forest which ranks no.1 in China. It has fostered the earliest modern bamboo processing industry in China. It has the largest bamboo product trading market in the world and has developed the first bamboo instrument manufacturer in China; it has established the first bamboo orchestra for the land of bamboo, and has the first Moso bamboo science and technology demonstration park and the first bamboo shoot food recipe in China; moreover, it has cultivated a group of entrepreneurs who are active in national bamboo producing areas.

The orchid culture in Lanxi. The orchid is the flower of Lanxi City. The city abounds in orchids, the custom of cultivating and loving orchids is prevalent in the city; the city is an area in which orchid production is relatively concentrated nationwide. In May, 1999, Lanxi was rated as the Land of the Orchid in China. The people of Lanxi have transformed the orchid culture into their characteristic economy, and they have also turned the annual orchid festival, which inherits the flower culture, into an emerging industry.

The camellia culture in Jinhua. Jinhua City enjoys a moderate climate and abundant rainfall; in particular, thanks to its acidic soil, it is suitable for growing camellia; it is the Land of the Camellia in China which is famous at home and abroad. Jinhua's camellia is famous for its beautiful tree form, large and gorgeous flower, varied flower postures and a long flowering season. Camellia is the flower of Jinhua City and is the symbol of the city's civilization. With economic development and the improvement in the living standard, the people have increasingly appreciated camellias with extremely high value. Camellia activities have involved economic and cultural fields, folk customs, and many aspects of daily life. Camellia culture has become an important part of traditional culture.

The tea culture in Kaihua. Kaihua County is situated in the "Golden Triangle" area of the Chinese green tea; it is the source of the Qiantang River. Kaihua Longding tea has an excellent quality and a long history. With the "ecological brand", the small Longding tea leaves in Kaihua have generated a large industry worth 400 million yuan in several years, providing a path of sustainable development for increasing the income of farmers in mountainous areas.

The mushroom culture in Qingyuan. Situated in southwestern Zhejiang, Qingyuan is one of the birthplaces of artificially cultivated mushrooms in the world. The county has a population of 190,000, more than half of them are engaged in mushroom production and relevant industries, so it is reputed as the Land of Mushroom Growers.

The income from mushrooms accounts for more than 50% of the per capita income of Qingyuan County's farmers and local fiscal revenue. Small mushrooms hold up half the sky for the material life of local people.

The gingko culture in Changxing. Gingko in Changxing is famous for the long tree age and their large number. More than 100,000 ginkgo trees grow in Changxing, and more than 10,000 ginkgo trees have existed for more than 100 years. In Xiaopu Town, Changxing County, there is a 12.5 km-long ancient ginkgo corridor, which contains 57,000 ancient ginkgo trees with luxuriant foliage, among which 2,320 ginkgo trees have existed for more than 100 years, presenting a spectacular landscape.

The wood flooring culture in Nanxun. Nanxun is an ancient town with a flat landscape which is an area south of the Yangtze River and has no abundant forest resources, but the people of Nanxun have successfully developed a large remarkable forestry area without forests. At present, there are more than 400 wood floor producers; it has become the wood floor production base with the largest scale of production, a large number of brands, the most concentrated areas and the longest industrial chain in China.

The wood carving culture in Dongyang. Wood carving in Dongyang has a history of more than 1,300 years. It is one of the six famous wood carving areas in China and ranks no.1 among three famous wood carvings in Zhejiang; it has been included in the first batch of the key traditional varieties of arts and crafts under protection in Zhejiang and the first batch of national intangible cultural heritages.

Zhejiang officially initiated the activity "Care for the Forests" in 2008. Zhejiang has vigorously advocated the philosophy of making the forests available in cities and letting cities embrace the forest; it has upheld the leading role of the Party and the government, orderly, philosophy-led and forest-focused development, the expansion of the common ground, and made great efforts to develop forests for making the people rich by putting the people first and it has carried out extensive publicity to create a good atmosphere, thus making active contributions to building a forest-loving Zhejiang, carrying forward the ecological civilization and promoting the harmonious development of the people and nature. In 2011, 15 billion yuan in a green development fund was utilized, 67.80 million trees were planted and 633,000 mu of afforestation renewal was completed in Zhejiang.

The activity of choosing and naming Zhejiang the Provincial Ecological Cultural Base was jointly organized by the Department of Forestry of Zhejiang Province and the Zhejiang Association of Ecological Culture; it is conducted once a year. Based on a preliminary recommendation by the forestry bureaus of cities, groups of experts carry out on-the-spot inspection, scoring, appraisal and review to make the final decision. At the next step, two organizers will continue, according to the requirement, put forward by the 18th National Congress of the Communist Party of China, of vigorously promoting ecological development and the overall requirement of greening plains and the region of rivers and lakes in five years and building a forest-loving Zhejiang in ten years, to further carry out the ecological culture, deeply tap a variety of ecological thoughts in Zhejiang's forest culture, mountain and river culture, marine culture, traditional farming culture, tea culture, flower culture, bamboo culture and stone culture, to expand the coverage of Zhejiang's ecological developmental activity, to ensure that the philosophy of carrying forward ecological culture, fostering ecological awareness and enhancing ecological responsibility is promoted in various fields and industries throughout the whole society, and that the activity produces a better demonstration effect on promoting ecological development and building a beautiful Zhejiang. The Department of Forestry of Zhejiang Province and the Zhejiang Association of Ecological Culture have jointly issued a document to name 258 ecological cultural bases, including Dipu Village in Jiangnan Town of Tonglu County.

In the process of spreading ecological culture, Zhejiang has proceeded from local conditions to carry out some influential, distinctive, effective and substantial ecological cultural publicity and educational activities, which have produced the driving, demonstration and stimulation effects on the sustainable economic and social development of the surrounding areas. The Meihuazhou Scenic Spot in Nanhu District, Jiaxing City integrates ancient Buddha culture, leisure farms, outdoor activities and 100 mu of orchards and covers a rich content; it has held the Jiaxing Ecological Cultural Tourism Festival and the Nanhu Peach Flower Festival for four consecutive years.

3. The development of ecological social culture

In parallel to vigorously carrying forward the ecological marine culture and the ecological forest culture, the Party Committee and the People's Government of Zhejiang Province have also made great efforts to develop the ecological culture in various fields of social life by promoting relevant work in a top-down way, transmitting the driving forces in a down-top manner and making joint efforts at the upper and lower levels, so Zhejiang has stayed ahead nationwide in ecological cultural development. Since Zhejiang appraised and identified the first batch of 32 green hotels in 2000, Zhejiang has identified 200 more green hotels in ten years and has become the province with the largest number of appraised and identified green hotels in China. The green hotel standard of Zhejiang Province, issued in September, 2001, is the first mature local standard for environmental protection in China's tourist hotel industry. In March, 2010, based on the local standard of Zhejiang Province for green hotels, the National Tourism Administration issued the national green tourist hotel standard for the tourism industry and promoted Zhejiang's experience in building green hotels throughout the whole industry.

In the meantime, Zhejiang has also carried out the development of green administrative villages and green enterprises. At present, there are 7 administrative villages. These administrative villages share two similarities: first, the forest coverage is relatively high, villages are clean, the environment is beautiful; second, the ecological culture is prosperous, there are unique features in residential buildings, courtyard facilities, cultural relics and historic sites, ecological landscapes and historical allusions. Dipu Village in Jiangnan Town of Tonglu County focuses on the culture of filial piety and righteousness and is dedicated to protecting ancient heritage, inheriting ancient culture and continuing close connections, it has become the model for the development of the ancient village culture in the whole county, even in the whole province. There are two green enterprises, the ecological industries thrive, actions have been taken to actively develop multistoried cropping and the breeding industry, and ecological industries including countryside tourism, sightseeing leisure, flowers, plants and nursery stock. Ancient trees and famous species of wood are protected and managed in a well-regulated way, and the regulations and systems for the management of ecological and environmental protection have been developed and effectively executed. Enterprises are required to, while obtaining economic returns, work on the philosophy of "advocating environmental protection and giving something back to the society" and the mission of "inheriting the Chinese civilization and carrying forward the silk culture" to energetically develop the ecological cultural bases which combine clean production with ecological culture, such as the ancient mulberry and silk museum.

In 2007, the People's Government of Zhejiang Province unveiled the master plan for building a conservation-minded society and stressed that Zhejiang would vigorously optimize the industrial structure, greatly develop the industries that are harmless to the ecological environment and strictly restrict the industries that are harmful to the ecological environment. As of 2010, the energy consumption per unit of GDP decreased by about 20%.

4. The development of ecological cultural parks

In 2013, Zhejiang unveiled the plan for the main functional areas and vowed to focus on building ten mountainous counties, while these counties are different in the quality of their ecological culture. Chun'an is the area where the famous national-level scenic spot Thousand-island Lake is located; it is hailed by tourists as one of the most beautiful, cleanest and safest towns in China. Pan'an boasts lush mountains and clear waters, a beautiful environment; it is reputed as the Large Potted Landscape in Central Zhejiang, the Natural Oxygen Bar City. Kaihua is famous within and outside the province for its tea culture, root carving culture, river source culture and weightlifting culture. Wencheng is the hometown of Liu Ji, a Chinese politician and writer from the Ming Dynasty (1368–1644), and Tongling Mountain Forest Park in Wencheng is hailed by tourism experts as a great pothole wonder in China. In Yunhe, the 1,000-layer terrace with a 1,000-year-old history and 1,000 m drop in height is praised by the photography circle as the most beautiful terrace in China. The Qinshun Wooden Arch Bridge is hailed as an amazing masterpiece in the history of the world's bridges and has been designated by the State Council as a major historical and cultural site protected at the national level. Qingyuan is the first hometown of the wooden arch bridge in China and the first mushroom city in China. Longquan is the famous capital of celadon ware and the land of treasured swords. Suichang is hailed as the source of the Qiantang River and the Ou'jiang River and the green sea south of the Yangtze River; it is the birthplace of the cultural treasure, the *Peony Pavilion*, created by Tang Xianzu, a famous dramatist of the Ming Dynasty, and also the land of Chinese folk culture and art. The style of the She Ethnic Group in Jingning is unique in Eastern China and is representative in China. With the development of various types of national ecological parks, Zhejiang's ecological industry has begun to take shape, and ecotourism has become the pillar industry in Zhejiang's mountainous counties. In 2012, ten key counties in ecological functional areas received tourists for about 26,308,600 person-times, and their comprehensive tourist income was about 11,077 million yuan. With regard to ecological agriculture, such industries as edible mushrooms, medicinal materials and tea in different areas have been roughly put into shape. The silicon industry in Kaihua, wooden toys in Yunhe and NongFu Spring in Chun'an are well-known nationwide. Regarding the development of the national ecological parks, Zhejiang's green development line of thought has become increasingly clear; in 2003, Zhejiang worked out the *Plan of Zhejiang Province for Building an Ecological Province*; in 2010, Zhejiang made the *Decision on Promoting Ecological Development*; in 2013, Zhejiang developed the *Plan of Zhejiang Province for the Main Functional Areas*, thus building a planning and policy system for ecological development. During the period of the 12th Five-Year Plan, ten counties in Zhejiang identified the building of a county (county-level city) through ecological development as a long-term developmental strategy; five counties put forward the strategy of building some national-level ecological counties, famous ecological counties and famous Chinese green counties; many cities and counties set the goals of building the ecologically liable counties, China's best ecological tourist counties, "the Homeland of Landscape, the Fairytale World". As of late 2014, 16 cities and counties in Zhejiang had been designated by the Ministry of Environmental Protection as national ecological counties. Kaihua County has become the first in the province to vow to build the National Eastern Park and has put in practice the building of the national ecological park.

5. The integrative development of ecological culture and industries

Developing the ecological industries is the primary task for ecological development. With a focus on clean production, Zhejiang has advocated the philosophy of a green GDP adjusted for environmental pollution and ecological damage, and promoted the vigorous development of the ecological industries which are circular and feature symbiosis and stable ecology. Zhejiang has actively highlighted its ecological cultural advantages to promote the integrative development of ecological culture and industries, especially speeding up the integrative development of the agricultural industry and the tourist industry. Zhejiang has made great efforts to develop modern agriculture, it has relied upon scientific and technological innovations to increase the industrial level, and pushed forward the integration between agriculture and tourism, developed agricultural industrial belts and tourism scenery lines in coordination with the construction of the beautiful countryside, its efforts in "cleaning, greening and beautifying the areas beside highways, railways, rivers, mountains, cities and villages", "three renovations and one demolition" and "governing five waters in tandem". The circulation of agricultural land has been accelerated to speed up the base-oriented, standardized and large-scale development of agricultural industries; the leading agricultural enterprises should play the leading role; modern agricultural parks should be turned into sightseeing agricultural parks with the functions of both agricultural industrial park and tourism sightseeing park. The ecological and cultural advantages should be converted into economic advantages. The ecological economy should be earnestly developed; efforts should be made to achieve ecological economization, economic ecologicalization, cultural economization and economic culturalization. Agricultural enterprises, tourist enterprises and cultural enterprises should work together and interdependently to achieve integrative development, they should develop, become large and strong amidst efforts in boosting the integration between agriculture and tourism. With such integration, more people can seek development from the ecological, tourism and cultural fields; agricultural enterprises should seize opportunities, capture business opportunities, fully leverage Zhejiang's ecological and cultural platforms to cooperate with tourist enterprises and cultural enterprises, set up stores, and greatly move towards the e-commerce field while continuing to improve the traditional sales modes and channels, so that edible mushrooms, tea, bamboo shoots and other special green and ecological agricultural product markets in Zhejiang can develop nationwide and worldwide amidst Zhejiang's ecological and cultural development.

4.3 Future Measures for Developing the Ecological Culture

Bringing about an ecological society and ecological social customs is the important task for building a harmonious society. It is essential to take ecological education as universal education, whole-process education and lifelong education, turn ecological awareness into universal awareness and global awareness, advocate ecological ethics and ecological behaviors, uphold the outlook on ecological improvement, ecological conscience, ecological justice and ecological obligations, and build ecological and cultural communities. In a society, only when the people have ecological ethics and ecological behaviors and there is the participation of the entire population and the entire society can the foundation for ensuring environmental and ecological safety be laid. Zhejiang is an ecologically sensitive region and also the eastern coastal region which has become the first to speed up industrialization, thus its ecological problems are especially severe, and its ecological development is particularly arduous and complicated. In response to severe and widespread traditional industrial pollution, Zhejiang has taken feasible measures and carried out a variety of ecological and cultural publicity activities year by year; in particular, Zhejiang has experienced the process of arranging points, connecting lines and forming areas in the campaign of greening Zhejiang. In Zhejiang, the philosophy of development has evolved from the building of a green Zhejiang, an ecological Zhejiang to the building of a beautiful Zhejiang, and the developmental mode has been transformed from the one in which the "gold and silver mountains" is pursued at the expense of clean, clear waters and lush mountains, and the one in which equal emphasis is placed on the "gold and silver mountains" and clean, clear waters and lush mountains to the one in which clean, clear waters and lush mountains are the "gold and silver mountains"; there has been a shift from the initial point-based work—voluntary tree planting and village greening—to area-based work—the development of the Yangtze River protection forest and coastal protection forest, thus systematic urban ecological development focusing on building forest-loving cities has been fostered gradually. Zhejiang has vigorously promoted urban ecological development to enhance the quality of urban life, and make great improvements in forest coverage, ecological functions, green spaces and urban-rural integration, with efforts made to lay a foundation for achieving the goal of harmonious development of urban ecology that the sky becomes bluer, mountains become greener, water becomes clearer and the landscape becomes more beautiful. The ecological culture has become increasingly prosperous, the ecological cultural research and ecological education have been strengthened continuously, the activities of green development have been extensively carried out, the philosophy of making ecological progress has been deeply rooted among the people, the healthily and culturally advanced lifestyle has preliminarily taken shape, and the cultural and ethical support for promoting ecological development has become more vigorous.

4.3.1 Actively Carrying Forward the Ecological Culture

Great efforts have been made to develop the advanced ecological culture and make sure that the notion of making ecological progress is deeply rooted among the people and the philosophy of making ecological progress becomes the mainstream value outlook in the society. First, fostering the philosophy of making ecological progress. Arranging and promoting ecological development by taking it as a priority in the building of a culturally strong province, by extensively carrying out the ecological and cultural publicity activities, including Zhejiang Ecology Day, China Low-carbon Day, World Environment Day and World Earth Day. Persistently conducting the publicity regarding provincial conditions, which are a shortage of resources and a weak environment, and guiding the people to consciously save water, electricity, paper and grain. Second, advocating a green lifestyle. Carrying out the guiding principles adopted by the Central Committee of the Communist Party of China and the Party Committee of Zhejiang Province, practicing strict economy, and resolutely opposing hedonism and extravagance. Guiding and standardizing the production of green products, smoothing out the channels for the circulation of green products, and expanding the consumption of the energy-saving, low-carbon and environmentally friendly green products. Strengthening the building up of the capacity for public service, adopting various means, including prices, to regulate and guide residents' green living and green travel. Third, actively giving play to the public's roles. Guiding the whole society to participate in ecological development, increasing the participation of the public in local environmental legislation, environmental impact assessment, the making of major environmental policies and plans, environmental law enforcement, and expanding the public opinion foundation for environmental decision-making and law enforcement. Promptly releasing environmental information, putting in place the people's rights to know and supervise.

4.3.2 Energetically Developing the Ecological Culture

First, it is necessary to greatly improve the mechanism for ecological development, explore the new ecological developmental mode, and establish the new-type harmonious relationship between people and nature. The issues concerning ecological safety and environmental governance should be incorporated into ecological development; the new ecological and cultural developmental mode should be improved in terms of the system of factor allocation, that of investment and financing and that of social responsibility. The ecological and cultural mechanism entails the following process of management: subject to respecting nature, observing the objective law, safeguarding the overall interests of the social, economic and natural systems, the Scientific Outlook on Development is taken as the basic principle, and emphasis is placed on not destroying the ecological environment or mitigating the impact on the ecological environment, the government, enterprises, social organizations and the public make joint efforts, a legal system and measures are developed and implemented to ensure that ecology is not destroyed and the natural environment is restored, so that ecology becomes sound, the environment becomes optimal, coordinated and sustained, and that there is harmonious coexistence between people and nature. The marine ecological mechanism is a system which consists of various elements, including public participation, regional cooperation, environmental early warning, ecological compensation, scientific and technological support, ecological restoration and intensive use, and an organic whole with certain functions which results from interconnection and interaction among elements. In a certain future period of time, it is essential to improve the system for the evaluation of the government, to study and establish a system of indicators for the evaluation of ecological development, really strengthen the communication and information sharing among areas and departments, build the cross-regional and interdepartmental coordination and cooperation mechanism, achieve co-building and sharing of the regional ecological environment, and it is necessary to establish a system of scientific and technological innovation, actively reinforce the scientific and technological support for ecological and cultural development and build a platform for scientific and technological innovation. It is also necessary to improve the system of ecological compensation, strengthen the design of various systems, including those relating to the patent for financial transfers for ecological and environmental protection, comprehensive evaluation, reward and punishment involving the ecological and environmental quality; moreover, it is necessary to establish and improve the mechanism for market-oriented factor allocation in such fields as land, water and emission rights, build a good pattern with the driving roles played by both government and the market. It is necessary to improve the investment and financing system, the fiscal and tax support policy and strengthen the legal guarantee for ecological development. Actions should be taken to intensify the basic survey on Zhejiang's biological resources and ecological environment, scientifically carry out restoration projects, adopt more ecological restoration means—enhancement and releasing, artificial fish reefs, algal cultivation, grazing and cutting prohibition—in important sea areas for fishery, coastal mud flats, forestland and other ecologically sensitive sea areas, and initiate the pilot projects for nature reserves and ecological restoration, gradually achieve sustainable development and the utilization of ecological resources and the benign circulation of the natural ecosystem as well as enhancing the capacity for resource and environmental support for Zhejiang's sustainable economic and social development.

Second, it is necessary to cultivate a healthy lifestyle and carry forward ecological moral and cultural concepts. Actions should be taken to actively conduct publicity and education on ecological development, popularize and embody, at multiple levels and from various perspectives, the value outlook, the outlook on officials' performance, wealth and life relating to ecological development; actions should also be taken to give importance to cultural protection and research, combine the protection of characteristic cultural resources with the development of various ecological demonstration areas so that they become the important platforms for carrying and promoting ecological cultures. To establish an awareness of ecological development, it is necessary to assume responsibilities for the ecological and resource environment while developing the economy, show care for and kindly treat the ecological environment, integrate ecological development into the urban innovation system and daily lifestyle, foster an atmosphere for ecological development with Zhejiang's regional cultural characteristics, enhance public awareness about participation in ecological development, establish a culturally advanced and healthy lifestyle, combine the improvement of the living environment with the development of the ecological environment, exquisitely leverage the natural landscape on islands and at sea, build a safe, comfortable, ecological and culturally advanced living environment, advocate a green and low-carbon production mode and lifestyle and build livable, business-friendly garden cities suitable for tourism so that the people live ideally.

Third, it is necessary to improve the quality of the people in promoting ecological development. Ecological development means the improvement in the people's awareness about natural ecological protection, the popularization of ecological ethical knowledge and the formation of a high quality of ecological culture. Actions should be taken, based on ecological cultural development, to promote publicity and education regarding ecological and environmental protection, turn ecological development into the conscious actions of the entire population, strengthen the philosophy of ecological development, and give full scope to the moral and ethical support of ecological cultures for ecological development. It is essential to vigorously advocate ecological ethics, reinforce ecological education among the leading cadres within various groups of cadres, primary and secondary school and university students, the people at the primary level, strengthen the moral rules relating to ecological development, improve the quality of the entire population in promoting ecological development and foster new social customs for ecological development. It is necessary to speed up the building of green organs, guide the primary-level units and urban and rural residents to extensively carry out popular green developmental activities, including the building of green schools, green communities and green households. It is necessary to actively guide and encourage green consumption, advocate green transportation and traveling and a green and culturally advanced lifestyle. Full scope should be given to the important roles of enterprises, industrial associations, news media, non-governmental environmental protection organizations and all of the citizens in promoting ecological development; great synergy should be built by ensuring that the whole society cares for, supports, participates in and supervises ecological development. As the important units in the whole society, enterprises should play a very important role in ecological development, entrepreneurs should develop an awareness of their social responsibility for ecological development; enterprises should carry out production, operations and management in an ecological, economical and green way, and should not seek wealth by polluting the environment and harming public interests.

4.3.3 Greatly Promoting the Development of the Ecological Creative Cultural Industry

Prominence should be given to the ecological and cultural characteristics of different areas in Zhejiang, so that the creative ecological and cultural industry in different areas is consistent with the characteristics of their resource endowment. The creative ecological and cultural industry results from the integration between the creative cultural industry and the ecological cultural industry; it is necessary to identify the ecological and cultural characteristics of different areas in Zhejiang and build a creative ecological and cultural industrial system which adapts to the local conditions and can fully leverage comparative advantages. It is necessary to carry forward the ecological and cultural characteristics of different areas, especially remote areas, in Zhejiang, effectively guide the optimization of the economic structure and the transformation of the economic growth mode in different areas in Zhejiang, tap the ecological cultures in different areas to integrate the existing industrial and market resources and build a creative cultural industry with competitive comparative advantages. It is essential to rationally choose the developmental mode for the creative ecological and cultural industry, strengthen the planning, development and cultivation of that industry, bring innovations to notions and to the industrial structure, fully tap the ecological and cultural resource characteristics in different areas, and achieve the overall coordinated development of Zhejiang's ecological cultural industry. It is necessary to further cultivate creative ecological and cultural talents in backward areas, more rapidly improve the mechanism for the flow of creative ecological and cultural talents, continuously innovate the operational mode for the regional creative ecological and cultural industry, and promote leapfrog development of that industry. It is necessary to attract high-level creative cultural talents in central cities to the remote areas with abundant ecological and cultural resources for carrying out short-term creative and entrepreneurial activities.

4.3.4 More Rapidly Promoting the Development of National Parks

Zhejiang should, with the goal of building a beautiful Zhejiang, effectively protect ecological and natural resources, popularize and expand ecological education and scientific research, ensure that the ecological and recreational environment is clean and beautiful, that ecological tourism is rich, comfortable and convenient, that the ecological economy is characteristic and environmentally friendly, that ecological resources are capitalized, ecological brands present a high value, ecological input is socialized, ecological contributions become diverse, the ecological price is low, that people and nature coexist harmoniously, mountainous areas and cities develop in a coordinated way, ecological reduction and ecological production interact with each other to achieve a balance and that global greening, landscape-oriented green development, park-based landscape development and modern management are carried out. Speeding up national industrial development is an important and effective means for achieving the above goal.

Zhejiang is one of the provinces with the largest number of national parks in China. At present, Zhejiang has been home to a number of national parks, including Xixi National Wetland Park and those in Linhai, Suichang, Zhuajiajian Daqingshan, Changshan and Xianju. In the future, the development of national parks will remain a priority in Zhejiang's ecological cultural development. Zhejiang should proceed from regional ecological endowment characteristics and ecological developmental needs to take national forest parks and nature reserves as the ecological protection sphere, take historical cultural relics and remains as the ecological tourism sphere, take green industries as the ecological economic sphere, take towns and villages as the ecological living sphere to build a number of national ecological parks. There are four criteria for building national ecological parks: (1) natural ecological resources are unique and intact; (2) historical cultural relics and remains are valuable; (3) ecological economic development is sustainable; (4) the ecological living environment is beautiful and people-oriented.

To build national parks, it is necessary to uphold the leading role of ecological culture, more rapidly build the regional mechanisms for ecological compensation, ecological migration and community participation, while the core breakthrough in building the national park system lies in building a mechanism for ecological compensation, arousing the vitality of the residents of the community for participating in the development and management of national parks, entrepreneurship and employment relating to the ecological economy, and increasing the level of the income of local farmers and the public service level through ecological development; this can also be summarized as a new mode of local urbanization and of the urbanization of nearby areas with Zhejiang's characteristics. The developmental strategy for national parks is a regional coordinated economic developmental strategy. The connotation of ecological cultural development in Zhejiang's areas which abound in ecological resources is that ecological migration and community participation are promoted to combine national parks with the ecological agricultural economy, the rural tourism economy, the ecotourism economy, to integrate and inherit an ecological culture, a farming culture, folk culture and ethnic culture, to promote the symbiotic development of people, nature and cultures, making sure that natural and cultural heritages are intact. The ability of an ecological resource management mode to properly deal with the relation between effective protection and rational utilization will certainly become the focus of ecological development. The national parks must be built by upholding ecological protection as the primary goal. In the process of promoting regional ecological economic transformation and upgrading, it is essential to identify the attribute of ecological natural resources as public goods and that of ecological economic resources as private goods, and refrain from abusing the market economic mechanism and carrying out excessive commercial development to break the boundary of national parks as a public service. Therefore, it is necessary to draw upon international experience and practices in government-led efforts, multiparty participation, regional coordination, area-specific management, separation between management and operation and franchising, to actively explore the operational mode for the integrative development of national parks and a county-level ecological economy which conforms to Zhejiang's conditions and characteristics, to become the first to make the policies, regulations, technical standards and specifications for the development and management of national parks, keep an alignment with the world's advanced standards for ecological protection, protect the ecological protective screen and ecological cultural sites in the country's eastern region, and ultimately integrate multiple ecological resource protection modes to establish the national park system and organizational structure with Zhejiang's characteristics.

Chapter 5
The Construction of a Beautiful Countryside with Harmony Between People and Nature

Lei Zhang

The No.1 Central Document in 2013 made the work arrangements for strengthening rural ecological development, environmental protection and comprehensive improvement and building a beautiful countryside. In the Central Rural Work Conference, Xi Jinping stressed, "To make China strong, agriculture must be strong; to make China beautiful, rural areas must be beautiful; to make China rich, farmers must be rich".[1] It is necessary to continue promoting the construction of a new socialist countryside and build a homeland of happiness and a beautiful countryside for farmers. The construction of a beautiful countryside is an important part of ecological development and of the building of a beautiful China. As the forerunner for building a beautiful countryside in China, Zhejiang has, since 2003, started with the project of "1,000-village demonstrations and 10,000-village improvements" to continuously improve and enrich its connotation, follow the sequence of "the development of the ecological environment—the building of a green Zhejiang—the building of an ecological Zhejiang—the building of a beautiful Zhejiang", make unremitting efforts, and take a path with the features of the times and Zhejiang's characteristics for building a beautiful countryside. As a result, the urban-rural relationship, the relationship between people and nature, the relationship between tradition and modernity have developed well. Zhejiang's building of a beautiful countryside is Zhejiang's practice in building a beautiful China and also the upgrading of the construction of a new socialist countryside.

[1] The Central Rural Work Conference Was Held in Beijing, Xi Jinping and Li Keqiang Delivered Important Speeches, *People's Daily*, December 25, 2013.

L. Zhang (✉)
Zhejiang Ecological Culture Research Center, Zhejiang Sci-Tech University, Hangzhou, China

J. Pan and M. Shen (eds.), *Chinese Dream and Practice in Zhejiang – Ecology*, Research Series on the Chinese Dream and China's Development Path,
https://doi.org/10.1007/978-981-13-7209-4_5

5.1 From the "10-Million Project" to "a Beautiful Countryside"

After the reform and opening-up was initiated, Zhejiang witnessed rapid economic development, it rapidly came out in front in the per capita income of farmers, became the first nationwide to leap from the state of ample food and clothing to the state of being well-off. However, the urban-rural gap increasingly widened, rural areas were disorderly and unsystematic, the ecological environment tended to worsen, infrastructure was outdated and backward. In the Zhejiang Rural Work Conference in January, 2003, the then Secretary of the Party Committee of Zhejiang Province, Xi Jinping, vowed to comprehensively push forward the environmental improvement of villages and towns and ecological development. In June, 2003, Zhejiang held an on-the-spot meeting regarding the project of "1,000-village demonstrations and 10,000-village improvements". Xi Jinping decided to take five years to comprehensively improve about 10,000 administrative villages among the 40,000 villages across the province, and turn about 1,000 central villages among them into model villages for building a moderately prosperous society in all respects. Since then, Zhejiang has begun its journey down the path of building a beautiful countryside. Such projects as the 10-Million Project and rural house renovations are the main means for building a beautiful countryside[2]; the 10-Million Project has narrowed the urban-rural gap and optimized the urban-rural relationship. The construction of a beautiful countryside based on the 10-Million Project has optimized the relationship between urban and rural areas, between people and nature; the construction of a new socialist countryside has been integrated into ecological development. As shown by more than 10 years of work, Zhejiang's construction of a beautiful countryside has broadly been divided into three stages.[3]

5.1.1 The Stage of Producing the Demonstration and Guiding Effects (2003–2007)

In 2003, the Party Committee and the People's Government of Zhejiang Province made the great decision of carrying out the project of "1,000-village demonstrations and 10,000-village improvements" according to the requirement of achieving coordinated urban and rural development put forward by the 16th National Congress of the Communist Party of China in response to the new expectations of farmers. Zhejiang started with the environmental problems—dirty, disorderly and bad—which prompted the strongest reaction from farmers to promote the building of 5,000 km of

[2]Zhang Bin, Really Putting in Place the Construction of a Beautiful Countryside, *Zhejiang Today*, 2010(21).

[3]Li Qiang, Carrying out the Project of "1,000-village Demonstrations and 10,000-village Improvements", Comprehensively Promoting the Construction of a Beautiful Countryside, *Newsletter About Work in Rural Areas*, 2013(22).

clear water riverways, make comprehensive improvements to the rural environment focusing on garbage disposal, sewage treatment, latrine improvement, village road hardening and village greening. As of 2007, with five years of efforts, 10,303 villages were preliminarily improved, 1,181 model villages for building a moderately prosperous society in all respects and 10,303 villages for environmental improvement were built in Zhejiang. This is the key segment of Zhejiang's decisive efforts at starting the "beautiful relay". It has helped get off to a good start and has better led the way in building a beautiful countryside; it has further made villages clean and culturally advanced; it has promoted production development, increased the income of farmers, stimulated the implementation of a series of projects, including those conducive to coordinated urban and rural development and the improvement of the people's well-being. Great changes have taken place in some rural areas in Zhejiang. The Party Committee and the People's Government of Zhejiang Province have held an on-the-spot meeting of the 10-Million Project with a focus on one priority each year. The main leaders of the Party Committee of Zhejiang Province have delivered reports, conducted inspections, made efforts to put in place and promote relevant work. Zhejiang has promoted the mode of urban-rural co-building, sharing and assistance under which emphasis is placed on government-led efforts and the farmers' leading position, and the input mechanism has improved continuously.

5.1.2 The Stage of Universal Promotion (2008–2010)

As from 2008, under the 10-Million Project, according to the requirement of ensuring equal access to basic public services in urban and rural areas, Zhejiang deepened and extended the successful experience from model villages for building a moderately prosperous society in all respects in all villages across the province. In 2008, Anji County, Zhejiang officially unveiled the plan of building a beautiful countryside in China, and issued the *Action Plan of Anji County for Building a Beautiful Countryside in China*. Thus, the concept of a beautiful countryside officially came onto the historical stage. This process consolidated the achievements made during the first stage. In this process, comprehensive improvement in the rural environment was made on a wider scope; the work was extended to the control of non-point source pollution and renovation of rural houses, including domestic sewage, livestock excrements, chemical fertilizers and pesticides; work priorities included the collection of domestic garbage and the treatment of domestic sewage; comprehensive improvement in the rural environment was promoted at the source, thus gradually giving rise to a pattern in which farmers obtained benefits extensively, village points were fully covered, the operational mechanism was well-functioning, the living conditions for rural people and the ecological environment were improved at the same time. As from 2010, Zhejiang has held a working conference on plains afforestation each year, and initiated the provincial "1818" action for plains afforestation to boost the construction of a beautiful countryside.

According to the "Eight-Eight Strategies" and the overall strategy of making the people rich by starting businesses and of building a strong province through innovation, Zhejiang actively implemented the strategy of coordinated urban and rural development, promoted the construction of a new countryside across the board, achieved transformation from comprehensively overcoming urban-rural segmentation to comprehensively promoting urban-rural integration. In 2008, the People's Government of Zhejiang Province carried out the project of making low-income rural households become well-off, and took eight major actions—assistance based on industrial development, assistance through training and employment, assistance based on migration from mountains, infrastructure construction, coverage of social assistance, regional cooperation for promotion, financial service support, social aid and care—and broadened the channels for increasing the income of low-income rural households, encouraged migration from mountains and improved rural infrastructures. In 2010, the Party Committee and the People's Government of Zhejiang Province further decided to promote the construction of a beautiful countryside, so the construction of a beautiful countryside unfolded fully across the province.

5.1.3 The Stage of Intensification and Improvement (Since 2011)

Since 2011, Zhejiang's "beautiful relay" has entered the stage of intensification and improvement. Zhejiang has worked on enhancing the quality and expanding the coverage, made comprehensive improvements to entire towns, and developed the *Action Plan of Zhejiang Province for Building a Beautiful Countryside (2011–2015)* according to the requirement of making ecological progress and building a moderately prosperous society in all respects. Zhejiang has put forward the development requirement of enhancing the connotation of a beautiful countryside by scientifically planning and arranging efforts to enhance the beauty, by making villages clean and the environment beautiful, by starting businesses to increase the income and create a good life, and by making villages culturally advanced, suitable for living, doing business and tourism. Zhejiang initiated the work on protecting and utilizing historical and cultural villages, so the relations between urban and rural areas, between the people and nature have improved continually, historical and cultural inheritance is combined with the development of a modern civilization. In 2012, the Party Committee and the People's Government of Zhejiang Province called for comprehensively promoting the construction of a beautiful countryside,[4] a beautiful countryside has become the developmental pulse of Zhejiang's practice in building a beautiful China. In an effort to promote the construction of a beautiful Zhejiang, in 2013, the Party Committee and the People's Government of Zhejiang Province made "three renova-

[4]Zhao Hongzhu, Comprehensively Promoting the Construction of a Beautiful Countryside, *Zhejiang Today*, 2012(21).

tions and one demolition", took the action of governing "five waters" in tandem, and carried out a number of key projects relating to the people's well-being.

From initiation to intensification, development and improvement, in more than ten years, Zhejiang pushed forward the 10-Million Project by taking a path of intensifying the work from points to areas and making local and then comprehensive improvements. Zhejiang has gradually established a mechanism in which successive Party committees and governments at various levels undertake the specified tasks for building a beautiful countryside year by year. Each year since 2003, Zhejiang has held an on-the-spot meeting in a typical county or country-level city, with participation of the main leaders from the Party Committee and the People's Government of Zhejiang Province. Each year the work has been different, but the determination about the 10-Million Project, name and theme have not changed, and it has been persistently carried out. Relevant supporting projects are being carried out in coordination with the 10-Million Project. There have been the broad road project, connected roads, 5,000 km of clear water riverways, drinking water for farmers, the greening of model villages, comprehensive improvements to rural land, the renovation of old and dilapidated rural houses, rural electrification, modern commercial and trade service model villages and well-off sports villages, thus a strong atmosphere for village improvement has been fostered. From the perspectives of connotation and essence, the project of "1,000-village demonstrations and 10,000-village improvements" is a work carrier, designed and implemented by the Party Committee and the People's Government of Zhejiang Province, which is based on the improvement of the rural living environment and extends urban infrastructure, urban public services and modern urban civilization to rural areas. It started from coordinating the relationship between urban and rural areas and has gradually extended to coordinate the relationship between the people and nature, between tradition and modernity; it integrates the construction of a new countryside, ecological development and rural cultural development.[5] According to the timeline, Zhejiang started relevant work early, it is the forerunner and pathfinder in building a beautiful China; in terms of a spatial line, it is also an important measure for ecological development and a concrete action for building a beautiful China in rural areas.

Over ten years, with unified planning, step-by-step implementation and gradual intensification, the Party Committee and the People's Government of Zhejiang Province have started from the improvement of the living environment, infrastructure and public services, and continuously expanded the content of construction, thus promoting the construction of a beautiful countryside across the board. Zhejiang has incessantly pushed forward coordinated urban-rural integration and the development of relevant rural systems and mechanisms, allocated more public financial resources to rural areas, and improved the rural public service system. With integration between garden cities and a beautiful countryside, a benign relationship of mutual consumption and service providers has been established between urban and rural areas. In Zhejiang, the fast-growing rural tourism industry, a good ecological environment

[5]Kang Jinchang, Xu Xueya, Zhejiang's Path of Building a Beautiful Countryside, *Newsletter About Work in Rural Areas*, 2013(22).

and authentic rustic style have attracted a large number of urban residents to rural areas, stimulating the development of the local economy and increasing the farmers' income. Other activities in urban and rural areas, such as shopping, cultural promotion, leisure and entertainment, have boosted consumption and domestic demand, and Zhejiang has fulfilled the great aspiration of building a better life in urban areas and making rural areas more attractive for urban residents.[6]

5.2 The Line of Thought and Measures Concerning the Construction of a Beautiful Countryside

5.2.1 The Line of Thought Regarding the Construction of a Beautiful Countryside

1. Combining scientific planning with plan implementation

In an effort to carry out the project of "1,000-village demonstrations and 10,000-village improvements" and build a beautiful countryside, the People's Government of Zhejiang Province has proceeded from the reality, developed plans according to local conditions, scientifically managed the positioning and depth of plans, ensured that the master plan specifies the direction, special plans coordinate with each other, key plans are profound and construction plans can be put in place, thus there has been a planning system in which the master plan for the construction of a beautiful countryside plays the leading role and special plans coordinate with each other. First, developing the unified plan for urban and rural areas, combining village layout planning with urban system planning to constitute an urban and rural planning system with "central cities—county towns—central towns—central villages" as the framework. Second, developing plans according to local conditions, rationally determining the layout of villages and the size of the population of each type of village, their functional orientation and direction of development, avoiding unnecessary repeated construction, extensive demolition and construction, rationally designating the living, production, ecological and other functional areas and properly arranging service facilities in villages. Third, coordinating the supporting plans, ensuring coordination between village layout planning and village construction planning at the county level, promoting a connection between the county-level village layout planning and the master plan for land use, urban system planning, infrastructure construction planning, achieving full coverage of urban and rural areas in the plans, full coordination among factors and overall construction at the county level.

[6]Gu Yikang, A Beautiful Zhejiang Cannot Be Built without a Beautiful Countryside, *Newsletter About Work in Rural Areas*, 2013(16).

2. Combining point-based improvement with area expansion

With efforts in pushing forward the project of "1,000-village demonstrations and 10,000-village improvements" and the construction of a beautiful countryside, point-based improvement serves as the foundation, while area improvement is the goal. During the initial stage of village improvement, emphasis is placed on garbage collection, sewage treatment, latrine renovation, river and ditch clearing, road hardening and village greening, priority is given to improving the villages with better conditions and foundations. The model of centralized collection and disposal of rural garbage, characterized by collection at households and villages, transportation to towns and treatment at counties, has been fully promoted. Open cesspits have been completely cleared out. Simple household latrines have been fully renovated; a long-term rural cleaning mechanism has been established, a routine cleaning system under which the village collective plays the leading role, cleaning personnel are responsible for relevant work and the tasks for specified areas are assigned to rural households has been adopted, with a view to keeping villages clean. The five-year action plan for the construction of a beautiful countryside has been fully carried out since 2011, with a focus on solving the fundamental and regional rural environmental problems, promoting the development of an ecological living environment, ecological environment, ecological economy and ecological culture in a coordinated way, pushing forward the integrated construction of regional road networks, pipeline networks, a network of forests, a network of rivers, a network of garbage disposal network and one of sewage treatment in a coordinated manner, speeding up village improvements by taking points as the base, connecting points to form lines and linking lines to form areas. Actions have been taken to comprehensively improve the areas along expressways, national highways, surrounding scenic spots and towns, and the environment in all of the towns. In 2013, the treatment of rural sewage was identified as the main carrier and extension of the construction of a beautiful countryside. According to the requirement of scientifically and resolutely governing waters according to laws and by everyone, efforts have been made to make the water in rural areas clean and ponds clear, to reshape the landscape in the region of rivers and lakes south of the Yangtze River, and force the transformation and upgrading of rural production mode, lifestyle and construction mode, elevating the construction of a beautiful countryside to a new level.

3. Combining production with life

Zhejiang has taken the construction of a beautiful countryside as an effective way to foster the new points for increasing the rural economy; Zhejiang has leveraged location conditions, ecological resources, humanistic deposits and other advantages, strengthened the philosophy of village operations, encouraged farmers to carry out innovations and start businesses, endeavored to shape a good pattern in which environmental beautification and economic development reinforce each other, the construction of a beautiful countryside and the work on making farmers rich are promoted at the same time. First, transforming the rural economic developmental mode. Pushing forward the reform of the property rights system in the rural collective economy, further expanding the rural collective economy; the income of the collective economy per village across the province was about 1 million yuan in 2013. More rapidly promoting agricultural scientific and technological advancement and innovation in the farming system, continuously optimizing the layout, structure, scale and mode of livestock and poultry breeding, and greatly reducing agricultural non-point source pollution. Earnestly developing the economy relating to health, sports, creative cultural development, property and labor services, turning the potential advantages of beautiful villages with a good ecology into realistic advantages for industrial development, and working on ecologicalization of the rural economy and ecological economization. Second, transforming the farmers' lifestyle. Improving facilities and services, at the same time carrying out education, training and publicity among farmers and guiding them, helping farmers gain more consciousness and a higher sense of responsibility for safeguarding rural environmental hygiene, stimulating farmers to change their notions, way of acting and their lifestyle. Third, changing the manner of rural construction. Resolutely dismantling the illegal buildings in rural areas, energetically promoting the construction of beautiful and livable towns and villages and the renovation of old and dilapidated rural houses, more quickly making traditional buildings modern, modern buildings localized and living conditions people-oriented, and promoting coordination between rural appearance, rustic buildings and the natural landscape.

4. Combining the building of modern rural communities with the protection of local culture

Zhejiang has combined the construction of a beautiful countryside with new-type urbanization. Zhejiang has solidly promoted pilot cultivation of 200 central towns and 27 small cities, allocated more resources and factors to rural areas, especially to central towns and central villages, made industrial layout more rational and the population more concentrated and speeded up the transformation of traditional rural communities into modern rural communities. According to the requirement of ensuring that the supporting projects and funds of relevant departments are made available in the areas designated for village improvement, Zhejiang has greatly extended the work and services in various respects to rural areas. First, extending urban infrastructure to rural areas. Arranging infrastructure including water, electricity and gas in a coordinated way, extending classified highways, post offices, telephones, broadband, radio and television to all villages, making electricity available to all rural households at the same price in both urban and rural areas, connecting passenger buses to 95% of the villages, and making the coverage of safe drinking water reach 97%. Second, extending urban public services to rural areas. Building an integrated urban-rural public service system in which county towns play the leading role, central towns serve as nodes and central villages are the foundation, extending the services concerning education, science and technology, training, employment, medical treatment, health, social security, culture, entertainment, commercial and trade finance to rural areas, and basically forming a 30-min public service circle with central villages as the main platforms.[7] Third, extending modern urban civilization to rural areas. Taking the action of delivering cultural products including drama, films and books, carrying out the project for the sharing of cultural information resources, building a network of distance education for the Party members and farmers' mailbox service platforms, more quickly developing the agritainment leisure tourism industry, and promoting direct interaction between farmers and citizens, and the transformation of farmers' notions and ways of behaving.

In the meantime, Zhejiang has brought about a combination between construction and protection, between cultivation and inheritance to protect rural culture and give expression to the rustic characteristics in the construction of beautiful villages. In village improvement, Zhejiang has tapped the farming-reading culture and ethnic customs which reflect village characteristics, it has drawn upon the industrial cultures, folk skills and arts which embody regional characteristics, Zhejiang has built farming-reading culture exhibition halls, characteristic skills and arts experience halls and halls for the tasting of special delicacies, distinctive cultural villages which feature one product at one village, one industry at one village, one rhyme at one village and one landscape at one village, and the regional cultures which fully display different characteristics of mountainous areas, hills, plains, the region of rivers and lakes and islands. In 2012, Zhejiang fully initiated the work on protecting and utilizing the historical and cultural villages, holistically promoted comprehensive protection of ancient buildings and villages' ecological environment, discovery and inheritance of excellent traditional cultures, the scientific improvement of villages' living environment and the orderly development of rural leisure, with efforts made to educate farmers to treasure the heritage from their ancestors, carry forward excellent traditional cultures, improve village customs and appearance, ensure that the memory of "nostalgia" brings together the mobile people and that cultural heritage is passed on to the future generations.

5. Combining government guidance with respect for farmers' willingness

The construction of a beautiful countryside is a concrete action for carrying out the Party's rural policy and is a beautiful undertaking bringing benefits to farmers. Since 2003, several Party Committees and the People's Governments of Zhejiang Province have given great importance to village improvement and to the construction of a beautiful countryside; each year an on-the-spot meeting has been held and such work has been identified as one of the priorities for doing substantial work for the people. The Party committees and governments at various levels have set up coordinating groups headed by the Party and government chiefs. Zhejiang has further integrated the developmental forces and resources, developed an action plan, a development plan and construction standards for the construction of a beautiful countryside, a series of relatively independent and unified institutional systems covering supervision, inspection and evaluation, and Zhejiang has made active experiments

[7]Liu Mei, Vivid Practice in Building a Beautiful Countryside and a Beautiful Zhejiang—A Review of Zhejiang Province's Implementation of the Project of "1,000-village Demonstrations and 10,000-village Improvements", *Zhejiang Daily*, November 20, 2013.

in improving the rural governance system. The key to the construction of a beautiful countryside lies in arousing the enthusiasm and initiative of various social forces, including the farmers. Thus, the Party committees and governments at various levels, especially the primary-level organizations, have always respected farmers' willingness in various respects and segments of the work, and made farmers take on the initiative in village improvement and the construction of a beautiful countryside. They have always respected the will of the people, safeguarded the people's rights and relied on and reinforced private capital.

5.2.2 *Measures for the Construction of a Beautiful Countryside*

In the construction of a new socialist countryside, Zhejiang has started with village improvements to focus on industrial development, build brands and seek support from cultural development and uphold the leading role of standardized management; it has become the beautiful model for the construction of a new socialist countryside nationwide.

1. Starting with village improvements

Different areas in Zhejiang differ greatly in natural conditions and are at different levels of economic development. Moreover, planning and renovation involve heavy workloads and extensive fields, so the tasks are arduous; Zhejiang has become the first to carry out such work, thus it is not easy to make village improvements. In the first on-the-spot meeting regarding the 10-Million Project, Zhejiang stressed that Zhejiang would not launch extensive demolition and construction, it would not impose uniformity in various respects, and Zhejiang would emphasize the role of planning in guiding the way and act under classified guidance, according to local conditions and abilities.

(1) Developing a village plan as the guide, determining the positioning of village improvements in a classified way. Zhejiang has first developed the village layout plans and construction plans, and has built a planning system which is coordinated with the urban and rural system, land use and with the plans for the construction of infrastructures. First, scientifically developing the village layout plans. Zhejiang has first surveyed rural areas, put forward the line of thought of "reducing natural villages, dismantling "hollow villages", renovating the villages within cities, migrating from villages on high mountains, protecting cultural villages and cultivating central villages"; it has optimized the layout of towns and villages, identified 200 central towns, 4,000 central villages, 16,000 retained villages and 971 historical and cultural villages across the province, and developed a relatively scientific urban-rural spatial layout plan (see Table 5.1). Second, developing the village construction plans in a classified way. Zhejiang has developed construction plans according to the conditions of

the villages. For central villages, Zhejiang plans to build the five-in-one public service centers to aggregate the population and stimulate the development of the surrounding areas; for the historical and cultural villages, Zhejiang has determined the protection and construction plans according to the requirement of integrating historical sites and natural ecology into the living environment; for the remote villages on high mountains and "hollow villages", Zhejiang plans to carry out non-local relocation, and relocate farmers to county towns or central towns by making farmers gather in towns, switching to other types of production, other work and identities; for the villages within cities and in suburbs, Zhejiang plans to carry out "three renovations and one demolition" and turn them into new urban communities; for the improvement plan for general villages, the key point consists in determining the plans for environmental and village improvement. For 85% of the planned retained villages and central villages, the construction plans have been developed. Third, building a coordinated planning system. The village layout and construction plans serve as an important basis for editing and revising the master plan for land use, the master plans for counties and county-level cities and relevant special plans; efforts are being made to achieve coordination between the village layout and construction plans and the plans in various relevant respects. This practice of seeking truth from facts and acting according to local conditions ensures scientific village improvement.

(2) Starting from garbage disposal and sewage treatment, continuously expanding the work on village improvement. Zhejiang has started with the environmental problems—dirty, disorderly and bad—which prompted the strongest reaction from the people, incessantly enriched the work on improvement, gradually built a pattern of improvement in which farmers obtain benefits extensively, village points are fully covered and the operational mechanism is well-functioning. First, pushing forward comprehensive improvement of the village environment. Zhejiang has adopted the mode of centralized collection and disposal of rural garbage, characterized by collection at households and villages, transporting it to towns and treatment within the counties; Zhejiang has piloted classified sorting of garbage, tracing at the source and collection of garbage at fixed points, and has made garbage reduced, harmless and recycled. Zhejiang has cleared open cesspits, renovated simple household latrines, and adopted a number of sewage treatment methods, including unified treatment at plants and village-level ecological treatment, to make more rural sewage discharged according to relevant standards. During the period 2003–2014, 1.12 million garbage cans were added, 23,000 public latrines were reconstructed, and domestic sewage from 3.11 million households was treated. Second, promoting environmental improvement along lines and in areas. Zhejiang has improved the areas along expressways, national highways, surrounding scenic spots and towns, and the environment in all of the towns; 60% of the towns have seen the completion of improvement tasks. Zhejiang has taken special actions, including "three renovations and one demolition", "cleaning, greening and beautifying the areas beside highways, railways, rivers and mountains", "clearing riverways

Table 5.1 The village layout planning of Zhejiang province

Type	Measure	Scale (village)	Action
Central village	Cultivation	4,000	The five-in-one public service center
Cultural village	Protection	971	Integration between historical sites and village environment
Village within cities	Renovation	16,000	"Three renovations and one demolition"
"Hollow village"	Demolition		
Natural village	Reduction	/	Non-local relocation, relocation to county towns or central towns
Village on a high mountain	Relocation		

and cleaning the countryside", and has promoted the building of networks of pipelines, forests, rivers, garbage disposal and sewage treatment. Third, carrying out the construction of a beautiful countryside. Zhejiang has endeavored to build a beautiful countryside at the county, town, village and household levels, promote the development of ecological habitats, ecological environments, an ecological economy and ecological culture, to make planning and layout scientific and beautiful, villages clean and culturally advanced and the environment beautiful, to increase the income by starting businesses and to create a good life. Fourth, establishing a mechanism for long-term cleaning. Zhejiang has adopted a routine cleaning system under which the village collective plays the leading role, cleaning personnel are responsible for relevant work and the tasks for specified areas are assigned to rural households; Zhejiang has built a cleaning funding mechanism led by the government in which funds are shared by the collective and farmers, and it has carried out comprehensive cleaning work at points, along lines and in areas, including courtyards, villages, roads and riverways.

2. Centering on industrial development

Based on village improvement, Zhejiang has put to use rural resources and ventured down the path of operations suited to the conditions of villages. Zhejiang has taken the construction of a beautiful countryside as the new point for increasing the rural economy, it has encouraged farmers to carry out innovations and start businesses, and it has made sure that environmental beautification and economic development reinforce each other, the construction of a beautiful countryside and the work on making farmers rich are promoted at the same time. In the traditional planting industry or breeding industry, there is only income from the primary industry. Zhejiang has accelerated the integrative development of the primary, secondary and tertiary industries, and has given expression to the added value in agriculture. The countryside is "operated", the direction, line of thought and measures for rural development have been identified.

(1) Taking advantage of beauty to develop rural leisure tourism. Zhejiang has leveraged the beautiful environment from the construction of a beautiful countryside to speed up the development of the agritainment leisure tourism industry. In 2014, Zhejiang's agritainment industry received tourists for 180 million person-times, and the direct operating income was 14.12 billion yuan. The agritainment operators have mostly developed on the basis of achievements in village improvements, the protection of historical and cultural villages and the construction of a beautiful countryside. In the process of gradually turning ecological and rural resources into developmental advantages, it has become a new growth point for Zhejiang's tourism economy and an important source for increasing farmers' income.

(2) Seizing opportunities to cultivate new forms of business in rural areas. Zhejiang has turned potential ecological advantages into realistic advantages for industrial development, while the old-age care industry, holiday economy, creative cultural industry and experience-oriented agriculture have emerged in beautiful villages. In the industrial development of Zhejiang's beautiful villages, the tourism economy dominated by leisure sightseeing and holiday experiences, the health preservation economy focusing on homestay-based holidays in summer, the health preservation economy mainly involving old-age care and heart-nourishing, sports competition focusing on expedition and outdoor training, the creative cultural economy primarily involving root seeking, sketching and creation have greatly enriched new forms of business in Zhejiang's rural areas; they have activated the economy of flowers and fruits, the nursery stock economy and they have also given rise to some forward-looking philosophies, including the "beautiful industry" and the "beautiful economic pilot area".

3. Building brands

The construction of a new countryside in Zhejiang is characterized by building brands and benefiting the people. Under the overall brand of a beautiful countryside, local authorities have, in light of local, industrial and humanistic characteristics, created a number of shining brands with regional rural characteristics which cater to the people's tastes, are inspiring, highbrow and easy to remember.[8] Brand names reflect characteristics and are impressive, memorable. As a result, the construction of a beautiful countryside is combined with localized construction. Names exquisitely correspond to geographical names or are associated with geography and a topographical feature or with humanity and folk customs. As they are highly readable and show lasting appeal, they promote the development of rural characteristic industries and an increase in the income of farmers to some extent. Brand operation is the most valuable operational mode among all the operational modes, such as Sunny Wenling, Golden Pinghu, Untrammeled Zhoushan, Unrestrained Tonglu, Worldly Xianju, Beatific

[8]Tang Yuanjie, Shen Baozhen, and Jiang Wenlong et al, The Path of Building a Beautiful Countryside with Chinese Characteristics—Focusing on Decade-long Transformation of Zhejiang's Countryside, *Farmers' Daily*, September 16, 2013.

Changxing, Mountainously Gorgeous Lishui, Idyllic Xiuzhou, Dreamy Fenghua, and Blissful Jiangshan. Sunny Wenling means that the first sunlight in the new century shone on Shitang in Wenling; Untrammeled Zhoushan is called that because of the attitude of the Buddhists in the Putuo Temple about standing aloof from worldly success; Unrestrained Tonglu is famous thanks to the *Ten Poetic Proses Praising Tonglu*, written by Fan Zhongyan, a great writer of the Song Dynasty. Accordingly, Zhejiang has carried out the “5525” Project—5 exotic belts in a beautiful countryside, 5 great rural festival activities and 25 characteristic villages. Brands represent positioning, features and competitiveness. Under the guidance of brands, the local authorities of counties and county-level cities have further made strategic decisions about industrial development in light of respective characteristic resources. From product operations, resource operations to brand operations, branding has been widely applied in the construction of a beautiful countryside in Zhejiang.

The overall brand of Hangzhou City—a city of quality life—is built on the basis of county, town and village brands. With improvement on both banks of the Qiantang River, the Fuchun River and the Xin’an River, Hangzhou City has witnessed great improvements in its rural areas, with fabulous scenery everywhere and at every village. Hangzhou has built the brand of a beautiful countryside according to local conditions. In light of historical and cultural or industrial characteristics of development, counties have creatively carried out brand positioning. For instance, the brand of the West Lake District is “The Most Beautiful West Lake, A Charming Countryside”; Xiaoshan District: “Culturally Advanced Xiaoshan, A Homeland of Happiness”; Yuhang District: “A Livable Idyllic City, A Beautiful Urban Countryside”; Tonglu County: “Unrestrained Tonglu, A Scenic Countryside”; Chun’an: “A Homeland of Clean, Clear Water, A Beautiful Countryside”; Lin’an City: “A Green Homeland, Splendid Mountain Villages”; Jiande City: “Spectacular Mountain Villages, Livable Jiande”; Fuyang City: “A Luxuriant Mountain Refuge, A Beautiful Countryside”. Some of these brands are distinguished by one or two different words, but they very accurately reflect respective characteristics, goals and developmental strategies. Counties and county-level cities have their own positioning, towns have their own brands; some excellent villages, characteristic villages and central villages have adopted respective brand names for publicity. Twenty-eight excellent villages in Lin’an City have different names, for instance, Shangtian Village in Banqiao Town is called “Civil and Martial Shangtian with Tea Fragrances and a Sea of Bamboo”; Tianmushan Village in Tianmushan Town is positioned as “A Kingdom of Towering Trees, Tianmu Cozy Households”, and these names are distinctive. With an increase in the building of county, county-level city, town and village brands, interaction and complementation in brand building, Hangzhou, the overall brand of Hangzhou City—a city of quality life—has become increasingly rich.

Huanxi Village in Tonglu County has built the brand of the “lotus flower” by displaying the environment against the background of historical culture. The residents of Huanxi Village are the 14th descendants of Zhou Dunyi, a great philosopher of the Northern Song Dynasty; Huanxi Village has built its brand on the basis of the *Ode to the Lotus Flower* written by Zhou Dunyi; the symbol of Huanxi Village is derived from the family name of Zhou Dunyi and features the lotus flower, suggesting that the origin of this village lies in the *Ode to the Lotus Flower*. There are a lotus flower square, a lotus flower fountain and a lotus flower theme park in this village. This village is surrounded by 100-mu of lotus flowers. Since the project "Cleaning Tonglu" was initiated, 25 million yuan have been used to improve Huanxi Village. The riverways and streets in this village have been improved; electric lines, broadband and digital television have been made available; 9 domestic sewage treatment tanks have been built, and more than 600 rural households in this village now have sewage pipes available to them. There are flowers and grasses on these sewage treatment tanks, making them unidentifiable by their appearance. There are more than 100 villages operated by collecting admission fees internally and externally, including Huanxi Village, in Zhejiang; they have built their brands on the basis of the fruit industry, such as tangerines, strawberries and peaches and plums, or the aquaculture industry, or the natural landscape, or profound historical and cultural deposits, so characteristic development has been fostered.

4. Relying on cultural development

There are a great variety of historical and cultural relics in Zhejiang. Thirty out of 350 officially announced Chinese famous historical and cultural towns and villages are located in Zhejiang, ranking Zhejiang no.1 nationwide; there are 8 international intangible cultural heritages and 189 national intangible cultural heritages in Zhejiang; Zhejiang ranks no.1 nationwide in the number and ranking of intangible cultural heritages. Historical culture is Zhejiang's soul and is also fundamental to Zhejiang. With the thriving rural culture as its soul, Zhejiang has given expression to the characteristics from the construction of a beautiful countryside. Zhejiang has deeply tapped the farming tradition, ethnic customs, folk skills and arts, other rural cultures to cultivate characteristic cultural villages, promote the protection of historical cultural villages and the construction of cultural auditoriums, enhance the appeal, cohesion and soft power in rural development.

(1) Protecting historical cultural villages. With a focus on restoring elegant traditional buildings, carrying forward time-honored historical culture, building a beautiful living environment and fostering a lifestyle of leisure, actions have been taken to protect, restore and utilize historical cultural villages. Zhejiang has holistically pushed forward the comprehensive protection of ancient buildings and the villages' ecological environment, the discovery and inheritance of excellent traditional cultures, the scientific improvement of the villages' living environment and the orderly development of rural leisure. The integrity and historical authenticity of historical cultural villages have been retained, while the continuity and sustainability of their life are also embodied.

With the advantages of ancient buildings, Cangpo Village in Yongjia, Zhejiang has built a cultural tourism exhibition hall for earnestly publicizing Yongjia's ancient village culture, its farming-reading culture, folk culture, ancestral temple culture, Kun Opera culture and other traditional cultures and it is promoting the deep integration of local tourism and culture. In Tonglu, Hangzhou, there are 1,400 ancient buildings with a certain historical cultural value which were constructed before the Republic of China. As from 2010, 119 ancient buildings were intensively repaired, 463 of them underwent general repair, and 36 million yuan were allocated to repair them. Historical buildings have taken on an entirely new look; they, in conjunction with the mountains and waters in Fuchun, constitute a rural tourist route of "humanity + history + natural mountains and waters", they are distinctive and unique. In coordination with five main rural festivals including the native cuisine festival, the rural carnival, the mountain flower festival, the outdoor leisure sports festival and the water summer festival, actions have been taken to build five exotic belts in a beautiful countryside, covering poetic mountains and waters, ancient customs, industrial bizarrerie, sports leisure, and ecological health preservation. Ancient villages and an ecological rural environment bring characteristic competitive power and offer valuable experience for the construction of a new countryside in Zhejiang.

(2) Building rural cultural auditoriums. Based on the ancestral temples, meeting halls, unused school houses and other venues, actions have been taken to build cultural auditoriums integrating learning, teaching, etiquette and entertainment and create a cultural homeland for farmers. In 2013, cultural auditoriums were built across the province; the People's Government of Zhejiang Province identified the project as one of the ten projects that were to do substantial work. Zhejiang plans to build 1,000 cultural auditoriums, which can serve as models. In principle, each of the counties, county-level cities and districts should build more than 12 cultural auditoriums. Cultural auditoriums are mainly based on the integration of resources and are built by utilizing old ancestral temples, meeting halls and unused school houses; the list of students, the list of the god of longevity and the list of contributions are arranged within cultural auditoriums, in which the content is native and vivid.

5. Carrying out a standardized manner of management to guide the way

As the forerunner in the construction of a beautiful countryside, Zhejiang has become the first to issue the *Standard for the Construction of a Beautiful Countryside*, providing a model for the construction of a beautiful countryside. Standardization is an important part and a sign of agricultural modernization and is also the vigorous means for promoting the construction of a new countryside. With standardization, the strategy for the construction of a new countryside whose philosophy is ambitious but whose connotation is abstract becomes a system of working which adopts explicit means and is easy to refine and operate, and the construction of a new countryside with rich content and a large complicated system becomes a logical, scientific and well-regulated system of responsibility.

The *Standard for the Construction of a Beautiful Countryside*, officially implemented as from April 6, 2014, is aimed at standardizing and promoting the experience and achievements in the construction of a beautiful countryside, and exploring the Zhejiang Model for the construction of a new socialist countryside. The standard brings about unification, normalization and quantification on the basis of summarizing Anji County's successful experience in the construction of a beautiful countryside. Based on Anji County's successful experience in the construction of a new countryside, the *Standard for the Construction of a Beautiful Countryside*, adopted by Zhejiang, cites the 21 existing national, industrial and local standards concerning the construction of a new countryside and provides operable practical guidance for that construction in terms of 36 indicators involving 7 aspects, including village construction, ecological environment, economic development, development of social programs, cultural and ethical development. The standard only serves as an operational guide and does not require uniformity in the construction of a beautiful countryside; it gives expression to the characteristics of villages and calls for retaining, to the greatest extent, the authentic rural culture and rural characteristics in line with rural natural endowment, historical tradition and future developmental needs, and thus catering to the developmental needs of different villages.

5.3 The Achievements and Modes in the Construction of a Beautiful Countryside

Agricultural resources can be turned into agricultural capital, while ecology, the environment and culture, as important resources, can also be converted into capital. The Secretary of the Party Committee of Zhejiang Province, Xia Baolong, pointed out that Zhejiang did not compete with others in GDP, Zhejiang should compete with others in building a beautiful countryside and in making the countryside more beautiful, affluent and culturally advanced. During the national working conference on improving the rural living environment on October 9, 2013, Vice Premier Wang Yang vowed to earnestly study and implement the guiding principles from the important written instructions made by Xi Jinping and Li Keqiang on comprehensively improving the rural production and living conditions, summarize and promote Zhejiang's experience in carrying out the project of "1,000-village demonstrations and 10,000-village improvements", and more quickly promote the comprehensive improvement of the rural living environment. Zhejiang's experience and practices deserve to be learnt and drawn upon. The authorities in the areas where conditions permit should endeavor to push forward the construction of a beautiful countryside. A beautiful countryside has become a shining symbol in the building of a beautiful Zhejiang.

5.3.1 Main Achievements in the Construction of a Beautiful Countryside

1. The ecological economy has developed rapidly, clean, clear waters and lush mountains are becoming the "gold and silver mountains"

The successive Party Committees and People's Governments of Zhejiang Province have attached great importance to agricultural development in the construction of a beautiful countryside and have actively pressed ahead with market-oriented reform in agriculture. Highly-efficient ecological agriculture has shown evident characteristics. In 2013, the output value of ten leading industries—including vegetables, tea, animal husbandry, fruits, edible mushrooms, silkworm breeding and mulberry growing, traditional Chinese medicinal materials, flowers and nursery stock, freshwater aquaculture and bamboo—accounted for nearly 80% of the total output value of agriculture, forestry, animal husbandry and fishery. Tea, silkworm cocoons, edible mushrooms, bee products, flowers and nursery stock occupied an important position nationwide. In late 2013, the planting bases for pollution-free agricultural products, green food and organic products certified in Zhejiang covered an area of 14,536,000 mu. In the meantime, Zhejiang enjoys a good foundation for developing the processing industry for agricultural products; leisure and sightseeing agriculture has developed rapidly and has become a new bright spot for increasing the farmers' income. In 2013, there were 638 processing-type leading agricultural enterprises with an annual sales revenue of 100 million yuan in Zhejiang. Forty-eight agricultural product logistics distribution centers in Zhejiang completed distribution worth 2.48 billion yuan. At present, 55 provincial pilot enterprises for the chain operation of agricultural products have 8,510 outlets; they have connected 5.06 million mu of production bases and realized 7.8 billion yuan in sales volume. In Zhejiang, there are 2,334 leisure and sightseeing agriculture areas (points), delivering 14 billion yuan in output value. With the three-year action plan for agricultural modernization ("8810" action) as the overall means, based on the goals in 8 respects including industrial development, operational benefits, quality safety, material equipment, scientific and technological support, ecological environment, policy input and the level of professional services, the Party Committee and the People's Government of Zhejiang Province organized 8 special actions involving industrial agglomeration, the high quality of products, facility-based production, operational organization, technology integration, IT-based means, environmental ecologicalization and service socialization. They also carried out ten key projects for improving the quality of farmland, promoting agricultural standardization and agricultural mechanization, developing the modern seed industry, cultivating new-type operators, building circular ecological agriculture, enhancing agricultural services, boosting intelligent agriculture, increasing the village-level collective economy and the creative agricultural cultural industry, with a view to building a strong province of highly-efficient ecological agriculture and a large province of excellent characteristic agriculture and speeding up industrial transformation and upgrading.

In Zhejiang, clean, clear waters and lush mountains are becoming the "gold and silver mountains". The farmers' income and standard of living improved significantly in ten years, the transfer and property income grew rapidly (see Fig. 5.1). In 2013, the total value of agriculture, forestry, animal husbandry, sideline occupations and fishery reached 283,739 million yuan and the per capita net income of rural residents stood at 16,106 yuan in Zhejiang. The economy of health preservation, the sports economy, the creative cultural economy, the property economy, including rural house leasing and participation in meeting halls as shareholders, the labor service economy, including the processing of materials supplied by clients and the processing of tourism products and a large number of "beautiful industries" have emerged in Zhe-

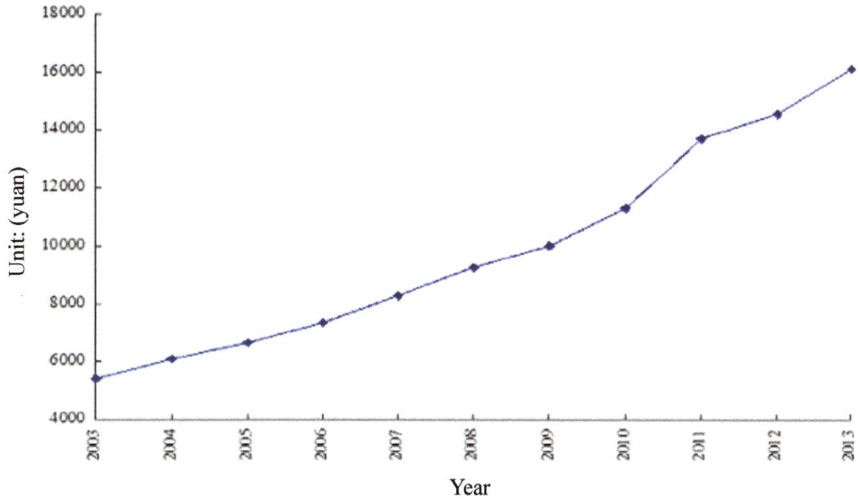

Fig. 5.1 The level of per capita net income of rural residents in Zhejiang province

jiang's countryside; a pattern in which environmental beautification and economic development benignly reinforce each other to make villages beautiful and farmers rich is in the making. Zhejiang has ranked no.1 among provinces and autonomous regions across the country in the per capita net income of farmers for 29 consecutive years.

In 2013, the income of the village-level collective economy exceeded 30 billion yuan in Zhejiang. A total of 8,844 village economic cooperatives underwent reform in the system of joint stock cooperatives, accounting for 28.7% of all cooperatives in Zhejiang. Forty-four counties, 739 towns and 15,316 village economic cooperatives were built and recognized as those with a well-regulated financial management. New-type players, including farmers' specialized cooperatives, have developed rapidly, industrial and commercial enterprises and private capital have been active in the agricultural field, and the level of industrialized operations has been relatively high. At present, there are 7,492 leading agricultural enterprises, 9,401 family farms, 37,428 farmers' specialized cooperatives, and 1,146,000 rural households as members of the cooperatives, stimulating the development of 4,664,000 non-member rural households; the area of the circulated land was 8.65 million mu, accounting for 45% of all contracted cultivated land. In 2013, 15 billion yuan from industrial and commercial capital was invested in agriculture, and there were more than 600 enterprises with an annual sales revenue above 100 million yuan in Zhejiang. In the meantime, the rural social security has improved continuously and the people's happiness has been guaranteed in Zhejiang. In late 2013, 28,551,000 people were covered by new rural cooperative medical care, accounting for 97.8%; the per capita funding standard was 557 yuan, the maximum payment limit in all areas within the scope of fund pooling exceeded 8 times the national per capita net income of rural residents and was not less than 80,000 yuan.

2. The ecological environment has become increasingly beautiful, an environment with a blue sky, clear water, green land and lush mountains has been fostered

Since 2003, Zhejiang has carried out the project of "1,000-village demonstrations and 10,000-village improvements" under the theme of comprehensive improvement of the rural environment for two consecutive rounds. In 2010, with Zhejiang included in the first batch of the 8 national demonstration provinces for contiguous improvement of the rural environment, the Ministry of Finance and the Ministry of Environmental Protection signed the *Agreement on Demonstrations in the Contiguous Improvement of the Rural Environment* with the People's Government of Zhejiang Province. Under the principle of "one-time planning and three-year implementation" and according to the requirement of making regional integration and concentrated contiguous improvement, with a focus on the Qiantang River and Lake Taihu basins and important protected areas for the sources of drinking water, Zhejiang has scientifically developed a plan for the implementation of improvements, it has adopted the technically advanced and economically rational pollution control mode according to local conditions, and built the mechanism for the operational maintenance of pollution control facilities, thus achieving the goal of "improving one contiguous area, changing one contiguous area and consolidating one contiguous area", really improving the quality of the rural living environment and the ecological environment. As of late 2014, 6,120 villages had undergone rural domestic sewage treatment in Zhejiang. A total of 1.50 million rural households obtained benefits; 1,901 villages underwent garbage reduction and recycling, and 97% of the villages in Zhejiang realized the centralized collection and disposal of domestic garbage.

During the period 2003–2014, Zhejiang promoted the construction of a new countryside at specific points, along lines and in areas, and the countryside presented an entirely new look. The environment in all of the towns has been improved; actions have been taken to holistically boost the integration of a regional road network, a network of pipelines, networks of forests, rivers, garbage disposal and sewage treatment. Zhejiang has, with the county as the unit, built the model areas for a beautiful countryside; Zhejiang has built scenic corridors along transportation routes; with villages developed into characteristic scenic spots, Zhejiang has built a large number of new homelands with a well-arranged layout, a beautiful environment, complete facilities and full services; farmers have generally improved their health habits, their philosophy of environmental protection, their democratic awareness and their outlook on social morality; the work on village improvement and its effectiveness serve as an important yardstick for the people to evaluate the primary-level cadres and an important basis for primary-level democratic elections, and so the ability of village organizations and the primary-level cadres to handle affairs has generally improved; the pattern of the overall planning of urban and rural areas by government departments has fully taken shape, the overall synergy in promoting urban and rural development in a coordinated way and making agriculture, rural areas and farmers better has generally increased.

3. Rural ecological culture has become increasingly rich so that the native style is deeply rooted among villagers

In the construction of a beautiful countryside, with the thriving rural culture, Zhejiang has kept carrying forward local culture and has worked on meeting villagers' cultural needs. A beautiful countryside cannot be built without a developed economy, a beautiful environment, the rich cultural life of rural people and the rural civilization combining tradition with modernity in the new period. Protecting and inheriting an excellent rural culture is essential in the construction of a beautiful countryside; rural areas have the characteristics and inherent laws that are different from those of urban areas, so it is necessary to respect and observe the typical context of rural areas, get rid of the stale and bring forth the fresh, and combine tradition with modernity; only in this way can the enduring vitality be nurtured. Zhejiang has vigorously protected historical and cultural villages, and initiated the protection of 260 historical and cultural villages. In 2013, Zhejiang initiated the construction of 1,000 rural cultural auditoriums. In an effort to improve the rural environment, local authorities have paid particular attention to tapping the clan culture, ethnic customs, folk arts and the anecdotes of famous people which reflect the characteristics of villages, and have built the farming culture exhibition halls, halls for experiences of characteristic skills and arts and halls for the tasting of special delicacies. In some villages, rural cultural auditoriums integrating learning, teaching, etiquette and entertainment have been built by utilizing old ancestral temples, old academies of classical learning, meeting halls, unused school houses and cultural activity centers; there are village rules, village songs and annals of village history.[9]

In Zhejiang, ancient villages abound in cultural resources. Regarding the protection and inheritance of traditional villages and rural traditional cultural resources, Zhejiang has fully utilized these cultural resources and promoted the prosperity and development of cultural heritage. Diverse village cultures, such as traditional drama and folk activities with local characteristics, have enriched the people's cultural and ethical life, and reshaped the farmers' cultural beliefs. The combination between rural intangible cultural heritage and tourism has boosted the development of the ecological tourism industry. In May, 2014, in the beautiful countryside of Dipu Village, Zhejiang, a rural tea bar with a strong sense of reminiscence was built by an old house with mudstone walls, timber beams, black tiles, wooden casks and stone cisterns, the originally dirty and stinking stone houses for raising pigs were turned into tea bars,[10] innovative techniques have attracted a large number of tourists and it has become the typical model for the construction of a beautiful countryside. In the construction of a beautiful countryside, Dipu Village has taken the innovative perspective of restoring the characteristics of an area, so that the native elements, unsophisticated and profound customs, memorable past events attract urban residents. Zhejiang has moved faster to protect traditional villages, making sure that

---

[9]Fang Yibo, Wang Zheng, and Shang Yiying, Clean, Clear Waters and Lush Mountains Are the "gold and silver mountains"—Demystifying the Zhejiang Model for a Beautiful Countryside, *Economic Information Daily*, November 11, 2013.

[10]Xu Yuwen, Turning an Ancient Pigsty into a Rural Tea Bar, *Farmers' Daily*, May 14, 2014.

there are venues for retaining everyone's nostalgia, and that culture can be inherited and carried forward.

Zhejiang's vivid practice in the construction of a beautiful countryside has presented the beautiful appearance of a new socialist countryside featuring increased production, higher living standards, a high cultural level, clean villages and democratic management. This is closely associated with Zhejiang's efforts in earnestly carrying out the Scientific Outlook on Development and the guiding principles, adopted by the Central Committee of the Communist Party of China, relating to agriculture, rural areas and farmers. According to the "Eight-Eight Strategies" and the developmental strategies of making the people rich by starting businesses and of building a strong province through innovation, building a materially affluent and culturally advanced Zhejiang and a beautiful Zhejiang for a good life, Zhejiang has vigorously carried out the strategy of promoting coordinated urban and rural development, comprehensively pushed forward the construction of a new countryside, achieved the transformation from comprehensively overcoming urban-rural segmentation to boosting urban-rural integration in an all-round way; Zhejiang has stayed ahead of the times in the construction of a new socialist countryside.[11]

5.3.2 The Typical Mode for the Construction of a Beautiful Countryside

In the construction of a new countryside and urban-rural integration, Zhejiang has made experiments to establish all-inclusive theories and practice. The villages with different national endowments and at different levels of economic development have proceeded from their respective unique advantages to seek out a variety of modes for the construction of a beautiful countryside.

1. The mode of ecological economization

Under the mode of ecological economization, ecological agricultural products, the processing industry for ecological agricultural products and the ecological service industry are under economic operations, the protection of the ecological environment is turned into ecological economic products. This mode is a relatively common mode for the construction of a beautiful countryside. The mode of ecological economization is suitable for the counties with an abundant endowment of natural resources, in which institutional and technical innovations effectively convert natural resources into ecological capital. Based on the initial funds, arising out of the reform of forest rights, for the development of ecological agriculture and the service industry, the government of Suichang County has actively experimented the path of e-commerce for agricultural products and developed a characteristic marketing path for agricultural products which caters to the Internet era. Kaihua County has strictly followed

[11]The Office of Agriculture and Rural Work of Zhejiang Province: Grand Spectacle in the Construction of a Beautiful Countryside in Zhejiang, *Zhejiang Daily*, February 10, 2011.

the philosophy of building the county through ecological development to develop the green industries, speed up the development of green agricultural products and embark on the developmental path for an ecological economy.

Suichang County is part of Lishui City. In Suichang County, there are 3,277,000 mu of forest land, the per capita area of forest land is 17 mu, the volume of forest stock is 5.92 million m^3 and the forest coverage is 82.3%; it is a large county of forests in Zhejiang. In the construction of a beautiful countryside, Suichang County has stayed ahead nationwide in the reform of forest rights and e-commerce marketing of agricultural products and has developed the Suichang Model for blazing new trails.

Experience One from Suichang Model is the reform of the system of forest rights. The reform brings about a fund guarantee for the development of the ecological economy. Forest resources have always been the largest asset owned by Suichang's farmers. In Suichang County, the mountain forest covers an area of 3.32 million mu, the per capita volume of forest stock reaches 40 m^3; 865,000 mu of mountain forests in Suichang County were incorporated into the key ecological public-welfare forests for protection in 2003; Suichang County is hailed as the source of the Qiantang River and the Ou'jiang River and the green sea south of the Yangtze River. In early 2006, Suichang County became the first in the province to launch the forest rights mortgage loan, making it possible to turn "leaves into money". In March, 2008, the leaders of the National Forest Reform Office fully recognized Suichang's achievements in the reform of forest rights, held that Suichang had stayed ahead nationwide in the reform of forest rights, and used the Suichang Model to encapsulate Suichang's innovation in the reform of forest rights.[12] In the reform of the forest rights, Suichang County has made beneficial experiments in three modes—small mortgage loans based on forest rights, mortgage guaranteed loans based on forest rights and direct mortgage loans based on forest rights.[13] The reform of the system of forest rights under the Suichang Model has put to use resources, turned mountain forests into banks from which cash can be withdrawn, and has also provided active guidance to avert the investment risks for forest workers so that the green wealth for forest workers is on the increase.

Experience Two from the Suichang Model is the construction of agricultural e-commerce. Agricultural e-commerce has stimulated the expansion of the marketing network for ecological agricultural products. Based on IT applications, the Party Committee and the People's Government of Suichang County have built a good e-commerce developmental ecosphere, and have made all-out efforts to upgrade the scientific and technological development demonstration area in Suichang County. In 2012, there were more than 1,500 e-commerce traders and more than 6,000 people engaged in e-commerce, the total volume of online retail sales based on e-commerce reached 200 million yuan in the county. The development of e-commerce

[12]Hua Wenli, The Suichang Model: A Model of the Reform of the Forest Rights, *Zhejiang Forestry*, 2009(3).

[13]Jiang Guohai, The Suichang Model in the Reform of the Forest Rights, *Zhejiang Economy*, 2009(23).

has stimulated the development of a large number of micro, small and medium-sized enterprises and farmers' specialized cooperatives in the county; there has been an e-commerce ecosphere in the county, which has an online store association as platform and is supported by government departments and features win-win cooperation among e-commerce traders, suppliers and the third-party service providers.

Kaihua County is situated at the source of the Qiantang River. Its forest coverage reaches 80.4%. In 2002, Kaihua County was designated as a national ecological demonstration area. In 2003, Kaihua County became the first nationwide to adopt the master plan for building an ecological county. In 2011, it was officially named as the National Ecological County by the Ministry of Environmental Protection and became the National Pilot Area for Ecological Development. Kaihua County is the national model county for greening, the national comprehensive demonstration county for developing agriculture through science and technology and achieving sustainable development, the national production demonstration base for pollution-free agricultural products, the birthplace of the famous Chinese tea Longding, the birthplace of the Chinese root carving art. The Qiantang River Source National Forest Park and the Gutianshan National Natural Reserve are located in Kaihua County. With a focus on the development of the experimental area for the scientific development of mountainous areas and the Qiantang River source tourist resort area, Kaihua County has fully leveraged the ecological advantages from its location at the source of the Qiantang River to cultivate and develop an ecological economy, take a new path of transformation and upgrading which features ecological resources, ecological capital, an ecological economy and ecological advantages and create a new highland for ecological development.[14]

Experience One from Kaihua County lies in building the county through ecological development and making industries green. Kaihua County has strictly regulated functional areas, and has designated 95.7% of its areas as the areas to which access is prohibited and the areas to which access is restricted. In the "991" action plan for the circular economy taken by Zhejiang, Kaihua County has issued the guiding catalogue for industrial development to strictly control the ecological threshold, and it has massively carried out the project for governing mountains and waters and treating sewage, it has made strategic adjustments of the industrial structure, shut down the enterprises with high energy and material consumption and heavy pollution, focused on developing the "2+6" leading industries, the highly-efficient ecological agriculture and the ecological leisure tourism industry. With regard to green industries, there has been a pattern of development covering new and hi-tech industries in the Kaihua and Huabu industrial functional areas, the green food processing industry in the Majin industrial functional area and the light processing industry in the Chihuai industrial functional area. The electric light source and green food processing industries have been included in the ten provincial advanced manufacturing bases planned and cultivated by Zhejiang. Huakang Pharmaceutical has ranked No.1 in

[14]Lu Ning, The Experience in the Construction of National Ecological Counties, Realistic Difficulties and the Direction of Development—A Field Survey from Kaihua, Zhejiang, *Contemporary Social Science Perspective*, 2014(2).

Asia and No.2 in the world in xylitol production and volume of sales. There is the most complete system for the recycling of synthetic material chlorine in China; two enterprises, including Qinghua Chemical and Synthetic Materials, have become the models in ammonia nitrogen control and environmental protection in Zhejiang; 18 enterprises have passed the provincial acceptance check on clean production.

Experience Two from Kaihua County consists of developing an ecological economy and achieving ecological economization. Sixty-three agricultural products in Kaihua County have passed the certification of organic foods and green pollution-free agricultural products and the base area accounts for more than 60% of the county's crop planting area. Kaihua Longding has been identified and recognized as a well-known Chinese brand, and it has been included in the list of China's top 100 regional public brands of agricultural products released by the Ministry of Agriculture. In Kaihua, there are 5,400 ponds, covering an area of 1,550 mu, in 17 towns and 4,600 rural households for the farming of freshwater fish. Farmers' specialized cooperatives have built 20,000 mu of ecological rice bases, where a one-mu area with milled rice with embryos delivers returns equivalent to that from a four-mu rice field, while the price of organic rice is 80 yuan/kg.[15] The ecological agricultural project of "pigs—marsh gas—crops and cash crops" has been carried out; Zhejiang's demonstration county for circular ecological agriculture has been built, a centralized supply of marsh gas has been promoted in all of the villages, the mode of circular ecological industrial development has been realized. Kaihua County has called for turning the comparative advantages of ecological resources into the competitive advantage of an ecological economy, and has taken a developmental path for concurrently promoting economic development and ecological development, and enhancing industrial competitiveness and environmental competitiveness at the same time.

2. The mode of resource recycling

The mode of resource recycling is an important mode for the construction of a beautiful countryside. Under this mode, the circular economic principle of reduction, reuse and recycling is followed to achieve efficient recycling of agricultural resources; this saves resources and also protects the environment and promotes sustainable economic and social development. In terms of the mode, natural resources are relatively lacking, it is necessary to bring about innovations to the line of thought and methods under the condition of limited resources and explore the manner of resource recycling. Tengtou Village in Fenghua City started early in resource recycling and has accumulated rich experience. In 1990, it highly emphasized resource recycling at the time of renovating its rural land and water treatment system, and has gained extensive experience in multistory development of agricultural products and resource recycling. In the development of the highly-efficient ecological agriculture, Lin'an City has relied on its own characteristics to intensify technical innovations and achieve resource recycling.

[15] Yang Junxiong, Zhu Jianping, and Xu Zhu'an, Kaihua Has Vigorously Developed an Ecological Economy, *Zhejiang Daily*, December 9, 2012.

Tengtou Village in Fenghua City is part of Ningbo City, it is listed among the global top 500 ecological villages and the global top 10 harmonious villages; it is the only village selected worldwide for the Shanghai World Expo; it is a national Grade 5A scenic spot village. It covers an area of 2 km^2 and has 344 households, with 854 villagers. Originally, Tengtou Village was a poor village with uneven farmland and rugged roads, its yield per mu was only 200 yuan; at present, it has become rich, the villagers have become smart, and every household has become well-off; it has embarked on the path of sustainable development guided by a green and ecological philosophy.[16]

Tengtou Village has made active experiments in the circular agricultural mode of development and started early in this regard, great effects have been produced, which are mainly reflected in the following respects. First, combining soil improvement with the scientific application of fertilizers, building a circular system in which farmland is sustained by farmland. In the village, organic fertilizers are mainly applied, with chemical fertilizers as a supplement; fertilizers are determined according to the soil, fertilizers are used to make up for the deficiency; scientific application of fertilizers is promoted steadily, green manure has been planted extensively for a long time; straw is returned to the farmland; farmyard manure, such as river mud and animal excrement, is used, the content of organic soil matter is increased. Second, establishing a multilevel and multistory cropping and raising circulation system. Tengtou Village has 7 multistory cropping and raising modes, including crop rotation, rotation of grain with green-manure crop, inter-planting of fruits and cereals, multistory cropping and raising, multistory cultivation in fishponds, indoor multistory rearing in pens and multistory development of mountainous areas. The development of multistoried agriculture has enhanced the efficiency of agricultural resource production and has brought about more economic returns, it has beautified the rural environment and generated more ecological benefits. Third, building a circular system which integrates planting, breeding, processing and marsh gas. Since 1978, Tengtou Village has studied waste processing and conversion for an efficient utilization of bioenergy. With the large-scale development of the planting and breading industries, the village has planned and built feed mills and marsh gas stations for addressing the issue of domestic fuel for all villagers, and has also built a system for the benign circulation of waste and old materials, mitigated the damage from firewood and straw burning to the living environment and increased the rate of waste circulation and conversion.[17]

Lin'an City has 13 towns, 5 sub-districts and 298 administrative villages; it has been rated as a national sanitary city, a national forest city, a national model city for environmental protection and a national ecological city; it has been listed as the only national pilot city for ecological civilization in Zhejiang. In 2010, Lin'an City made the strategic arrangement for relying on ecological advantages, scientific and tech-

[16]Liu Songjiang, The Practice and Line of Thought of Tengtou Village in Building a Green Homeland, *Garden*, 2012(12).

[17]Zhou Yayue, Xu Lei, Regional Agricultural Sustainable Development under the Guidance of the Ecological Culture—Take Tengtou Village in Ningbo as an Example, *Agricultural Economy*, 2008(2).

nological innovations to build a green homeland and wealthy and splendid mountain villages. Lin'an City has emphasized both "wealthy" and "splendid", followed the requirement of "one product in one village" and "one industry in one village", and has carried out the project of enhancing the new green industries to promote the coordinated development of the primary, secondary and tertiary industries in rural areas[18]; the city has obtained good economic returns in boosting resource recycling and has built the ecological agricultural developmental mode with the city's characteristics.

Typical Case One involving resource recycling in Lin'an City is the recycling of hickory hulls. After hickory raw materials in Lin'an and from other areas are shucked, about 70,000t of hulls are generated each year. In the past, they were discarded casually in open spaces and streams, thus polluting rivers and affecting crop growth. With the recycling and harmless disposal of hickory hulls, more than 90% of the hickory hulls are recycled.[19]

Typical Case Two involving resource recycling in Lin'an City is the circular ecological breeding chain. The ecological farm at Qianchuan Town has actively performed circular ecological agricultural production; it has controlled pollution in livestock and poultry breeding while making pollution-free organic agricultural products. With a focus on raising chickens, the farm has conducted multistory agricultural development relating to forests, fruit and fish, and has built a benign agricultural ecological chain. The ecological farm has adopted the ecological circulation mode of "pork pigs → pig manure → earthworm → layer chickens, broiler chickens → chicken manure → marsh gas → biogas slurry → freshwater fish → fruit, bamboo and vegetables, grass → chickens, pigs" and "pork pig → pig urine → duckweed, other aquatic forage grass → freshwater fish → sludge → fruit, bamboo and vegetables, grass → chickens, pigs", so resource recycling has been realized scientifically.

3. The mode of making scenic spots universal throughout the county

This mode stresses the notion of building universal scenic spots throughout the county and fosters the development characterized by coordination among villages and universal scenic spots across the county. The mode is suitable for the construction of a beautiful countryside in the counties with certain quantities and on a certain scale and level of development. Under this mode, the scattered beautiful villages are connected, the returns at points involving the beautiful villages are delivered in a contiguous way, producing a scale effect and driving forward the economic development of the whole county. Anji County has become the first nationwide to initiate the construction of a beautiful countryside. Anji County's standard setting in integrative development of industries and a beautiful countryside has produced a demonstration effect on the rest of the country. Making scenic spots universal throughout the county has become a new front in the construction of a new countryside. Chun'an County enjoys the

---

[18]Qian Xiafang, Fang Xuefeng, A Green Homeland, Wealthy and Splendid Mountain Villages—Lin'an City Has Made All-out Efforts to Build the Characteristic Brand of a Beautiful Countryside, *Daily Business*, September 16, 2012.

[19]Li Zhong, A Dream, Which the Farmers in Lin'an Have, of Turning Waste into Wealth Has Stimulated Industrial Recycling, *Hangzhou Daily*, June 12, 2013.

resources from the world-famous Thousand-island Lake. The developmental orientation of building the county on the basis of the lake tallies up exactly with the developmental mode of making scenic spots universal throughout the county.

Anji is part of Huzhou City, and it is famous for bamboo. It has become the first to initiate the construction of a beautiful countryside. The *Standard for the Construction of a Beautiful Countryside*, developed on the basis of Anji's practice, has been promoted throughout the province and has produced a demonstration effect nationwide. Anji County has been granted a number of titles, including national ecological county, national demonstration county for the ecological development of forestry, national ecological demonstration area, the birthplace of Chinese folk art (calligraphy and painting), the China Habitat Environment Example Prize, national ecological film and television shooting base, national hygienic city. Anji has ventured down the path of building a moderately prosperous society with integration among the primary, secondary and tertiary industries and coordinated urban and rural development, with the realization of the people's affluence and happiness; Anji has made agriculture strong, the countryside beautiful, farmers rich, and urban and rural areas develop in a harmonious way. The Anji Model and the Anji Spirit present the trajectory of coordinated development among agriculture, rural ecology and industries in China in the new period. Anji County has planted, developed and utilized Moso bamboo, for example by making a series of bamboo products including bamboo flooring, bamboo fiber, bamboo carbon and bamboo leaf flavonoid, so that the value of Moso bamboo has soared, and bamboo products are made available all over the world. In the meantime, priority has been given to the building of a well-known Chinese brand of agricultural product—the Anji White Tea; it has become an excellence among high-grade green tea, its highest price reaches 12,000 yuan/kg. As other advantageous agricultural products, including the silkworm and mulberry products, tobacco and bamboo shoots, have made more and more farmers obtain more income and become rich, the primary industry has developed rapidly. Anji has leveraged its beautiful environment to actively develop leisure agriculture and the rural tourism industry, build a leisure and holiday paradise, and they have become Anji's pillar industries. Under the guidance of the Standardization Administration of China, China's indicator system for the construction of a beautiful countryside, developed by Anji County, has become the national standard for the construction of a new countryside. In 2010, Anji County was identified as China's National Standardization Demonstration County for a Beautiful Countryside by the Standardization Administration of China.[20]

Based on great achievements in the construction of a beautiful countryside, making scenic spots universal throughout Anji County is a new direction for Anji County to start experiments. On November 20, 2013, the National Tourism Administration identified Anji County as the National Experimental Area for Rural Tourism Holidays. Anji County has turned the universal development of scenic spots into the universal development of resorts, and more rapidly developed the leisure tourism industry and promoted the development of extensive scenic spots at the county level.

[20]Yang Xiaowei, Anji County's Practice and Inspirations in the Construction of China's Beautiful Countryside, *Zhengce Liaowang*, 2012(9).

The plan for making scenic spots universal, developed by Anji County, covers the central leisure area with Lingfeng Resort as the core, the leisure tourism circle supported by a beautiful countryside, two main ecological tourism functional areas, including the Bamboo Sea and the Huangpu River Source, two cultural leisure areas, including Changshuo's Hometown—Changshuo is a master of art—and the White Tea Fragrance, and the modern rural sightseeing leisure area with the rim of Bijia Mountain as the core. Anji County has organized county-wide efforts to build China's beautiful countryside; it has planned the county as a scenic spot and designed a village as a scenic spot. Making scenic spots universal throughout the county has represented a new direction for the construction of a new countryside in Anji County.

Chun'an County is part of Hangzhou. It boasts the Thousand-island Lake with national Grade I quality of the water and a number of famous brands, including NongFu Spring, Chunpai Organic Fish, and Thousand-island Lake Tourism. The Thousand-island Lake is reputed as the best water in the world; the Thousand-island Lake is a national scenic spot; Chun'an County is a national strong county of tourism. In Chun'an County, there are multiple ecosystems, including forest, streams and reservoirs, species resources are very abundant, which lay a good ecological environmental foundation for building an ecological county. The Thousand-island Lake has become the shining brand and developmental support for Chun'an. The 12th Party Congress of Chun'an County fully identified the developmental orientation of developing the county on the basis of the lake, and vowed to comprehensively turn Chun'an into a leisure resort, a China water industry base, an ecological livable area of happiness, an urban agricultural garden and thus a culturally advanced and harmonious area. Under the guidance of the orientation of developing the county on the basis of the lake, the county has witnessed the rapid development of ecological agriculture, an ecological industry, ecotourism, an ecological habitat, an ecological environment and ecological culture.[21] In 2012, Chun'an County passed the evaluation and acceptance check concerning national ecological counties, the Thousand-island Lake became the first across the province to be listed as a national pilot area for protecting the good ecological environment of the lake. Regarding policy-making, Chun'an County has made continuous innovations to the mechanism of lake management, and has vowed to build the pilot area for joint ecological development and ecological progress sharing in the trans-provincial river basin. In 2013, Chun'an County comprehensively carried out the evaluation of the quality of the water at the cross sections between towns, strictly implemented a reward and punishment system for towns, and effectively promoted the protection of the Thousand-island Lake's source. As Chun'an County is the experimental area for building a beautiful Hangzhou, it is necessary for Chun'an County to build a full range of new modes covering ecological and environmental protection, ecological economic development, guarantee and improvement of the people's well-being which can be copied, drawn upon and promoted in other similar areas of the country.

[21] Xu Changsheng, A Study of Chun'an's Practical Mechanism for Building a National Ecological County, *Reform & Opening*, 2009(7).

Making scenic spots universal throughout the county coincides with Chun'an County's positioning of building the county on the basis of the lake. At present, with the goal of building the world mountain and water park and the oriental famous lake for leisure, Chun'an County has made all-out efforts to push forward the development of the first county-level universal scenic area in China, making sure that there are scenic areas throughout the county, that landscape can be found everywhere, that every village is a scenic spot and that everyone is a tour guide. Actions have been taken to ensure that 4,427 km^2 of clean, clear waters and lush mountains, urban and rural areas present the look, landscape and environment of scenic spots, the green roads, resources, products, business, services and culture available in scenic spots. Chun'an County should build a pattern of county-level universal scenic spots consisting of one city, one lake, one circle and multiple points, and develop multiple tourism modes, including an urban tour shared by the host and the guest, a lake area tour for excellent leisure, a tour around the lake and a distinctive country tour. Efforts have been made to promote tourism from the lake area to the surrounding areas, extend scenic spots from county town to rural areas, and turn the traditional lake area tour into a county-wide tour, so that the whole county becomes a large scenic area of the Thousand-island Lake.[22]

[22]Hu Yuehua, Chun'an County Has Made Scenic Spots Universal throughout the County, *Qianjiang Evening News*, June 22, 2012.

Chapter 6
The Green and Low-Carbon Development of Ecological Cities

Shouxian Zhu

The development of ecological cities is an important sign that Zhejiang's urbanization is changing from a quantitative increase to qualitative improvement; it is also one of the priorities for strengthening ecological development. Urbanization is one of the two main goals of modernization; in the new period, Zhejiang's urban development cannot be pursued merely by controlling environmental pollution like that in the middle and later periods of the 20th century, Zhejiang should elevate urban development to sustainable development towards urban ecological progress. It is necessary to draw upon domestic and foreign experience in urbanization, build and renovate cities according to high standards and really achieve green and low-carbon development of ecological cities, and thus build a beautiful Zhejiang.

6.1 Urban-Rural Integration

Urban-rural integration and coordinated urban-rural development mean that urban and rural areas are considered holistically rather than separately in terms of development. Urban-rural integration represents more a long-term goal and stresses the outcome of integrating urban and rural areas. Coordinated urban-rural development mainly entails operational means and a realization path, it means that the line of thought of coordination is adopted to guide the practice of development, and it stresses a process. Urban-rural integration calls for eliminating the gap between urban and rural areas to achieve the common development of urban and rural areas; coordinated urban-rural development emphasizes more the efforts to develop urban and rural areas in a coordinated way while retaining the natural characteristics of both areas.

S. Zhu (✉)
Chinese Academy of Social Sciences, Beijing, China

J. Pan and M. Shen (eds.), *Chinese Dream and Practice in Zhejiang – Ecology*,
Research Series on the Chinese Dream and China's Development Path,
https://doi.org/10.1007/978-981-13-7209-4_6

6.1.1 The Background of Economic Development for Zhejiang's Urban-Rural Integration

In late 2014, Zhejiang had a population of 55.08 million, it lacked natural resources and did not enjoy the advantages of natural resource conditions for economic development. Before the reform and opening-up, Zhejiang had been at the middle level nationwide in economic development for a long time. In recent years, Zhejiang has still kept rapid economic growth and has become the focus of attention at home and abroad. Industrial development has played a crucial role in Zhejiang's drive towards modernization, which is marked by the growth of its economic aggregate and the increasing proportion of its industry in the GDP. Accordingly, Zhejiang has stayed ahead nationwide regarding the economy.

During the period 2000–2013, the evolution of Zhejiang's industrial structure tended to be mature, and the secondary industry accounted for 49.1–54.1%. As from 2012, the tertiary industry accounted for 45.2%, and it surpassed the industry to become the leading industry for the first time in 40 years, and the adjustment of its industrial structure entered a new normal; in other words, the optimization and upgrading of the industrial structure was promoted steadily, energy conservation, carbon and emission reduction and ecological development were deepened, Zhejiang led the rest of the country in building ecological provinces and ecological cities.

6.1.2 Urbanization

In 2013, the rate of urbanization based on permanent resident population reached 64% in Zhejiang. Zhejiang's urbanization has entered the later period of rapid development, namely, the mature stage.

The city group around Hangzhou Bay, the Wenzhou-Taizhou city group and the city group in central Zhejiang are forming more rapidly, while the Hangzhou, Ningbo, Wenzhou metropolitan areas have preliminarily taken shape. According to relevant data, the land for urban construction, which accounts for 2.55% of Zhejiang's land area, carries more than 79% of Zhejiang's GDP, while the urban areas in three main central cities including Hangzhou, Ningbo and Wenzhou have 16.7% of Zhejiang's population and create 1/3 of Zhejiang's GDP.

City group and metropolitan area are the main forms of Zhejiang's urbanization; they have enhanced the ability of cities to gather high-end factors and stimulate the development of rural areas, and they have also made cities more suitable for living and doing business.

New-type urbanization provides new spaces for spatial agglomeration and regional reorganization of small and medium-sized enterprises and urban industries at the county level. Zhejiang has 312 massive economies with an annual sales revenue above 1 billion yuan and 72 massive economies with an annual sales revenue above

10 billion yuan. In Zhejiang, 42 massive economies are more quickly undergoing transformation and upgrading towards modern industrial clusters.

Urbanization has brought about the fusion of resource factors and the rapid growth of the headquarters economy, the creative economy, the cybereconomy, the building economy, the port economy and other new-type urban economies. The economic aggregate in cities and county towns has accounted for more than 60% of Zhejiang's economic aggregate.

In the process of promoting new-type urbanization, Zhejiang has always focused on putting the people first, and has vigorously promoted the reform involving the system of unified management of household registration in urban and rural areas and carried out the residence permit system across the board.

For the quality of urbanization in Zhejiang, as shown by the *Annual Report on the Development of Ecological Cities in China (2013)*, Hangzhou City is listed among the top ten ecological cities, environmentally friendly cities, resource-conserving cities, circular economic cities, landscape leisure cities, green consumption-oriented cities and comprehensive innovative cities, and Ningbo is listed among the top ten environmentally friendly cities.[1]

In 2013, Zhenhai District in Ningbo City, Dongtou County, Changxing County, Jiashan County, Chun'an County, Taishun County, Tiantai County, Yunhe County, Suichang County and Xihu District in Hangzhou City passed the evaluation and acceptance check concerning national ecological cities (counties, districts). Hangzhou is included in the second batch of national pilot areas for ecological development, while Jiashan County, Chun'an County, Xihu District in Hangzhou City, Zhenhai District in Ningbo City, Dongtou County, Tiantai County, Changxing County, Yunhe County, Suichang County, Taishun County, Zhoushan City, Lishui City and Deqing County are included in the sixth batch of national pilot areas for ecological development. Hangzhou City, Lishui City and Huzhou City are included in the first batch of national pilot demonstration areas for ecological development. The institutional innovations in Hangzhou's ecological development mainly include the development of energy saving and the environmental protection market, trading in emission rights, third-party control of environmental pollution, in the experimentation on and establishment of the monitoring and early-warning mechanism involving the carrying capacity of the resources in the environment. The institutional innovations in Huzhou's ecological development mainly include the experimentation on and establishment of a system for the evaluation of ecological development, the experimentation on and the preparation of a natural resource balance sheet, the experimentation on and establishment of a system for the natural resource asset property rights. The institutional innovations in Lishui's ecological development mainly include experimentation on and establishment of a system for the evaluation of the leading cadre which reflects ecological requirements, improvements to the systems of natural resource property rights, asset management and supervision.

---

[1] Sun Weiping, Liu Juke, *Annual Report on the Development of Ecological Cities in China (2013)*, Social Sciences Academic Press, 2013.

Since the 18th National Congress of the Communist Party of China and the 3rd and 4th Plenary Sessions of the 18th Central Committee of the Communist Party of China, Zhejiang has speeded up the development of ecological institutions and the promotion of the rule of law in the ecological field. It is necessary for Zhejiang to become the first to make breakthroughs in improving the system of ecological institutions, including improving the system of natural resource asset property rights and that of the regulation of usage, identifying the red line for ecological protection, exploring and developing the natural resource balance sheet, carrying out the systems concerning the compensated use of resources and ecological compensation, and reforming the system for the management of ecological environmental protection. In the meantime, Zhejiang should become the first to adopt a strict legal system to protect the ecological environment, and more rapidly establish the legal institutions for ecological development which impose effective constraints on developmental behaviors and promote green, circular and low-carbon development.

6.1.3 The Progress in Zhejiang's Coordinated Urban-Rural Development

Since the middle and later periods of the 1990s, Zhejiang has, in an effort to accelerate urbanization, put forward the goal and requirement for promoting coordinated urban and rural development. The *Plan of Zhejiang Province for Urbanization*, developed by the Party Committee and the People's Government of Zhejiang Province in 1991, stated, "It is necessary to comprehensively improve the overall quality of cities, enhance the ability of cities to accumulate factors and stimulate the economic development of others, give full scope to the central role of cities in regional economic and social development and achieve coordinated urban and rural development."

Based on summing up Zhejiang's experience in economic development over the years, the 4th Plenary (Enlarged) Session of the 11th Party Committee of Zhejiang Province, held in July, 2003, comprehensively and systematically summarized eight advantages for Zhejiang's development and put forward eight measures for future development—the "Eight-Eight Strategies", namely, further giving scope to eight advantages and pushing forward eight measures. Specifically: (1) further giving play to Zhejiang's advantages regarding institutions and mechanisms, actively promoting the common development of the economy with different types of ownership while ensuring that public ownership remains dominant and continuously improving the socialist system of a market economy; (2) further drawing on Zhejiang's locational advantage, proactively keeping in line with Shanghai, actively engaging in cooperation and communication with the Yangtze River Delta and constantly increasing the level of opening up internally and externally; (3) further leveraging the advantages of massive characteristic industries in Zhejiang, speeding up the building of advanced manufacturing bases, venturing down the path of new industrialization; (4) further giving play to Zhejiang's advantages in coordinated urban and rural development, promoting coordinated urban and rural economic and social development and quickening urban and rural integration; (5) further utilizing Zhejiang's ecological advantage, turning Zhejiang into an ecological province and building a green Zhejiang; (6) further exploiting the advantages of mountain and marine resources, energetically developing the marine economy, boosting the leapfrog development of the underdeveloped areas, and making the development of the marine economy and the underdeveloped areas become the new economic growth point of Zhejiang Province; (7) further utilizing Zhejiang's environmental advantages, actively pushing forward the construction of infrastructures, practically strengthening legal construction, credit construction and improving the efficiency of organs; (8) further leveraging Zhejiang's humanistic advantage, actively rejuvenating the province through science, technology and talent and stepping up efforts to turn Zhejiang into a culturally strong province. The decisions and arrangements, in the "Eight-Eight Strategies", concerning coordinated urban and rural development and accelerated urban and urban integration in Zhejiang have laid the foundation for comprehensively evaluating the level of coordinated urban and rural development, which Zhejiang has become the first to promote.

In April, 2007, the Zhejiang Provincial Development and Reform Commission and the Zhejiang Provincial Bureau of Statistics jointly released the *Annual Report on the Comprehensive Evaluation of the Level of Coordinated Urban and Rural Development in Zhejiang Province 2005*, making Zhejiang become the first nationwide to conduct a quantitative analysis and evaluation of the level of provincial coordinated urban and rural development. The system of indicators for comprehensive evaluation for the level of coordinated urban and rural development which Zhejiang became the first to build consists of 20 indicators concerning four fields including coordinated urban and rural economic development, coordinated urban and rural development of social programs and infrastructure, coordinated urban and rural development of the people's living standard and social security, and the coordinated urban and rural improvement of the ecological environment (see Table 6.1). The system of indicators is designed to evaluate the level of coordinated urban and rural development by the year 2020 according to the standard 45–60 points, 60–75 points, 75–90 points and above 90 points, and it covers four stages including preliminarily coordinated development, basically coordinated development, holistically coordinated development and full integration.

Compared with the target values by 2020, the comprehensive score in the level of coordinated urban and rural development in Zhejiang was 61.87 points in 2005, Zhejiang was at the stage of basically coordinated development; the comprehensive score in the level of coordinated urban and rural development in Zhejiang was 88.49 points in 2013, Zhejiang was at the stage of moving from holistically coordinated urban and rural development to full integration (see Fig. 6.1).

In 2013, Zhejiang saw the steady development of the level of coordinated urban and rural development in four main fields, including economic development, public services, the people's life standard and the ecological environment, the degrees of goal realization were 86.2, 94.5, 83.9 and 89.8%; except for a few indicators, including the growth rate of added value from the primary industry, whose degrees

Table 6.1 Comprehensive evaluation of the coordinated urban and rural development in Zhejiang province

Field	No.	Indicator name	Unit	Weight	Target value
Coordinated urban and rural economic development	1	Proportion of the employed people in the secondary and tertiary industries	%	6	90
	2	Labor productivity in the primary industry	Yuan/person	5	25,000
	3	Per capita GDP	Yuan/person	6	80,000
	4	Per capita local fiscal revenue	Yuan/person	7	6500
Coordinated urban and rural development of social programs and infrastructure	5	Proportion of fiscal expenditure in agriculture, rural areas and farmers and its increase	%	6	15
	6	Proportion of administrative villages connected by standard roads	%	4	100
	7	Rate of coverage by safe drinking water in rural areas	%	4	90
	8	Ratio of educational funds per student in urban areas to that in rural areas	%	5	100
	9	Number of medical personnel per 1000 people	Person	5	6
	10	Ratio of agricultural technicians to the people engaged in agriculture	‰	4	10
Coordinated urban and rural development of the people's living standard and social security	11	Per capita resident income gap between urban and rural areas	Tmes	10	2
	12	Ratio of per capita expenditure on the consumption of domestic electricity in urban areas to that in rural areas	%	3	70

(continued)

Table 6.1 (continued)

Field	No.	Indicator name	Unit	Weight	Target value
	13	Ratio of per capita expenditure on culture, entertainment, education, medical and health expenses in urban areas to that in rural areas	%	3	60
	14	Ratio of the level of IT application in urban areas to that in rural areas	%	3	80
	15	Degree of difference in the level of subsistence allowances between urban and rural areas	Point/Score	3	1
	16	Proportion of the people covered by social insurance among all employed people	%	10	80
Coordinated urban and rural improvement of the ecological environment	17	Comprehensive score in environmental quality	Point/Score	5	6
	18	Proportion of rural garbage collected and disposed of	%	3	100
	19	Rate of popularization of sanitary toilets in rural areas	%	4	95
	20	Proportion of villages that have been improved	%	4	100
Total				100	

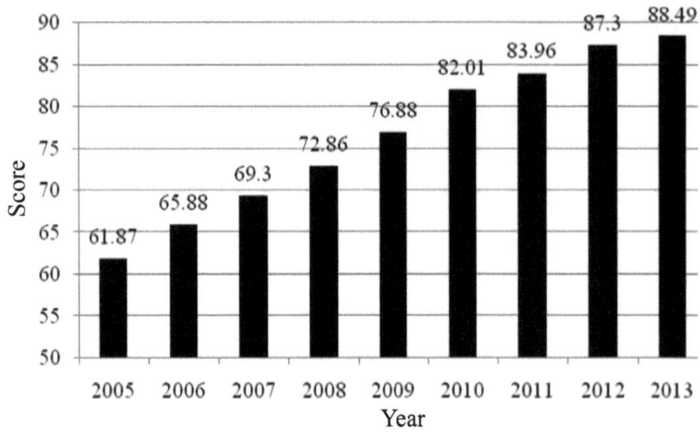

Fig. 6.1 The comprehensive score for the level of coordinated urban and rural development in Zhejiang Province, 2005–2013

of realization decreased to some extent, the majority of the 33 indicators showed rapid growth. There were 16 indicators whose degrees of realization reached or exceeded 90%, accounting for 48.5%; 11 indicators whose degrees of realization stood at 75–90%, accounting for 33.3%; 4 indicators whose degrees of realization were 60–75%, accounting for 12.1%; 2 indicators whose degrees of realization were lower than 60%, accounting for 6.1%.

In 2013, 11 cities divided into districts intensified their efforts to promote coordinated urban and rural development and further increased the level of that development. For the overall level of development, the score of five cities, including Jiaxing, Ningbo, Zhoushan, Hangzhou and Huzhou, exceeded 90 points, they were at the stage of full integration, among which Zhoushan and Huzhou were new entrants; the score of Shaoxing, Taizhou, Jinhua and Wenzhou exceeded 80 points, and they remained at the stage of holistically coordinated development; the score of Quzhou and Lishui was lower than 75 points, so they were at the stage of basically coordinated development.

In 2013, 61 counties (county-level cities, districts) with settlement under the first-level fiscal system in Zhejiang generally maintained a steady increase in the level of coordinated urban and rural development. Eleven counties (county-level cities, districts), including Keqiao District, were at the stage of full integration; 31 counties (county-level cities, districts), including Haiyan County, were at the stage of holistically coordinated development, while 19 counties (county-level cities, districts), including Yunhe County, were at the stage of basically coordinated development.

According to the results of the evaluation, in 2013, Zhejiang's coordinated urban and rural development generally showed three main improvements: the level of coordinated urban and rural development in key fields increased significantly, marked effects were produced in ecological environmental development in four main fields, in particular, the quality of the water environment improved significantly, it ranked no.1 among four main fields in increasing the level of coordinated urban and rural development. The level of coordinated urban and rural development increased more quickly in weak segments, the gap between the field of the people's life standard with the lowest degree of goal realization and the field of public services with the highest degree of goal realization narrowed gradually. The level of coordinated urban and rural development in the underdeveloped areas generally increased, in Zhejiang, the average increase in the score concerning the level of coordinated urban and rural development in 26 underdeveloped areas, especially 11 key underdeveloped counties (county-level cities), was larger than that in 61 counties (county-level cities, districts).

In 2013, the level of coordinated urban and rural development in Zhejiang increased steadily, but some new developments and new problems occurred. First, there was heavy pressure to more quickly increase the farmers' income and arduous tasks for improving agricultural output. In Zhejiang, the ability to resist risks in agriculture was relatively low, agricultural output was vulnerable to a number of factors including the natural environment and the market environment. In 2013, the degree of realizing the goal concerning the growth rate of added value from the primary industry was only 13.3% and reverse changes occurred in Zhejiang; the growth rate of added value from the primary industry in 9 cities divided into districts, 41 counties (county-level cities, districts) was lower than that in the previous year; 5 cities divided into districts, 30 counties (county-level cities, districts) saw a decrease in the comparative labor productivity in the primary industry and difficulties in maintaining steady improvement in the efficiency of agricultural output. Second, there was great pressure for protecting the ecological environment and tough tasks for improving the quality of the ecological environment. Although the improvement of the water environment was comprehensively strengthened and the quality of the water environment improved significantly in Zhejiang, the air quality was not optimistic and haze often occurred in various parts of the province. In 2013, among 11 cities divided into districts and 61 counties (county-level cities, districts) in Zhejiang, only Yunhe, Longquan and Tonglu did not see a decrease in the rate of reaching the standard in terms of environmental air and the air pollution was very severe. Third, there was high pressure for continuously increasing fiscal revenue and arduous tasks for increasing input in promoting coordinated urban and rural development. Zhejiang's total fiscal revenue grew by 7.8% in 2013, down 0.4 percentage points compared with 2012. As economic development has entered a new normal, it is difficult to sustain the rapid growth of the fiscal revenue and there are difficulties in maintaining a high level of input in promoting coordinated urban and rural development.

Given the above developments and new problems, according to relevant requirements for building a materially affluent and culturally advanced province and a beautiful province for a good life, the following measures should be taken in order to achieve integrative development of urban and rural areas at a higher level, and make the people of Zhejiang become richer, the environment more beautiful and the society more harmonious.

- Improving the quality and efficiency in agriculture, increasing the income of the farmers. With a focus on "enhancing the labor productivity and continuously increasing the income of the farmers", comprehensively carrying out the project of "four changes and three famous actors" (vacating the cage to change birds, replacing the people with machines, optimizing the space to save land, expanding the market through e-commerce; cultivating well-known enterprises, famous products and eminent entrepreneurs) in agriculture, applying the philosophy of "vacating the cage to change birds, replacing the people with machines, optimizing the space to save land, expanding the market through e-commerce" in agricultural development, cultivating well-known enterprises, well-known products and eminent entrepreneurs in agriculture. More quickly phasing out the backward production modes, increasing the level of agricultural operators, promoting IT application in agriculture, increasing agricultural production benefits, pushing forward agricultural scientific and technological innovations, promoting agglomeration and efficient development and actively innovating the production and marketing modes concerning agricultural products, thus expanding the channels for increasing the income of the farmers.
- Reinforcing the protection of the ecological environment, beautifying the urban and rural environment. With a focus on "global coordination, whole-process control and participation by all the people", vigorously pushing forward special actions, including governing "five waters" in tandem, making "three renovations and one demolition", "cleaning, greening and beautifying the areas beside highways, railways, rivers and mountains", controlling the haze, saving energy and reducing emissions, thus building a beautiful countryside as well as further improving the look of urban and rural areas, building a beautiful living environment in urban and rural areas through global coverage, all-round promotion and all-weather monitoring.
- Intensifying the reform of the urban and rural system to narrow the gap between urban and rural areas. With a focus on "promoting an orderly flow of factors between urban and rural areas and enhancing the endogenous vitality in rural areas", deeply pushing forward the reform involving the right to contracted use of land, the usufruct of homesteads and equity in the rural collective assets in rural areas, actively pressing ahead with the reform of the household registration system, encouraging and guiding the migrant agricultural population to settle in urban areas by obtaining urban registered permanent residence, summarizing and promoting the experience from the pilot work in Jiashan, Deqing, Haining and other areas with respect to scientific development, the reform of the urban and rural system, the reform involving market-oriented allocation of resource factors, arousing the endogenous vitality in rural areas to promote urban and rural integration while boosting agricultural development through industry and stimulating rural development through urban areas.

6.2 The Building of Low-Carbon Cities

In order to mitigate global warming, it is essential to develop all-round policies, proceed from adjusting the energy structure, enhancing the efficiency of energy and improving urban planning to reduce greenhouse gas emissions. To develop all-round strategies for addressing climate change, it is necessary to integrate and coordinate the work of different policy departments, and the key lies in the political will of local governments. Some large cities in foreign countries, such as London, Tokyo and New York, Nordic cities, such as Copenhagen, Malmo, Oslo and Sonderborg, Japanese Kitakyushu City, Kyoto City and Yokohama City are the forerunners in the building of low-carbon cities. In order to strengthen cooperation among large cities throughout the world and reduce greenhouse gas emissions, in 2015, more than 40 large cities around the world established the C40 Cities Climate Leadership Group (C40), which held three conferences in London, Bali and Seoul in 2005, 2007 and 2009, respectively, confirming that the key to addressing climate change lies in cities.

As the developed countries and international metropolises actively took actions to move towards low-carbon economic transformation, in 2010, the National Development and Reform Commission of China identified five provinces, including Guangdong, Liaoning, Hubei, Shaanxi and Yunnan, and eight cities, including Tianjin, Chongqing, Shenzhen, Xiamen, Hangzhou, Nanchang, Guiyang and Baoding, as the first batch of national pilot areas for low-carbon development in China.

In November, 2012, the National Development and Reform Commission of China identified 29 provinces, municipalities, autonomous regions and cities, including Beijing, Shanghai, Hainan, Shijiazhuang, Qinhuangdao, Jincheng, Hulun Buir, Jilin, the Greater Khingan region, Suzhou, Huai'an, Zhenjiang, Ningbo, Wenzhou, Chizhou, Nanping, Jingdezhen, Ganzhou, Qingdao, Jiyuan, Wuhan, Guangzhou, Guilin, Guangyuan, Zunyi, Kunming, Yan'an, Jinchang and Urumqi, as the batch of national pilot areas for low-carbon development in China.

As of October, 2014, China had identified 6 pilot provinces and autonomous regions and 36 pilot cities for low-carbon development; each of the 31 provinces, autonomous regions and municipalities, except Hunan, Ningxia, Tibet and Qinghai, had at least one pilot city for low-carbon development; in other words, pilot areas for low-carbon development fully unfolded nationwide.

Six concrete tasks in pilot work on low-carbon development include: First, identifying the direction, principles and requirements of the work, taking "all-round, coordinated and sustainable" as the fundamental requirements for pilot work on low-carbon development, subject to comprehensively carrying out the general layout of making progress in the economic, political, cultural, social and ecological fields; second, developing the plan for low-carbon development; third, building a low-carbon industrial system characterized by low-carbon, green, environmentally-friendly and circular development; fourth, establishing a system of greenhouse gas emission data statistics and management; fifth, establishing a target responsibility system for controlling greenhouse gas emissions; sixth, actively advocating a low-carbon green lifestyle and consumption mode.

Hangzhou has been included in the first batch of national pilot cities for low-carbon development, while Ningbo and Wenzhou have been included in the second batch of national pilot cities for low-carbon development. The practice and experiments of low-carbon development in Hangzhou, Ningbo and Wenzhou have been comparatively analyzed to provide beneficial experience for Zhejiang to fully build low-carbon cities and promote ecological development.

6.2.1 Hangzhou Has Actively Promoted Low-Carbon Urban Development in Six Fields

In order to comprehensively carry out the Scientific Outlook on Development and thoroughly implement the strategy of building the city through environmental development, Hangzhou has seized the opportunity from being included in the first batch of national pilot cities for low-carbon development and the second batch of national pilot cities for ecological development, and has taken philosophy innovation as the guide, obtained support from technical innovation, derived a guarantee from institutional innovation, identified the reduction in the intensity of carbon dioxide emissions as the main hallmark to actively explore a low-carbon developmental path which ensures that low-carbon industries lead the economy, citizens live a low-carbon life and the low-carbon city is the blueprint for urban development and which reflects Chinese characteristics, the characteristics of the times and Hangzhou's characteristics, with a view to promoting low-carbon urban development which features a low-carbon economy, low-carbon buildings, low-carbon transportation, a low-carbon life, a low-carbon environment and a low-carbon society.

1. Cultivating low-carbon industries, building a low-carbon economy

Building a low-carbon industrial support system. Centering on "adjusting the economic structure and changing the developmental mode", more rapidly building a "3 + 1" modern industrial system ("3" means modern agriculture, modern industry and modern service industry, "1" means the creative cultural industry), establishing national creative cultural industrial centers and the world's capital of e-commerce, achieving a historical leap from Made in Hangzhou to Created in Hangzhou, Service in Hangzhou and Creative in Hangzhou, promoting the development of the industrial structure in the low-carbon direction. Developing low-carbon science and technology, strengthening cooperation with famous universities and large institutions, establishing low-carbon laboratories, building a system of low-carbon technical innovation and achievement conversion in which enterprises play the leading role and there is cooperation among industries, universities and research. Advocating a low-carbon design, proceeding from design to reduce material and energy consumption relating to products in manufacturing, storage, transportation, circulation, consumption, and even recovery, guiding the people's mindset regarding consumption from the perspective of design, stimulating the people to develop a low-carbon habit of consumption.

Developing the new energy industry. Establishing a special fund for supporting the new energy industry, making the special developmental plan for the new energy industry, planning and building the new energy industrial bases, intensifying the attraction and cultivation of investments on the part of the existing enterprises, more rapidly building an industrial chain focusing on solar photovoltaic and wind power industries, actively developing energy-saving and environmental protection, new energy vehicles, new light sources, biomass energy, nuclear power equipment manufacturing and related industries, and cultivating and developing low-carbon industrial clusters. Pushing forward the adjustment of the structure of energy, energetically promoting comprehensive utilization of clean energy, including solar energy, marsh gas, natural gas, terrestrial heat, in production and life, rationally adjusting the proportion of coal, electric power, oil and gas utilized, reducing the utilization of fossil fuels, such as coal and oil, to the greatest extent, promoting the development of the structure of electric power from single coal power generation to coal power, gas power generation, power generation based on renewable energy, such as solar energy, and biomass energy, such as garbage and straw.

Promoting industrial energy saving and emission and carbon reduction. These actions include: comprehensively adopting the economic, legal and environmental protection and necessary administrative means to resolutely shut down the outdated production facilities in light, textile and other industries, firmly prevent the expansion of production capacity and the duplication of similar projects in the industries with excess capacity, thus resolutely prohibiting unauthorized expansion and reconstruction and non-local transfer of the outdated equipment which have been ordered to be shut down in the specified time; promoting desulfurization and dust removal renovation in industrial enterprises, thus strictly ensuring that pollutant emission meets relevant standards; more quickly building logistics parks, thus effectively integrating logistics resources; actively promoting the energy, water and material-saving products and technologies to save energy and reduce carbon emissions to the greatest extent; actively building green enterprises, strictly carrying out a system for the evaluation and examination of energy savings for the fixed asset investment projects which annually consume energy equivalent to more than 3,000tce, thus building a number of circular industrial economic demonstration enterprises and parks; pushing forward material saving and utilization of renewable materials and alternative materials to reduce the level of raw material consumption; increasing the rate of comprehensive recovery and utilization of industrial wastewater, waste gas and waste heat, thus developing the circular economy, building renewable resource distribution and processing centers, thus cultivating a number of enterprises at a certain scale which comply with national industrial policies, adopt the latest technologies and are engaged in key venous industries including waste and old car processing and recovery, waste and old metal processing and recovery, waste and old plastics processing and recovery, in this way building and improving a system of a network of renewable energy recycling.

2. Promoting building energy conservation, constructing low-carbon buildings

Actions in this respect include: carrying out the sunshine rooftop demonstration project; more rapidly carrying out the national Building Integrated PV (BIPV) application demonstration project and the provincial one-million rooftop power generation program, thus vigorously pushing forward Hangzhou's sunshine rooftop demonstration project, by fully utilizing the surface (including rooftop and curtain walls) of public buildings, industrial buildings, residential buildings, public facilities, other buildings and structures, installing solar photovoltaic cell modules, a system of electric energy control and a grid-connected system, achieving popularization and application of photovoltaic power generation in the building field; becoming the first to initiate the pilot work on the application of building photovoltaic power generation in major projects including subways, the east station hub, the Olympic Sports Expo Center and the New Energy School of Hangzhou Normal University, turning 2 MW BIPV grid-connected power generation project of Hangzhou Energy and Environmental Industrial Park, China (Phase I) into the first MW-class rooftop photovoltaic grid-connected power generation project in Zhejiang, and even in China.

Carrying out the urban green rooftop program. This includes: promoting a vertical planting policy and technical standards; institutionalizing and standardizing vertical planting; actively exploring new ways and new varieties of vertical planting and vertical greening, continuing the work on inserting green elements in all possible spaces, making green penetration into walls, rationally conducting green sowing and encouraging everyone to participate in greening, first promoting rooftop greening on all skirt buildings where conditions permit, concurrently greening the elevated stand columns, the side slopes of municipal trunk roads, retaining walls, the side slopes of river banks where safety is guaranteed, increasing the green concentration of urban three-dimensional spaces, thus reducing the island effect of urban heat.

Promoting the renovation of saving on building energy. This includes: in coordination with courtyard improvements, the improvement of old and dilapidated houses and other urban organic renewal projects, renovating building energy saving and increasing heat preservation and energy saving effects in buildings; organizing efforts to carry out the low-energy and green building demonstration projects, vigorously promoting energy and land-saving and environmentally-friendly buildings, thus actively pushing forward integrative development of renewable energy and buildings; conducting green rating of various types of buildings in terms of energy saving and issuing the corresponding certificate of the level of energy saving; making local regulations on building energy saving, promoting the sustainable building standards, popularizing the building energy saving management mode with design and construction standardization and closed acceptance check, standardizing the energy-saving building design standard and atlas, technical regulations on construction, acceptance specification and operational management rules, thus promoting building energy saving according to laws.

3. Advocating green travel, developing low-carbon transportation

Developing the system of free bike service. These actions include: according to the requirement of "arranging points nearby, enabling renting and returning based on the all-purpose card, building a distribution system and giving expression to urban esthetics", further improving the system of free bike service, more rapidly building bike lanes, especially the urban riverfront slow traffic system, so that free bikes can really play the role of a large public transportation system as a bond, and become the greatest highlight and characteristic in Hangzhou's building a low-carbon city; -thoroughly carrying out the No Car Day activity, organizing a series of theme publicity activities relating to green travel, stimulating citizens to choose the low-carbon travel modes with low energy consumption and low emissions, driving their private cars less, taking buses and riding bikes more, making the No Car Day activity regular, mitigating the difficulties in driving on the road and parking, thus protecting the ecological environment and promoting the implementation of Fitness-for-All programs.

Building a city with five-in-one zero-transfer public transportation. This includes: giving priority to public transportation, more rapidly building a large five-in-one public transportation system covering subways, buses, taxis, water buses and free bikes, building the first city with zero transfer among subways, buses, taxis, water buses and free bikes in China, and even in the world, giving priority to urban public transportation in urban transportation, building a low-carbon urban transportation system. Energetically pushing forward intelligent transportation and facility construction, increasing the scientific and technological content in transportation, reducing roundabout transportation, duplicated transportation, empty vehicle transportation, thus mitigating carbon emissions and fostering a clean, quiet, healthy and orderly transportation environment.

4. Advocating green consumption, building a low-carbon life

Advocating a low-carbon lifestyle. Actions to promote this include: advocating a diet with carbon reduction, more intake of vegetables and less intake of meat, especially an intake of local, in-season, organic and natural foods; advocating low-carbon dress, guiding citizens to pay more attention to and reduce clothing carbon emissions; carrying out a universal campaign of energy saving and carbon emissions, guiding citizens to uphold saving, oppose consumption, practice rational consumption and moderate consumption; stimulating the people to change their lifestyle from the traditional one of a high-carbon mode to a low-carbon mode in clothing, food, housing, traveling and use in daily life, develop a healthy and low-carbon lifestyle and living habit, thus eliminating carbon dependence; introducing the philosophy of a carbon footprint, setting up a carbon emissions calculator on related websites, so that everyone and every organization can evaluate their impact on the environment, and set the baseline for evaluating the state of future emission reduction, calling upon enterprises to incorporate carbon footprint evaluation as an important part of their social responsibilities, thus guiding individuals and organizations to "erase their carbon footprint and make carbon compensation".

Building a healthy city. These actions include: proceeding from daily life, starting with a rational diet, moderate exercise and control on their smoking habit, intensifying the publicity regarding the building of a healthy city, organizing efforts to conduct health theme activities, designing and promoting oil control products, carrying out the activities of controlling smoking in public places, building a number of health education and health promotion venues, extensively arousing the people's initiative, enthusiasm and creativity about participation in building a healthy city and enhancing the people's health awareness as well as the quality of their health.

5. Strengthening ecological development, building a low-carbon environment

Building a national forest city. This includes: advocating the philosophy of "integrating the city into the forest and making the forest embrace the city", earnestly carrying out the universal tree planting activity, so that everyone makes efforts to keep Hangzhou green, shifting focus from the greening rate to the forest coverage rate, from the visual effect to the ecological effect, from the land area for greening to the spaces for greening, from the greening of the built-up areas to coordinated urban and rural greening; fostering the best living environment with "the city within the forest, the forest within the city and the people within the green landscape", developing the city's ability to make carbon neutral, thus bringing benefits to the city and the people; better protecting six ecological belts in the western, northwestern, northern, southwestern, southern and eastern regions, "leaving a blank" in urban development, refraining from breaking the ecological limits and casually adjusting the land for protecting ecological belts; strictly controlling the building density around ecological belts, strictly controlling new projects, intensifying the investigation and handling of illegal buildings within ecological belts and the improvement of pollution sources, preventing main city, sub-center city and group construction from spreading to ecological belts, building the mode of "active protection" of ecological belts with Hangzhou's characteristics, thus better protecting the carbon sink lifeline of the city; developing an ecological economy, continuously improving the ecological environment, actively carrying forward an ecological culture, promoting coordinated urban and rural environmental protection, establishing a mechanism for ecological compensation, significantly improving the quality of the urban and rural environment, especially the quality of the air and surface water, fostering a social atmosphere in which the entire population jointly makes and shares ecological progress, thus giving shape to the pattern in which urban and rural areas reinforce each other regarding carbon source and carbon sink.

6. Changing urban management, building a low-carbon society

Putting in practice the development philosophy of a compact city. These actions include: introducing the low carbon philosophy to guide the formulation of urban planning, adopting the guideline of "making the dense ones denser and the sparse ones sparser", more rapidly building new towns and multifunctional and efficient urban complexes, promoting the building of metropolises with municipal networking, intensifying economical and intensive use of land, lowering the living costs for urban residents and business costs for enterprises, reducing resource and energy wasting caused by extensive urban expansion, thus building a new pattern of low-carbon urban development; building the Hangzhou Low Carbon Science & Technology Museum, China, giving prominence to the theme of low carbon and life, creating a low carbon center for science and technology popularization, a green building exhibition center, a low carbon academic exchange center and a center for low carbon information data, thus turning the Hangzhou Low Carbon Science & Technology Museum, China, into the first low carbon museum on the theme of science and technology in China, and even in the world.

Building low-carbon communities. This includes: drawing upon the successful international experience in low-carbon communities and carbon neutral communities, actively exploring the low-carbon life modes which feature zero carbon, zero wastes, sustainable transportation, local materials, local food, low water consumption, animal and plant protection, cultural heritage protection, fair trading and a happy and healthy lifestyle; promoting the means for planning a low-carbon community, an architectural technology and management mode, thus building a number of benchmark low-carbon communities; more rapidly making low-carbon operational improvements to government buildings and commercial office buildings, installing energy-saving lamps and solar lighting systems, controlling the use of air conditioners, adopting energy-saving computers and office equipment, promoting e-commerce and e-government, making office work paperless or printing on both sides of the paper as much as possible, building an office building management system for energy demand and use; building the Hangzhou Mode for collection and direct transportation of garbage; focusing on direct transportation supplemented by transfer, mainly carrying out incineration supplemented by landfilling, centering on dispersion supplemented by centralization; summarizing and promoting the experience and practice in the mode of direct transportation through garbage can vehicles, piloting the mode of dynamic direct delivery by pipelines; renovating and expanding the existing garbage incineration plants and properly selecting the sites for building new garbage incineration plants; upgrading and renovating the existing garbage transfer stations in old urban areas, pushing forward historical adjustment and transformation of the garbage collection and distribution mode, the transfer mode, the end-of-pipeline disposal mode, thus building the leading domestic and world-class mode of collection and direct transportation of garbage and achieving low carbon garbage disposal.

6.2.2 *Action Plan of Ningbo City for Pilot Work on a Low-Carbon City*

Carrying out pilot work on national low-carbon cities is an important measure for boosting scientific development and making ecological progress. Since the reform and opening-up, as a relatively developed coastal port city, Ningbo has witnessed a continuous increase in both its level of industrialization and the scale of its city as well as steady improvement in the living standard of its urban and rural residents, but the constraints from energy, resources and environment have become increasingly prominent, and the coordinated and sustainable economic, social and ecological development has been subject to tests. Conforming to the international trends and building a low-carbon system of development with Ningbo's characteristics is essential for overcoming resource and environmental constraints, speeding up transformation and upgrading, opening up new spaces for development and promoting ecological development.

The main tasks relating to pilot work on low-carbon cities are as follows.

1. Pushing forward the low-carbon development of industries

(1) Optimizing and developing the port-based industries, stabilizing total carbon emissions in key industries. This action includes: pressing ahead with base-oriented, circular and high-end development of port-based industries, relying on the existing port-based industrial cluster areas to actively promote the integration and upgrading of port-based industries including petrochemical engineering, iron and steel, shipbuilding, automobiles, continuously improve the port-based circular industrial chain and build the port-based advanced manufacturing clusters with good market prospects, high output and returns and up-to-standard emissions; carrying out relevant port-based industrial plans, thus encouraging the development of large gas power generation units in the electric power industry; in principle, no additional coal-fired generating units were built during the period of the 12th Five-Year Plan; the arrangement of major equipment in the petrochemical industry was basically completed during the period of the 13th Five-Year Plan; efforts are being made to adjust and optimize the product structure, and the total energy consumption is no longer being extended to the iron and steel industry.

(2) More rapidly developing the modern service industry, achieving structured carbon reduction. These actions include: increasing the scale of the service industry; according to the requirement of "increasing the speed and energy level and improving the quality", building a three-in-one port and a system of shipping logistics services, more rapidly developing a headquarters economy, e-commerce, service outsourcing and other emerging services, comprehensively enhancing life service functions conducive to making the people safe, happy and healthy, thus building a structurally rational, distinctive, vigorously supported and highly competitive modern service industry system. As of 2015, the main operating income of the service industry exceeded 2000 billion yuan, the added value of the service industry reached 450 billion yuan.

(3) Actively developing low-carbon agriculture, promoting high-efficiency and ecological development. These actions include: vigorously promoting and applying agricultural ecological technologies and techniques for minimum tillage and no-tillage, enabling large-scale and ecological disposal and utilization of straw and livestock and poultry breeding excrement; by the end of the period of the 12th Five-Year Plan, 3.50 million mu of farmland underwent soil testing and formulated fertilization, the coverage of soil testing and formulated fertilization technique exceeded 85%, the rate of comprehensive utilization of crop straw exceeded 95%, while that of large-scale livestock and poultry breeding excrement exceeded 97%.

2. Optimizing and adjusting the energy structure

(1) Increasing the proportion of natural gas consumption. This includes: expanding natural gas sources through multiple channels, thus enhancing the capacity for natural gas supply; in 2015, the city's capacity for receiving natural gas reached 9 billion m^3; expanding the coverage of the network of natural gas pipelines; in 2015, the medium and high-pressure backbone network covered all the counties (county-level cities) and districts (urban areas), the network of gas supply pipelines fully covered central urban areas, and more than one million households used natural gas; actively shifting from coal to gas and from oil to gas in the boilers in central urban areas, encouraging the development of the distributed energy system, fully turning the three-river area into an area without coal-based fuel.

(2) Expanding the scale of non-fossil energy use. These actions include: intensifying the development of wind power, focusing on developing the wind farms at such islands as Tantoushan and Gaotang in Xiangshan, promoting the development of offshore wind farms in Xiangshan and Cixi; making the installed capacity of wind power exceed 700,000 KW by 2015; encouraging the development and utilization of solar energy, carrying out the sunshine rooftop program, developing distributed solar photovoltaic power generation demonstrations; making the scale of solar photovoltaic development and the heat collection area of solar water heaters reach 40,000 KW and 1.20 million m^2, respectively, by 2015; speeding up the utilization of biomass energy, carrying out the marsh gas power generation projects at large livestock breeding farms and garbage landfills in Cixi, Yinzhou, Ninghai and Xiangshan; energetically promoting and applying systems of ground source (water source) heat pump air-conditioning, air source heat pump hot water systems; making the area of ground source heat pump air-conditioning application and the installed power of air source heat pumps reach 3.50 million m^2 and 800,000 KW, respectively, by 2015.

3. Continuously enhancing energy efficiency

(1) Reinforcing the work on industrial energy saving. These actions include: strengthening the supervision and management of key energy-consuming units and projects, fully adopting means for energy auditing and clean production auditing, thus improving the work on energy saving and consumption reduction in key energy-consuming industries and actively shutting down the outdated production facilities; actively promoting and applying the energy-saving technologies relating to energy production, conversion and transmission; more rapidly building smart power grids, enhancing the capacity for electric power transmission, the adaptive capacity of urban and rural areas for power distribution and their capacity for receiving renewable energy and reducing the loss in power transmission, transformation and distribution.

(2) Improving the efficiency of energy use in buildings. This includes: conducting the study of energy saving standards for low-carbon buildings in Ningbo,

pushing forward the building of the renewable energy building application demonstration city, comprehensively promoting and applying energy-saving and renewable building materials and energy-saving equipment, actively piloting the application of the distributed energy system; vigorously promoting the transformation to energy saving at the existing buildings; adopting advanced heat pump hot water and air-conditioning systems in 20% of the new building area by the end of the period of the 12th Five-Year Plan.

(3) Building a low-carbon transportation system. These actions include: more rapidly developing low-carbon logistics, improving the platform for logistics information, bringing innovations to the mode of doing logistics business, promoting "dual transportation", actively pushing forward sea-railway combined transportation and river-sea combined transportation, thus building an efficient urban and rural distribution system; vigorously promoting low-carbon public transportation, comprehensively building a network of convenient and fast public transportation systems covering both urban and rural areas, thus increasing the proportion of citizens who travel via public transportation; accelerating the development of urban rail transit; completing and putting into operation no.1 rail transportation line and Phase I project of no.2 rail transportation line by 2015; improving the management of urban bike lanes, gradually building a network of public bike renting and use services; more rapidly increasing the energy saving level of vehicles, actively promoting new products and new technologies for saving energy in transportation; exploring the electrification of urban public transportation and green power supply; intensifying the work on shifting from oil to gas in the transportation system.

(4) Advocating a low-carbon lifestyle. These actions include: actively building low-carbon communities, schools and families; popularizing low carbon scientific knowledge, promoting the application of low carbon labels, improving the low-carbon consumption policy, developing the guiding rules for citizens' low-carbon behavior and the pact on energy and resource saving, thus actively guiding rational purchasing, moderate consumption, a simple life and other green consumption philosophies; building a system for recycling and comprehensive utilization of urban garbage; building the pilot points for the classification of garbage from 150,000 households in urban areas and making the recycling rate reach 70% by the end of the period of the 12th Five-Year Plan.

4. Increasing the level of ecological carbon sink

(1) Promoting the building of a "Forested Ningbo". These actions include: comprehensively pushing forward the building of forested city, towns and villages, and making the forest coverage within the city reach 51%, the greening rate in urban built-up areas exceed 35% and the per capita green park area be larger than 12 m^2 by 2020; carrying out the coastal protection forest improvement project, fully building the coastal green ecological protective screens; continuing to promote green development on both sides of the roads and river networks, thus building

urban ecological corridors; expanding the scale of ecological public-welfare forests; making the area of the city's ecological public-welfare forests reach 4 million mu by 2015; increasing the carbon sequestration effectiveness of greening works, carrying out the forest quality improvement project, speeding up the improvement of low production forests, accelerating the development of large-diameter timber and precious timber forests and guiding and promoting the manufacturing of bamboo and timber products.

(2) Strengthening the protection of ecological wetlands. This includes: pushing forward the demonstration project for protection and rational utilization of wetlands; taking the development of ecological wetlands in such areas as Hangzhou Bay and Sanmen Bay as models, identifying the city's natural wetland reserves at the seaside, beside bays, lakes, rivers and creeks, intensifying regulation and rational utilization of wetland areas, thus better giving scope to the ecological value of wetlands.

(3) Enhancing the capacity for marine carbon sequestration. These actions include: speeding up the development of marine ranching; adopting various means, including artificial fish reef and seaweed transplantation, to attract marine organisms to marine ranching areas, thus improving the environment of offshore areas with eutrophication and achieving carbon sequestration and nitrogen phosphorus reduction; building more than 6 artificial fish reef areas covering an area of 6000 hectares, transplanting 600 mu of seaweed to such sea areas as Xiangshan Harbor and Yushan and building a sea grass meadow by 2020.

5. Vigorously developing the emerging low-carbon industries

(1) Promoting the industrialization of low carbon technologies. These actions include: building a platform for low carbon technology research and development, encouraging cooperation among industries, universities and research groups, establishing specialized institutions for low carbon technology research dominated by enterprises, conducting special technology research for low carbon development, and building a repository of low carbon new and hi-tech projects; developing a catalogue of industrialization and promotion involving key fields, including industry, building, transportation, carbon sequestration and an increase in carbon sink, encouraging the introduction or research and development of key practical technologies, thus building a number of low-carbon industrial bases; making great efforts to develop new energy industries, focusing on developing new energy equipment manufacturing sectors, including solar photovoltaic equipment, complete equipment and supporting equipment relating to wind power, ground source (water source) heat pumps, power and energy storage batteries and intelligent power transmission and transformation equipment.

(2) Building a specialized service system for energy saving and environmental protection. These actions include: giving prominence to the guiding and leading role of standards; establishing and improving the system of standards covering

evaluation indicators, technical specifications, management guide and product marking; studying and developing the industrial standards relating to low carbon technologies which conform to Ningbo's industrial characteristics, thus improving the management capability of the standardization body.

Cultivating and developing the service industry for energy efficiency improvement. This includes: carrying out the project for the cultivation of a system for low-carbon market services, promoting contract energy management, supporting a number of integrated energy saving and low carbon service companies, encouraging the development of specialized low-carbon investment companies; cultivating and developing the bodies specializing in the amount of energy saving and carbon emission auditing, carbon sink metering, thus encouraging the provision of services relating to low-carbon development credit rating and low carbon technology evaluation.

6. Strengthening the building of the capacity for low-carbon support

Developing a low-carbon development plan and implementation scheme, enhancing the capability for greenhouse gas accounting. On this basis, it is necessary to improve the market guidance mechanism, more rapidly develop the energy saving and environmental protection service industry and explore the pilot energy trading market and carbon emission trading.

(1) Establishing the greenhouse gas statistical accounting system. This includes: developing the greenhouse gas list of Ningbo City; first initiating the formulation of the greenhouse gas list for the year 2005, concurrently developing the list of Ningbo City for 2006–2011 and preparing the annual report of the regular list; concurrently promoting the building of a platform for the list information database.

Establishing the basic statistical system for greenhouse gas emissions. This includes: incorporating the basic statistical indicators concerning greenhouse gas emissions into the government's system of statistical indicators, thus building and improving a statistical system suitable for greenhouse gas emission accounting; in such pilot fields as classification and disposal of garbage in central urban areas, green logistics, electrification of urban public transportation, establishing a verifiable carbon reduction account, improving the standing book and records concerning greenhouse gas emissions and energy consumption in key emitting units.

(2) Strengthening the guidance of a market mechanism. These actions include: establishing a green credit mechanism; building up a capital pool consisting of government guidance funds and supporting funds from commercial banks, ensuring that commercial banks follow the requirement of transformation, upgrading and low-carbon development to explore and establish a green credit mechanism and project evaluation criteria, proceed with closed operations, revolving use and targeted allocation of the capital pool and offer low-interest loans to enterprises involved; enterprises repay the principal and interest

by the gains they make through project income or carbon emission reduction trading.

Exploring and establishing a carbon emission trading market. This includes: studying and determining the carbon emission rights in key industries and key enterprises, building a trading platform for carbon emission rights integrating registration, bidding, trading, settlement, liquidation, certification and supervision, thus establishing a regional market system for carbon trading, and piloting carbon trading where conditions permit.

6.2.3 Wenzhou City's Practical Experience in Low-Carbon Development

Wenzhou City focuses low-carbon industrial development on increasing the added value per unit of input and output, and has made efforts to promote low-carbon transformation of its traditional industries and improvement along the value chain, to cultivate the emerging low-carbon industries and the modern service industry, and build a number of low-carbon model enterprises and a low-carbon industrial system with Wenzhou's characteristics. These efforts include:

(1) Pushing forward low-carbon transformation of traditional industries; encouraging enterprises to orderly transfer the traditional manufacturing segments to other areas, fostering the segments with high added value and low-carbon emission, including industrial design and brand marketing, along the value chain, achieving the low-carbon transformation of industries, thus building low-carbon model enterprises; strictly controlling the development of the industries with high energy consumption and high emissions, exercising strict control over industrial access, putting a stop to examination, approval and filing of the projects for production capacity expansion in the industries with heavy pollution and high energy consumption and excess production capacity, taking the action of shutting down the outdated production facilities and improving the industries with high energy consumption and heavy pollution, thus actively guiding enterprises to perform clean production.

(2) Energetically cultivating the emerging low-carbon industries. This includes: cultivating and developing the ten main emerging industries, including cybereconomy, tourism leisure, modern logistics, laser and photoelectricity, the port-based petrochemical industry, rail transportation, general aviation, new materials, the creative cultural industry and life health; earnestly developing complete equipment and supporting equipment relating to wind power, smart power grid equipment, nuclear power equipment, ocean energy equipment, garbage power generation equipment and new energy automobile equipment; more rapidly building modern industrial clusters, thus guiding and encouraging the development of the whole industrial chain.

(3) Actively developing the modern service industry. This includes: energetically developing the modern service industry covering finance, e-commerce, modern logistics, headquarters and commercial services, science and technology information service; actively cultivating the emerging fields in the modern service industry, such as fashion consumption, exhibition economy and service outsourcing; striving to build a number of low-carbon logistics parks and centers, ecotourism areas, and modern agricultural parks characterized by circular ecological agriculture, urban agriculture and leisure agriculture.

(4) Promoting the circular and low-carbon renovation of parks. This includes: under the principle of reduction, reuse and recycling, focusing on circular and low-carbon renovation of 11 industrial parks at and above the provincial level, cultivating and gathering a number of low-carbon enterprises, making the intensity of carbon emissions at parks reach the advanced level of domestic industries, thus guiding and stimulating the low-carbon development of industries in the city.

6.3 Full Implementation of the Strategy of New-Type Urbanization

6.3.1 The Background of Zhejiang's Strategy of New-Type Urbanization

6.3.1.1 Unswervingly Taking the Path Towards New-Type Urbanization

In August, 2006, the Zhejiang Urban Work Conference was held in Hangzhou. In the conference, Xi Jinping, the then Secretary of the Party Committee of Zhejiang Province and Chairman of the Standing Committee of the People's Congress of Zhejiang Province vowed to take the Deng Xiaoping Theory and the Important Thought of Three Represents as the guide, fully carry out the Scientific Outlook on Development, build a socialist harmonious society, follow a series of major strategic arrangements, made by the Party Committee of Zhejiang Province, including the "Eight-Eight Strategies", and focus on promoting coordinated urban and rural economic and social development and the construction of a new socialist countryside to further optimize the urban system, improve urban and rural planning, enhance urban functions, strengthen urban management, innovate the mechanism of development, and resolutely venture down the new-type urbanization path that is energy-conserving, environmentally friendly, economically efficient and socially harmonious through the coordinated development of small, medium-sized and large cities and small towns, the mutual promotion and joint improvement of urban and rural areas. It is necessary to combine urban development with the construction of a new countryside, go down the urbanization path with mutual promotion and joint improvement of urban and rural areas; it is also necessary to combine urban development with the optimization of the distribution of the province's population and productive forces, take the urbanization path with coordinated development of small, medium-sized and large cities and small towns, combine urban development with the improvement of the efficiency of resource utilization, thus going down the resource-conserving urbanization path, combine urban development with environmental protection, ecological development and go down the environmentally friendly urbanization path, combine urban development with growth mode transformation and venture down the economical and efficient urbanization path and combine urban development with the building of a harmonious society, going down the urbanization path with social harmony.

6.3.1.2 National Plan for New-Type Urbanization

In March, 2014, the *National Plan for New-type Urbanization (2014–2020)* was unveiled. According to the report delivered during the 18th National Congress of the Communist Party of China, the *Decision of the Central Committee of the Communist Party of China on Some Major Issues Concerning Comprehensively Intensifying the Reform*, the guiding principles adopted during the Central Urbanization Work Conference, the *12th Five-Year Plan of the People's Republic of China for National Economic and Social Development*, the *National Plan for Main Function Areas*, and the new requirement of "venturing down the new-type urbanization path with Chinese characteristics and comprehensively enhancing the quality of urbanization", the developmental path, main goals and strategic tasks for future urbanization were identified, the institutional and policy innovations in relevant fields were coordinated, and the *National New-type Urbanization Plan (2014–2020)* was recognized as the macro, strategic and basic plan for guiding national healthy urbanization. It mainly covers the following aspects: turning agricultural migrant population into urban residents in an orderly fashion, optimizing the layout and form of urbanization, improving the cities' capacity for sustainable development, promoting integrated urban and rural development and reforming and improving the urbanization system and mechanism.

In July, 2014, the National Development and Reform Commission, the State Commission Office of Public Sectors Reform, the Ministry of Public Security, the Ministry of Civil Affairs, the Ministry of Finance, the Ministry of Human Resources and Social Security, the Ministry of Land and Resources, the Ministry of Housing and Urban-Rural Development, the Ministry of Agriculture, the People's Bank of China, the China Banking Regulatory Commission and other ministries jointly issued the *Circular Concerning National Comprehensive Pilot Work on New-type Urbanization* (Fa Gai Gui Hua (2014) No.1229), calling for choosing pilot areas at different levels including the cities specifically designated in the state plan, provincial capitals, prefecture-level cities, county-level cities and towns where conditions permit to the extent that the number of pilot areas proposed by each province (autonomous region, municipality) is controlled within 4 in principle (including cities and towns at different levels). The pilot work is mainly as follows: with a focus on "establishing a mechanism for sharing the costs incurred by turning the agricultural migrant population into urban residents, a diversified and sustainable mechanism for investment in and the financing of urbanization, a city establishment mode for "innovating administrative management and reducing administrative costs", reforming and improving the rural homestead system", in accordance with the requirement of development in entrepreneurship and innovation, public services, social governance, green and low-carbon development, making pilot experiments in both comprehensive and classified ways and providing the experience and modes which can be copied and promoted in the rest of the country.

6.3.2 Zhejiang's Strategic Path for New-Type Urbanization

With new-type urbanization as a major strategy of development, Zhejiang has introduced a number of policies, strengthened top-level design, made more efforts to promote it, and ventured down a path of new-type urbanization with Zhejiang's characteristics.

1. Upholding the guiding role of planning, optimizing the system of urban level

First, taking city groups as the main forms for boosting urbanization and the important carriers for promoting regional coordinated development. This includes: developing and implementing a new round of the plans for a provincial urban system and the planning of three main city groups, including the city group around Hangzhou Bay, the Wenzhou-Taizhou city group and the city group in central Zhejiang. These three city groups are forming rapidly, and the Hangzhou, Ningbo and Wenzhou metropolitan areas have begun to take shape.

Second, taking the master plan of counties (county-level cities) as the "main platform" for coordinated urban and rural development in counties (county-level cities). This means: developing and implementing the master plan of counties (county-level cities), making counties (county-level cities) fully covered by urban and rural planning, thus ensuring fully coordinated factor allocation and development.

Third, pushing forward the central town cultivation project and the reform of the expansion of the powers of strong towns. This includes: identifying 200 provincial central towns, choosing 27 of them to pilot the cultivation of small cities, improving the ability of central towns to carry population and industries and deliver public services in rural areas and making central towns become important nodes in coordinated urban and rural development.

Fourth, pressing ahead with the central town cultivation project. This includes: identifying the first batch of 4000 provincial central villages, improving community infrastructures and public services, actively and steadily stimulating the rural population to gather together by reducing natural villages, dismantling "hollow" villages, relocating from the villages on high mountains, merging small villages, thus making central villages the important platforms through which farmers can fully enjoy public services.

2. Upholding coordination between urban development and industrial development, continuously enhancing the function of urban areas as points of absorption

(1) Strengthening the function of urban areas as points of agglomeration. This includes: pushing forward urban development from scale expansion to functional improvement, promoting benign interaction between urban development and industrial structural adjustment, fully enhancing the overall quality and the functions of factor agglomeration and economic radiation.

(2) Promoting the development of the agglomeration of industries. This includes: carrying out the plans for three main industrial belts around Hangzhou Bay, along the Wenzhou-Taizhou coast and the Jinhua-Quzhou-Lishui Highway, giving great impetus to the development of large platforms, large industries, large projects and large enterprises, speeding up the building of 14 provincial industrial agglomeration areas; pushing forward population aggregation while accelerating the development of industrial agglomerations.

(3) Propelling the construction of infrastructure. These actions include: more rapidly upgrading regional transportation infrastructure, developing the circle of main cities from a 4 h road transportation circle into a 3 h road transportation circle; making the per capita road area in the cities above the county level reach 17.6 m^2.

3. Upholding interaction between urban and rural areas, further giving play to the driving role of urban areas

(1) Innovating the small city cultivation mechanism. Becoming the first nationwide to pilot the cultivation of small cities, choosing 27 provincial central towns with good developmental conditions and an excellent location to speed up transformation towards small cities by scientific positioning, guidance in planning, functional improvement, policy support and institutional innovations.

(2) Extending public services to rural areas. This means: vigorously developing rural social programs, establishing and improving more than 30,000 village-level service centers; more rapidly improving rural public service facilities, making electricity available to all households and ensuring the same price in both urban and rural areas in the case of the same power grid, extending radio and television coverage to all villages, achieving full coverage by county-level libraries and cultural centers, town-level comprehensive cultural stations, town chain supermarkets and trustworthy shops in administrative villages.

(3) Exploring and establishing the mechanism of linking for the increase or decrease in land for urban and rural construction; and carrying out the project of one-million land reclamation and guarantees, thus striking a balance between land occupation and compensation across the province.

(4) Making significant achievements in village improvement. This includes: initiating two rounds of work on carrying out the project of "1000-village demonstrations and 10,000-village improvements", cultivating 823 central villages, thus building a number of beautiful villages, clean villages and villages of happiness.

4. Upholding reform-driven actions, building equitable urban and rural institutions

These actions include: focusing on key fields and crucial segments in coordinated urban and rural development, carrying out pilot work to make breakthroughs, developing models to lead the way, intensifying the comprehensive urban and rural reform and overcoming the dual urban-rural system; turning rural land into assets by carrying out reforms involving the exchange of homesteads for urban housing, separation between rural production and domestic houses, land circulation and establishment of the endowment insurance system for rural households; promoting intensive use of spaces by carrying out "1 + X" layout plans of villages and towns; ensuring equal access to basic public services by building an integrated urban and rural system for fair education and equal employment, achieving full coverage of both urban and rural areas by the old-age care and medical systems, establishing a large five-in-one social security system covering both urban and rural areas; making rural financial innovations by actively developing village banks, small loan companies, rural fund mutual cooperatives and other new-type rural financial organizations, establishing a mechanism for cross-regional development of rural cooperative financial institutions; turning the agricultural migrant population into urban residents by proactively experimenting with and promoting a reform of the new-type household registration system suitable for new-type urbanization, carrying out the residence permit system across the board and fully removing the restrictions on migration of registered permanent residence in counties (county-level cities) and the areas at the lower levels.

6.3.3 Zhejiang's Practical Experience in the Strategy of New-Type Urbanization

In 2014, Zhejiang's rate of urbanization reached 64.87%, up about 10 percentage points compared with the national average level, the ratio of urban residents' income to that of rural residents in Zhejiang was 2.08:1, making Zhejiang one of the provinces with the smallest urban-rural gap in China and the comprehensive effect of urbanization became increasingly prominent.

Zhejiang has mainly acquired the following practical experience in carrying out the strategy for new-type urbanization.

First, it has acquired experience in upholding the driving role of the Party committee and the government, attaching importance to top-level design. Since Zhejiang carried out the urbanization strategy in 1998, the Party Committee and the People's Government of Zhejiang Province have unveiled the urbanization plan and the urban-rural integration plan, issued the *Opinions on Further Strengthening Urban Work and*

Going Down the Path of New-type Urbanization and the *Plan of Zhejiang Province for Expanding New-type Urbanization* and it has put forward the requirements of coordinated development, intensive development, characteristic development, interactive development, harmonious development and innovative development.

Second, it has experience in upholding the guiding role of planning, optimizing the spatial layout of urban and rural areas. This involves: paying attention to the planning of city groups, developing and implementing a new round of plans for provincial urban systems and the plans of three main city groups, including the city group around Hangzhou Bay, the Wenzhou-Taizhou city group and the city group in central Zhejiang—these three city groups are forming rapidly, and the Hangzhou, Ningbo and Wenzhou metropolitan areas have begun to take shape; attaching importance to the master plan of counties (county-level cities), developing and implementing the master plan of counties (county-level cities), making counties and county-level cities fully covered by urban and rural planning, thus ensuring fully coordinated factor allocation and development, which has been highly commended by the Ministry of Housing and Urban-Rural Development.

Third, Zhejiang has acquired experience in upholding coordination between urban development and industrial development, enhancing the function of cities for stimulating the development of others; this includes intensifying the development of urban agglomeration function areas; the urban construction land which accounts for 2.6% of the land areas carries more than 79% of the province's GDP; the urban areas of Hangzhou, Ningbo and Wenzhou have 16.7% of the province's population and have created 1/3 of the province's GDP; more rapidly promoting the development of industrial agglomerations, carrying out the plans of three main industrial belts around Hangzhou Bay, along the Wenzhou-Taizhou coast and the Jinhua-Quzhou-Lishui Highway, speeding up the building of 14 provincial industrial agglomeration areas; making the economic aggregate of cities and county towns account for more than 60% of the province's economic aggregate, and the total industrial output value, total export volume and actually realized foreign investments in the development zones (parks) above the provincial level account for 52.2, 43.5 and 63.2% of the province's levels.

Fourth, upholding coordinated urban and rural development, pushing forward the development of towns. This includes choosing 27 provincial central towns with good developmental conditions and excellent locations to become the first nationwide to pilot the cultivation of small cities. At present, preliminary results have been achieved in these pilot towns. More quickly improving rural public service facilities, making electricity available to all households and ensuring the same price in both urban and rural areas in the case of the same power grid, extending radio and television coverage to all villages, achieving full coverage by county-level libraries and cultural centers, town-level comprehensive cultural stations, town chain supermarkets and trustworthy shops in administrative villages. In late 2013, the improved villages accounted for 94% and the rural domestic garbage collected and disposed of in a centralized way made up 93% in Zhejiang.

6.3.4 *Lishui's Experience in New-Type Urbanization*

In the process of promoting new-type urbanization, Lishui City has deeply carried out the guiding principles adopted in the Central Urbanization Work Conference, it has resolutely taken the green ecological developmental path that clean, clear waters and lush mountains are the "gold and silver mountains", and worked out the blueprint of a beautiful Lishui according to the overall goal of building a mountainously gorgeous Lishui, a great land for health preservation and a land of long life.

First, the philosophy of urban planning fully reflects ecological characteristics. Planning is based on Lishui's characteristic urban resources and developmental features; Lishui has developed three main philosophies, including "depicting the city with lush mountains, revitalizing the city through clean, clear waters and prospering the city through culture". "Depicting the city with lush mountains" means that lush mountains serve as the background of Lishui and become the most important ecological and landscape space among the elements of urban development, and urban spaces are shaped by making mountains visible to residents and reflecting the essential characteristic of a mountainously gorgeous Lishui. "Revitalizing the city through clean, clear waters" means that the distributional characteristics of the water system at the sites are clarified, large bodies of water connect to small bodies of water, the water within the city connects to the water outside the city, rivers and lakes are connected to make the water flow, actions are taken to conduct an ecological restoration of the Ou'jiang River basin and lay the foundation for shaping the beautiful waterfront spaces in the city, make water visible to residents, build an urban ecological pattern in which the city is surrounded by mountains and bodies of water. "Prospering the city through culture" means that the cultural connotation of the city is fully tapped, prominence is given to the protection of historical and cultural blocks and historical, cultural relics as well as historic sites; great considerations are given to the texture and characteristics of old towns; the cultural continuity and fusion of historical Lishui, contemporary Lishui and modern Lishui are maintained so that residents keep fond memories, the city's cultural impetus is fully highlighted, offering support for sustaining urban characteristics and providing power for making the city prosperous.

Second, the nature of the city gives prominence to an ecological brand. The current nature of the city is a habitat with mountains and bodies of water and an ecotourism city in the southern part of the Yangtze River Delta Region, a characteristic manufacturing base and a central city in southwestern Zhejiang. A new master plan presents new requirements for Lishui's functions and developmental orientation, it defines Lishui City as the central city in southwestern Zhejiang, Zhejiang's famous historical and cultural city, and the international city for ecological leisure, health preservation and tourism.

Third, the optimization of the urban structure highlights ecological functions. According to the relevant plan, as Lishui City is located on the Hangzhou–Fujian–Guangdong urban development axis and the port—hinterland development axis, first priority is given to turning Lishui into a central city, and a structure of a municipal urban system which features "one core, two axes, four areas and four levels" thus takes shape, and the overall strategy of a "strong center, excellent towns and beautiful villages" is carried out to enhance the capability for enabling agglomeration and stimulating the development of others and drive regional coordinated development. Actions are being taken to integrate and promote the global factors and resources, highlight characteristic and differentiated development, develop a green and ecological new-type urbanization path. In the central urban areas, based on the spatial structure of "one river and two cities", the functional urban layout of "residential development in the northern part, leisure in the central part, industrial development in the southern part" is being further optimized and improved. In the northern part, residential development is mainly pursued, an administrative center, a commercial center and a cultural center are being built; in the central part, the tourism leisure, holiday and health preservation industries are mainly developed along Lake Nanming of the Ou'jiang River and in the Yangmingshan Scenic Spot; in the southern part, industrial development is mainly fostered, a logistics center, an industrial center and a technical innovation center are being built.

Fourth, urban functions give prominence to ecological connotations. The plan for more rapidly renovating and improving old urban areas is being carried out to promote functional improvement and structural upgrading in old urban areas. Actions are being taken to orderly boost the renewal and renovation of old urban areas, improve the quality and inherit historical development, accelerate the renovation and transformation of old urban areas, achieve benign complementation in block functions and sound interaction in the block effect between new and old urban areas, continuously improve the landscape image of urban areas, enrich urban cultural connotations and enhance overall competitiveness. Old residential quarters, roads, back streets, alleys, public service facilities, green land squares are being renovated, underground spaces are being developed and utilized, urban functional blocks are being optimized and adjusted to bring about a new landscape in old urban areas. Actions are being taken to speed up the functional improvement of southern blocks, build a highland for ecological industrial agglomeration, develop the Fuling block into a new southern ecological urban area which integrates characteristic logistics and commercial leisure; industrial upgrading is being carried out and supporting functions are being developed in the Qibaiyang block and it is being turned into a demonstration area for industrial-urban integration.

Since GDP-based evaluation was abandoned, Lishui has developed relevant policies for promoting ecological industrial and economic development and relevant industrial developmental plans, it has explored and practiced the new evaluation method which is more ecological, thus producing marked effects.

First, developing a philosophy of ecological and green industrial development. Since Lishui was administratively turned from an area into a city, the successive Party Committees and the People's Governments of Lishui City have identified the developmental strategies, goals and line of thought which better reflect urban ecological characteristics. In 2003, Lishui put forward the developmental strategy of building the city through ecological development, making the city strong through industry and developing the city through green development. In 2008, Lishui set the main strategic goals of making ecological progress and building a moderately prosperous society in all respects. In 2012, Lishui put forward the overall strategic goal of achieving a green rise and a scientific leap.

Second, developing the philosophy of industrial selection of economic ecologicalization and ecological economization. Lishui has issued the *Guiding Catalogue of Encouraged Industries and Products for Industrial Transformation and Upgrading* and the *Guiding Opinions on Industrial Layout of Urban Areas*, stating that projects are subject to evaluation involving decision-making about park entry before projects are filed and approved, investment attraction projects are signed, and emphasis is placed on introducing the projects with low energy consumption, less pollution, high technological content and high added value. At present, Lishui has basically built a leading industrial system in which the equipment manufacturing industry plays the leading role, eight main industrial clusters are pillars and six strategic emerging industries are forerunners.

Third, building an institutional system for ecological, intensive and environmentally friendly development. (1) Introducing the evaluation system of taking the yield per unit area as the yardstick. Preliminarily establishing the evaluation mechanism in which the efficiency of land use by enterprises serves as the basis, guiding enterprises to increase the level of intensive land use and the input-output ratio. (2) Establishing and carrying out a differentiated mechanism for regional development and environmental management. Innovating the mechanism for the policy management of the main functional areas and ecological functional areas in light of the characteristics of regional economic and social development and ecological environmental factors, ecological environmental sensitivity and the law which governs the spatial distribution of ecological service functions. (3) Establishing the mechanism involving the evaluation of project's entry into parks.

Fourth, bringing about an early form of the ecological industry which is green, circular and low-carbon. Lishui has promoted energy saving and emission reduction by strengthening control at the source, engineering construction, supervision measures and policy guidance. In 2012, the energy consumption per unit of GDP was 0.5315tce/10,000 yuan, ranking Lishui no.3 in the province. Under the principle of reduction, reuse and recycling, Lishui has also actively performed clean production, speeded up circular industrial economic development. As of 2012, there were 200 enterprises which had passed phased acceptance checks in clean production auditing in Lishui. Fifteen production enterprises and 34 operating enterprises have carried out the work on comprehensive utilization of resources, and 965,200t of waste resources are recycled each year in Lishui. Lishui has fully completed the new "811" three-year action plan for environmental protection and the ecological renovation of 10 provincial development zones (industrial parks). Lishui Economic Development Zone has been identified as the National Advanced Pilot Base for the Circular Economy in the Synthetic Leather Industry and Songyang's technology for centralized treatment of stainless steel wastewater has been the leader nationwide. Lishui has stayed ahead nationwide and ranked no.1 in Zhejiang for 9 consecutive years in ecological and environmental quality. Lishui has made a number of ecological industrial products including high-end equipment, cutting tools, high-voltage electrical apparatuses, energy-saving lamps, motors, down- and feather-filled products, daily chemical washing products, auto parts, wood and bamboo products, green food and biological medicine.

Chapter 7
Investigations on the Institutions for Ecological Development with Checks and Balances

Jian Liu

Zhejiang has gained much successful experience in improving the institutions for ecological development. When Xi Jinping took charge of Zhejiang, he attached great importance to the improvement of those institutions, while the subsequent successive Secretaries of the Party Committee of Zhejiang Province also gave a great deal of attention to ecological development and made great efforts to investigate the institutions for ecological development suitable for Zhejiang's reality. Thanks to the guarantee from those institutions, Zhejiang has made outstanding achievements in building an ecological province, a beautiful Zhejiang and promoting ecological development, and it has rapidly developed a green economy, a circular economy and a low-carbon economy.

7.1 Institutional Reform in Ecological Development

Under the new situation of ecological development, reforms must be carried out to remove the institutional barriers which do not adapt to the new situation, and they affect and restrict ecological development. The focus of institutional reform is the reform of the system of government administration; institutional reform centers on changing government functions and conducting comprehensive institutional reform to rationalize the administration system.

7.1.1 Transformation of Government Functions

The transformation of government functions is essentially changing the government from an omnipotent government to a non-omnipotent one, the building of a service-oriented government. This goal is achieved mainly by streamlining

J. Liu (✉)
Zhejiang Academy of Social Sciences, Hangzhou, China

J. Pan and M. Shen (eds.), *Chinese Dream and Practice in Zhejiang – Ecology*,
Research Series on the Chinese Dream and China's Development Path,
https://doi.org/10.1007/978-981-13-7209-4_7

administration and delegating powers, especially the administrative power of examination and approval, to the lower levels. "The powers which should be delegated are fully delegated, and the things which should be administered are better administered."[1]

Zhejiang has pushed forward the streamlining of the administration and delegating powers to the lower levels for many years; Zhejiang carried out three rounds of reform of the system of administrative examination and approval in 1999, 2002 and 2003, producing marked effects: in the first round of reform, the original 3,251 items subject to administrative examination and approval were reduced by 50.6%; in the second round of reform, the items subject to administrative examination and approval were reduced by 46.7%; after the third round of reform, only 718 items subject to administrative licensing and 243 items not subject to administrative licensing examination and approval were retained at the provincial level.

Since 2013, Zhejiang has carried out the fourth round of this m reform, with a focus on establishing a centralized system of examination and approval, more rapidly improving the system of joint examination and approval, building a system involving early examination and approval and standardized management of intermediary services, carrying out the system of agencies to take care of the whole process regarding the examination and approval services concerning the projects whose entry into parks is sought and large projects, establishing the access system for the items subject to examination and approval, and improving the system of responsibility for examination and approval.[2] When summing up the work of the year 2013, the Governor of Zhejiang Province, Li Qiang, said, "The current government will give top priority to intensifying reforms in the whole process, and will start with reforming the system of examination and approval to transform government functions.

Zhejiang's main experience and practice in streamlining administration and delegating powers with respect to ecological development are as follows.

1. Disclosing the list of powers, enhancing transparency

The Department of Environmental Protection, the Department of Water Conservation and other departments of Zhejiang Province have actively responded to the provincial government's call by disclosing the list of powers to make powers transparent and accept public supervision, delegating the administrative authority for examination and approval to the lower levels and promoting localized examination and approval.

2. Building the Zhejiang government service portal

On June 25, 2014, Zhejiang launched the Zhejiang government service portal to bring together the government service resources of more than 4,000 government institutions across the province, having achieved coordination among the provincial, municipal and county levels and improved the administrative efficiency. In the

[1]Liu Xiangrong, A Multi-pronged Line of Thought of the Reform of the System of Ecological Development, *Journal of Hunan Normal University*, 2014(2).

[2]Wen Hongyan, Gu Chun, and Xu Juan et al., Zhejiang Has Made the Best Efforts in Streamlining Administration and Delegating Powers, *People's Daily*, May 12, 2014.

meantime, the list of government powers has been posted online, forcing government departments to streamline their administration and delegate powers to the lower levels.

3. Simplifying the examination and approval procedures, reducing the examination and approval time

The Department of Environmental Protection of Zhejiang Province has reduced the number of examination and approval steps from the original four—preliminary review based on environmental impact assessment, evaluation, review, examination and approval—to two—review based on environmental impact assessment, examination and approval.

7.1.2 Comprehensive Institutional Reform

Ecological development is a systematic project involving the economic, political, cultural and social fields, so it needs to be promoted by a comprehensive department. Moreover, under the traditional system of government, there is severe fragmentation among the different departments, government functions are excessively distributed among many departments—water yield, water quality, shipping and landscape are in the charge of the departments of water conservation, of environmental protection, of shipping and of parks, respectively, the government function involving environmental protection is separated from that involving water conservation, different pollutant components in the air are in the charge of different competent departments. Therefore, it is essential to carry out a comprehensive reform of the institutions. Zhejiang's main practices in comprehensive institutional reform are as follows.

1. Setting up the leading group for the building of an ecological province

In 2003, Zhejiang put forward the goal of building an ecological province. To promote the building of an ecological province, on May 6, 2003, the Party Committee and the People's Government of Zhejiang Province jointly issued a circular to establish Zhejiang's Leading Group for the Building of an Ecological Province, in which the Secretary of the Party Committee of Zhejiang Province, Xi Jinping, served as the leader, the Governor of Zhejiang Province served as Executive Deputy Leader, members of the Standing Committee of the Party Committee of Zhejiang Province, the Secretary General of the Party Committee of Zhejiang Province and vice governors of Zhejiang Province served as deputy leaders, the people chiefly in charge of the Party Committee and the People's Government of Zhejiang Province and relevant departments of Zhejiang Province were members. The leading group had an office, which was located at the provincial environmental protection bureau. Afterwards, this organization was established. The cities and counties also set up their own leading groups accordingly, with offices (ecology offices) of the leading groups at various levels located at the environmental protection departments, thus making important contributions to building Zhejiang into an ecological province.

2. Establishing the Department of Environmental Protection of Zhejiang Province, expanding the authority of the department of environmental protection

The department of environmental protection is an important functional department for promoting ecological development. In 2009, Zhejiang established the Department of Environmental Protection in charge of environmental protection across the province. Besides the functions of the former provincial bureau of environmental protection, its functions also include strengthening the coordination on major issues, including environmental policies, planning and standards, reinforcing the organization and implementation in building an ecological province, guidance and coordination in ecological development, intensifying guidance, coordination and supervision in environmental improvement and ecological protection, and taking further actions in achieving the national and provincial emission reduction goals as well as environmental supervision.

The upgrading of the competent department for environmental protection from the provincial bureau of environmental protection to the provincial department of environmental protection has not only increased the administrative level of the department of environmental protection, but it has also expanded the functions and authority of that department; this is favorable to the department of environmental protection in performing the functions of environmental protection and is conducive to boosting ecological environmental improvement and ecological development.

7.2 The Building of Mechanisms for Ecological Development

The priority in ecological institutional improvement lies in building a structure of checks and balances among the market, the government and the society. At present, the problems of the government replacing the market and running the programs which should be run by social forces are severe, there is a failure of the market mechanism, the government mechanism and the social mechanism,[3] and it is necessary to bring about innovations in those mechanisms.

7.2.1 Innovations to the Market Mechanism

For a long time, market failure has occurred in the field of resource environment due to the nature of environmental resources as public goods and the externality of the environment, so the intervention of the government is necessary. With increasing scarcity of environmental resources and decreeing costs for defining the resource property rights, it is possible to introduce a market mechanism into the field of

[3]Shen Manhong, Leveraging Ecological Advantages to Build a Beautiful Zhejiang, *Zhejiang Daily*, December 27, 2013.

resource environment. Therefore, it is essential to renovate the market mechanism and reform the system of environmental resource property rights so that the market mechanism plays the decisive role in resource allocation. Zhejiang's innovations to the market mechanism mainly include the reform of forest rights, the trading of emission rights and the mechanism for the trading of water rights. These innovations are at the forefront nationwide. Specifically, Zhejiang has brought about innovations to the forestry-related financial services and has become the first nationwide to promote the compensated use of emission rights and inter-regional trading of water rights.

7.2.1.1 The Reform of Forest Rights

Zhejiang's reform of forest rights started in the 1980s. Forest rights were stabilized, and private plots and the system of responsibility for forestry production were determined to arouse forest workers' enthusiasm about production. In the 21st century, the contract period of much mountain land and forest land expired, some local governments introduced policies to pilot extension of the contract period; for instance, Lin'an City issued the *Circular Concerning Improving the Work on Extending the Contract Period of Mountain Forests, Ascertaining Forest Rights and Issuing the Certificate of Forest Rights* in 2001 and basically completed contract signing and issuance of the certificate of forest rights in late 2002. In 2006, based on the experience from pilot work in different areas, Zhejiang issued the *Circular of the General Office of the Party Committee of Zhejiang Province and the General Office of the People's Government of Zhejiang Province Concerning the Extension of the Contract Period of Mountain Forests*, stating that the responsibility for forests which has been contracted to households continues to be contracted by the original contractors and the contract period is extended for another 50 years.[4] In 2007, Zhejiang issued the *Several Opinions on Further Intensifying the Reform of the System of Collective Forest Rights*, further specifying the ownership of and the right to use forests on the collective mountain land and the ownership of the rights to the contracted management and use of forest land (collective forest rights) on the basis of consolidating the achievements in extending the right to contract mountain forests.[5] In 2008, the State Council issued the *Opinions of the Central Committee of the Communist Party of China and the State Council on Comprehensively Promoting the Reform of the System of Collective Forest Rights*, calling for further clarifying property rights, loosening control over the management rights, putting in place the right of disposal and the right to earnings. Local authorities in Zhejiang have issued documents to further push forward the reform of collective forest rights.

---

[4]Xu Xiuying, Shen Yueqin, and Li Lanying, Explorations in the Reform of the System of Collective Forest Rights in Zhejiang Province, *Forestry Economics*, 2007(11).

[5]Several Opinions of the Party Committee and the People's Government of Zhejiang Province on Further Intensifying the Reform of the System of Collective Forest Rights, *Forestry Economics*, 2008(1).

Under the support and guidance of the Party Committee and the People's Government of Zhejiang Province, Zhejiang has made outstanding achievements in the reform of collective forest rights and has explored new modes of forestry reform. Its main practices and achievements are as follows.

1. Ascertaining forest rights and issuing a certificate of rights, becoming the first to complete the main reform of the collective forest rights

Zhejiang has actively promoted the renewal (issuance) of the certificate of forest rights to stabilize mountain rights and forest rights. The area for which the certificate of forest rights was renewed (issued) was 86,545,000 mu, accounting for 96.8% of the area for which the certificate of forest rights should be renewed (issued); 4,259,000 certificates of forest rights were renewed (issued), accounting for 99.0% of the certificates of forest rights which should be renewed (issued); 1,439,000 contracts of responsibility for forests were signed, accounting for 97.5% of the contracts which should be signed. Zhejiang has become the first nationwide to basically finish the main reform tasks. In the meantime, Zhejiang has pushed forward IT application relating to forest rights so that all of the data on the attribution of certificates of forest rights are entered into the system and digital management of forest rights is achieved.

2. Promoting the circulation of collective forest rights

As early as the 1990s, Zhejiang pushed forward the transfer of the forest land management right in various ways, including the transfer linking remuneration to output, concurrent leasing for multiple households and reverse leasing and contracting. In the 21st century, in order to further promote the circulation of forest rights and standardize the circulation behavior, Zhejiang has issued a number of policy documents, including the *Administrative Measures of Zhejiang Province for the Circulation of Forests, Forest Trees and Forest Land* and the *Interim Measures of Zhejiang Province for the Administration of Forest Resource Asset Mortgages* and established the East China Forestry Exchange; bodies for the management of forest rights were established in 75 counties (county-level cities, districts) and centers for the trading of forest rights were built in 65 counties (county-level cities, districts). The average annual circulated area was nearly one million mu. The circulated area and the circulation amount have reached 13.66 million mu and 21.4 billion yuan in Zhejiang.

3. Bringing innovations to the forestry-related financial services

Zhejiang's forestry-related financial innovations mainly cover the following two aspects. First, offering forest rights mortgage loans. In order to liquidize forestry assets and increase forestry input and the people's income, Lishui City became the first to provide forest rights mortgage loans in 2006 and it has explored new ways to extend loans, including IC cards relating to forest rights. To ensure that a forest rights mortgage loan is offered smoothly, Lishui City issued the *Opinions on Promoting the Circulation of Forest Resources*, the *Opinions on Promoting the Development of the Forest Resource Asset Mortgage Loan Services* and other relevant policy documents. Second, launching forest fire insurance. To mitigate the economic loss incurred to forest workers by forest fire, the Forestry Bureau of Zhejiang Province and local forestry bureaus have contacted insurance companies for providing insurance coverage relating to forest fires. With consultation among multiple parties, the work on forest fire insurance has been carried out across the province. At present, in Zhejiang, the area covered by valid policies of forest fire insurance and the area covered by valid policies of comprehensive insurance reach 43.96 million mu and 1,222,100 mu, respectively, and the combined area accounts for 46% of Zhejiang's total forest area.[6]

4. Cultivating new-type forestry operators

In order to withstand market risks, increase the increase and the degree of intensive development, forest workers in different areas have established various types of forestry-related specialized cooperative organizations under the guidance of local forestry departments.

5. Intensifying the reform of state forest farms

The People's Government of Zhejiang Province has issued the *Several Opinions on Accelerating the Development of Modern State Agricultural, Forest and Fish Farms*; Zhejiang has become the first nationwide to push forward the reform of state forest farms, and has identified 79 forest farms as public institutions.

7.2.1.2 The Mechanism of the Compensated Use and Trading of Emission Rights

Emission rights refer to the right of pollutant emitting units to legally emit pollutants into the environment. As a market mechanism has been introduced to promote the compensated use and trading of emission rights, full scope can be given to the role of the market in resource allocation, so that the emission reduction costs are minimized and the goal of environmental protection is achieved subject to the total quantity control of pollutants.

Zhejiang is not the province where the trading of emission rights was initiated at first, but Zhejiang has become the first nationwide to put the compensated use of emission rights into practice. Zhejiang has experienced the process from local pilot work to province-wide promotion with respect to the compensated use and trading of emission rights. Jiaxing is the city which has become the first in Zhejiang to initiate the compensated use and trading of emission rights. As early as 2002, to solve the problem of water pollution, Xiuzhou District issued the *Trial Measures of Xiuzhou District for the Administration of the Total Quantity Control of Water Pollution Emissions and the Compensated Use of Emission Rights*, initiating experiments in the system of compensated use and trading of emission rights. With four years of practice, good effects were produced; in 2005 alone, the trading volume reached 7.20 million

[6]Fu Xinghui, Sun Lu, An Analysis and Line of Thought of the Reform of the System of Collective Forest Rights in Zhejiang Province, *Forest Resources Management*, 2012(6).

yuan.[7] In 2007, based on Xiuzhou District's experience, with many years of study and experimentation, Jiaxing City issued and implemented the *Measures of Jiaxing City for the Trading of the Main Pollution Emission Rights (Trial)*, and established the center for the trading of emission rights reserves, initiated full implementation of the system of trading of emission rights across the city. As of November, 2009, there were 890 enterprises participating in the compensated use and trading of emission rights and the volume of trading totaled 149 million yuan in Jiaxing City.[8]

In 2009, based on the experience gained from the pilot work in different areas, the Ministry of Finance and the Ministry of Environmental Protection approved Zhejiang Province for officially piloting the compensated use and the trading of emission rights, Zhejiang unveiled the *Work Plan of Zhejiang Province for Piloting the Compensated Use and Trading of Main Pollutant Emission Rights* and the *Guiding Opinions Concerning the Pilot Work on the Compensated Use and Trading of Emission Rights*, thus initiating the pilot work on the system of emission rights across the province. In 2010, the People's Government of Zhejiang Province issued the *Interim Measures of Zhejiang Province for the Administration of the Emission Rights Permit* and the *Interim Measures of Zhejiang Province for Pilot Work on the Compensated Use and Trading of Emission Rights*, thus fully piloting the system of the compensated use and trading of emission rights across the province. In 2012, Zhejiang started full compulsory implementation of the system of emission rights.[9]

Overall, Zhejiang's reform of the system of emission rights focuses on the following four aspects: the environmental protection efforts shift from concentration control to total quantity control; the open type of environmental property rights is turned into a closed one; uncompensated use of the environmental capacity is turned into a compensated use; the environmental property rights are no longer untradeable and have become tradable. Specifically, Zhejiang's main experience and practices in promoting the compensated use and trading of emission rights are as follows.

1. Putting policies first, improving the policy, law and regulation system

With close cooperation among various departments, in order to promote the compensated use and trading of emission rights in Zhejiang, Zhejiang has introduced 11 provincial policy documents.[10] These policy documents cover different contents and focus on different aspects, but they improve the policy, law and the framework of the regulation system involving the compensated use and trading of emission rights from different perspectives.

[7] Du Derong, Xu Wenxiang, and Sun Dongming, The Xiuzhou Model for Initiating the Compensated Use of Emission Rights – Report from Xiuzhou District, *Zhejiang Today*, 2006(3).

[8] Shen Manhong, Xie Huiming, An Empirical and Normative Analysis of Economic Economization – Take the Compensated Use of Emission Rights in Jiaxing City as an Example, *Journal of the China University of Geosciences* (Social Science Edition), 2010(6).

[9] Shen Manhong, Xie Huiming, and Zhou Nan, The Zhejiang Model in the Reform of the System of Emission Rights, *Journal of Zhejiang Party School of C.P.C*, 2013(6).

[10] Zhou Shuxun, Chen Qi, The Zhejiang Model for the Trading of Emission Rights, *Environmental Economy*, 2012(3).

2. Giving importance to the supporting role of environmental monitoring technologies

Zhejiang has recorded all things concerning the increment and the decrement of the source of pollution, the amount of excessive emission or the amount of secret emission by installing online monitoring equipment and the systems for online monitoring of the environmental quality in key polluting enterprises to be able to monitor the emission behaviors of the enterprises emitting pollutants. Zhejiang has made great efforts to build an online monitoring system and promote the networking of enterprises emitting pollutants. As of early 2013, nearly 3,000 enterprises had been connected to the monitoring platform managed by the provincial department. The emission system based on swiping cards has been built. Under such a system, once the enterprises' emission quota has been used up, the valves at sewage pipes will automatically be closed, thus effectively monitoring the emission behaviors of enterprises each month. Zhejiang has built a system of networks for the comprehensive management of environmental protection covering the trading of emission rights and the management of emission permits, the examination and approval of construction projects and environmental law enforcement, so data sharing has been achieved.

3. Paying attention to further institutional innovations relating to emission rights

The cities in Zhejiang have brought about innovations in light of the reality to avoid or overcome institutional barriers or resistance and to solve practical problems. For instance, the Jiaxing Model for the initial allocation of emission rights helps circumvent the issue of legislation by recognizing the emission rights obtained by enterprises before pilot work, thus guaranteeing fairness; the system of emission rights mortgage loans introduced by Shaoxing and the system of emission rights leasing adopted by Pinghu City has effectively overcome the shortage of resources and the difficulties in obtaining loans for small and medium-sized enterprises. The People's Government of Zhejiang Province has, based on the successful experience from different areas, improved institutional innovations and promoted them across the province and has actively pushed forward the compensated use and trading of emission rights.

4. Letting the market set the prices of emission rights

In Zhejiang, the prices of emission rights were once set by the government, but these prices often deviated from the equilibrium prices on the market and hindered trading. Many local authorities have attempted to conduct auctions of emission rights; as prices are set by the market, auction prices are not necessarily the equilibrium prices on the market, but they reflect the enterprises' willingness to pay for emission rights, and auction prices are representative to some extent. Nanhu District of Jiaxing City and Hangzhou City conducted auctions of emission rights many times, many enterprises participated in them and basically all of the emission rights were bought, thus promoting the trading and transfer of emission rights .

5. Upholding the down-top promotion mode

Zhejiang has progressively promoted the system of the compensated use and trading of emission rights in a down-top manner and from local pilot work to province-wide promotion, thus guaranteeing an orderly promotion of work. Zhejiang has fully drawn upon the experience from pilot work in different areas when developing its policies, making sure that the policies conform to the reality and are enforceable and operable.

7.2.1.3 The Mechanism for the Trading of Water Rights

In China, water resources are owned by the state. Therefore, the trading of water rights is essentially the transfer of the use of water resources right among different traders.

The first episode of trading water rights in Zhejiang occurred between Dongyang City and Yiwu City; this is also the first episode of trading water rights in China. In November, 2000, the two sides reached an agreement on the trading of water rights, under which Yiwu City would invest 200 million yuan to purchase the right to annually use 49,999,000 m^3 of water from the Hengmian Reservoir in Dongyang; after the transfer of the water use rights, the original ownership of the reservoir remains unchanged, reservoir operations and its engineering maintenance are still performed by Dongyang, while Yiwu pays the comprehensive management fee (including the water resource fee) on the basis of 0.1 yuan/m^3 and for an annual actual quantity of a supply of water; the water pipeline project from the reservoir to Yiwu was planned, designed, invested in and carried out by Yiwu City, the water diversion section and pipe project within Dongyang was carried out by Dongyang and costs were borne by Yiwu City.[11] In January, 2005, the project was completed to initiate the supply of water, and the people of Yiwu can now drink water from the Hengmian Reservoir.

Zhejiang's innovations to the trading of water rights are mainly as follows.

1. Defining water rights

The water involved in Zhejiang's cases of the trading of water rights is basically reservoir water, it is easy to define its ownership, so there is no difficulty in defining water rights caused by the mobility and inter-regional flow of water.

2. Determining local governments as the main traders of water rights

Local governments represent local public interests. The trading of water rights among local governments can be conducted at lower costs on the one hand and can effectively guarantee local long-term interests on the other hand .

[11] Shen Manhong, *A Study of the System of the Trading of Water Rights – An Analysis of China's Cases*, Zhejiang University Press, 2006, p. 57.

3. Determining the principle of limited water supply

Several cases of trading water rights in Zhejiang are based on surplus water, priority is given to guaranteeing water use in production, life and in the ecological field, thus effectively preventing a lack of water in the place from which water is supplied and the resulting contradiction between the place from which water is supplied and the place where water is used.

4. Determining the principle of the compensated use and trading of water rights

In several cases of the trading of water rights in Zhejiang, the supplier of the water has obtained a certain economic compensation in different ways, regardless of conducting a direct buyout or becoming a shareholder to participate in the construction. That compensation can be used to carry out water conservation projects and protect water resources and to guarantee the quantity and quality of the water supply.

5. Achieving a win-win outcome for both sides in the trading of water rights

All cases of the trading of water rights in Zhejiang are spontaneously carried out by local governments in light of the reality; both sides in the trading have actual needs. The trading in water rights can address the shortage of water resources on the one hand and the exchange of surplus water for capital on the other hand, so both sides obtain benefits and a win-win outcome is achieved.

7.2.2 *Innovation to the Governmental Mechanism*

Ecological development often meets with government failure, thus innovation to the governmental mechanism is very important. That innovation lies in defining government functions, effectively fulfilling government duties and scientifically evaluating government performance. In that innovation, top priority should be given to fostering a correct outlook on officials' performance and improving the system involving the evaluation of the officials' performance.

In the traditional evaluation of officials, the proportion of GDP was excessively large, officials were evaluated mainly by the local GDP, the GDP served as the yardstick for the evaluation of performance, so local governments placed undue emphasis on the speed of the growth of the GDP and on the scale of the GDP, so ecological development was thus affected.

As early as 2004, Xi Jinping pointed out, "It is necessary to scientifically develop indicators for evaluating the performance of officials and correct employment guidance and system. As the conditions in different areas are different, the requirements and key points for measuring the performance of the officials should also be different; the GDP should be considered, but GDP cannot be taken as the only criterion." "The performance of the officials should be measured in the future under the principle of being widely recognized by the people and valuing actual performance, and this should serve as an important yardstick for evaluating officials; it is necessary to improve the content of evaluation, and take "whether the line of thought regarding development and the strategy for development are correct, whether the relationship between quality and quality and between speed and benefit are properly handled" as the important considerations in evaluating the leading cadres as to whether they have fostered a correct outlook on performance. In evaluation, emphasis should be placed on economic, social, humanistic and environmental indicators; there is a shift from the mere pursuit of speed to a comprehensive evaluation of the speed of growth, the level of employment, the educational input and environmental quality."[12]

In October of 2004, under the guidance of this line of thought, the Party Committee and the People's Government of Zhejiang Province conducted a relevant survey on the improvement of the cadre evaluation system, and developed a system with indicators for the evaluation of cadres with Zhejiang's characteristics which conforms to Zhejiang's reality. Afterwards, it was piloted at the local level.

Huzhou was the first area in Zhejiang to pilot this system. In late 2003, the People's Government of Huzhou City issued the *Opinions on Improving the Annual Comprehensive Evaluation of Counties and Districts*, and decided not to consider the GDP in the evaluation of cadres as from 2004. In practice, Huzhou City has attempted to develop a measure of classified evaluation, adopt different evaluation indicators for different areas, fully leverage local characteristics, and tailor evaluation measures to industrial, agricultural and ecological development. In this way, Huzhou City has completely freed the areas unsuitable for developing the industry from the constraint of a GDP-based evaluation, and thus embarked on a new path for harmonious development of an ecological economy.

In 2006, based on the experience from the pilot work in different areas, the Department of Organization under the Party Committee of Zhejiang Province became the first nationwide to issue the *Implementation Measures of Zhejiang Province for the Comprehensive Evaluation of the Leading Groups and the Leading Cadres of the Party and the Government of Counties (County-level Cities, Districts) (Trial).* In September, 2007, they studied and developed the *Implementation Measures of Zhejiang Province for Comprehensive Evaluation of the Leading Groups and the Leading Cadres from the Working Departments of the Party and the Government (Trial).* The promulgation and implementation of both measures suggest that Zhejiang's work on the evaluation of cadres has reached a new level and vigorously supports ecological development in different areas. Afterwards, Zhejiang continuously improved the work on the evaluation of cadres. From 2006 to 2011, after the promulgation of a series of implementation opinions and measures, the Department of Organization under the Party Committee of Zhejiang Province made some innovations and breakthroughs in the evaluation and appointment of cadres.

1. Placing less emphasis on the GDP, giving greater importance to ecological and environmental protection

In the long-term economic development, Zhejiang's ecological environment was damaged to some extent. In order to protect the ecological environment, build an

[12]Xi Jinping, *Zhijiang Xinyu*, Zhejiang People's Publishing House, 2007, p. 30, 73.

ecological Zhejiang and a beautiful Zhejiang, the Department of Organization under the Party Committee of Zhejiang Province has incorporated the indicators of an ecological environment into the system of indicators for the evaluation of cadres, carried out the accountability system and the one-vote veto system. For the local governments which fail to meet the requirements of the indicators of ecological environment assigned by the Party Committee and the People's Government of Zhejiang Province, a notice of criticism is circulated, and the granting of relevant honors and titles under evaluation is voted down; for the projects which violate regulations or exceed relevant standards, regular rectification should be made within the specified time; if rectification is not duly put in place, relevant responsible personnel will be held to account. In key ecological functional areas, including Lishui, Chun'an and Kaihua, GDP-based evaluation has been canceled, and priority is given to ecological protection, ecological economy and the people's living standard in carrying out the evaluation, with a view to promoting ecological development in these areas.

2. Giving importance to scientific guidance

Under the guidance of the incorrect outlook on development and the performance of the officials, some vanity projects occurred in different areas, where the rate of economic development was valued but the indicators relating to actual performance were overlooked. In order to reverse this problem, Zhejiang has designed the system of indicators for the evaluation of cadres which gives prominence to the evaluation of actual performance in economic transformation and upgrading, ecological development, improvement of the people's well-being, social harmony and stability, and it no longer takes the rate of economic development as the important basis for evaluation.

3. Paying attention to classified evaluation, avoiding the adoption of uniform indicators in all cases

The reality and the level of economic development vary with different areas in Zhejiang, so the same evaluation indicators cannot be used. For some areas and sectors, the characteristic systems of indicators for the evaluation which focus on different aspects should be built in light of their characteristics and according to the requirements of the Party Committee and the People's Government of Zhejiang Province for central work. For the evaluation of the cities and counties, 11 prefecture-level cities in Zhejiang are classified into three categories in light of the local reality and levels and rates of economic development. For the areas at a high level of economic development, such as Hangzhou and Ningbo, the evaluation focuses on economic structural adjustment, scientific and technological innovations and social services; for the areas unsuitable for industrial and economic development, such as Quzhou and Lishui, the evaluation should be conducted by giving more consideration to ecological development and the quality of the environment.

4. Giving importance to public evaluation

In the past, cadres were mainly under official evaluation. Even if there was public evaluation, its proportion was small; moreover, the degree of public participation was low and the public's understanding of evaluation was insufficient, and the evaluation covered less content. Zhejiang's new system of indicators for the evaluation of cadres gives prominence to the importance of public evaluation and expands the weights of relevant indicators. As the degree of public participation has been increased, more indicators concerning public satisfaction have been added and the popularity among the public has become stronger, evaluation reflects the voice of the people at the primary level and the people's will more.[13]

5. Giving importance to evaluating the moral conduct of officials

Zhejiang's system of indicators for the evaluation of cadres stresses the evaluation of officials' life circle and social circle, their personal conduct, social interaction and family life. Emphasis is placed on evaluating the moral conduct of the officials, and specific evaluation indicators have been adopted to quantitatively analyze their moral conduct. Prudence is exercised when appointing the officials who are subject to much negative evaluation by the public and whose conduct is hardly clarified in a short time; all of the officials who fail to reach the standard relating to moral conduct are subject to the one-vote veto system.

7.2.3 Innovation to the Social Mechanism

The failure of the social mechanism is mainly embodied in a lack of channels for the public's access to environmental information and the channels for public participation in environmental protection. Therefore, it is necessary to proceed from these two aspects to bring about innovations to the social mechanism. Zhejiang has mainly made the following contributions to innovations to the social mechanism.

7.2.3.1 The Mechanism for the Disclosure of Environmental Information

Environmental information includes the government's environmental information and enterprises' environmental information. The government's environmental information refers to the information which is developed or obtained by the environmental protection department in fulfilling the environmental protection duties, and it is recorded and kept in certain forms; the enterprises' environmental information refers to the information which is recorded and kept by enterprises and is related to the environmental impact exerted by the operating activities of the enterprises and their environmental behavior. The disclosure of environmental information can effectively safeguard the citizens' right to know and promote public participation in environmental protection. China has been improving the system for the disclosure

[13]Ma Yueming, Giving More Prominence to the Explicit Orientation of Scientific Development, Explaining One Opinion and Five Measures in the Evaluation of Cadres, *Zhejiang Today*, September 21, 2011.

of environmental information and has carried out the *Measures for the Disclosure of Environmental Information (Trial)* since 2008.

Zhejiang has made outstanding achievements in the disclosure of environmental information. According to the latest scoring and ranking of 120 cities in respect of the Pollution Information Transparency Index (PITI) during the period 2013–2014 published by non-governmental environmental protection organizations, including The Institute of Public & Environmental Affairs, The Natural Resources Defense Council, Green Anhui, Green Hunan, Green Qilu, the Fujian Green Home, and the Nanjing Green Stone, Ningbo City in Zhejiang ranked no.1, Wenzhou City and Hangzhou City ranked no.5 and no.6.[14] Zhejiang's successful experience and innovations to its mechanism regarding the disclosure of environmental information are as follows.

1. Ensuring a high degree of government attention to the disclosure of environmental information

In 2013, the People's Government of Zhejiang Province issued the *Current Key Work Arrangements of Zhejiang Province for Government's Disclosure of Information*, identifying the disclosure of environmental information as one of the tasks of information disclosure in nine key fields, requiring concentration on disclosing the information concerning the quality of the air and of the water, the assessment of the environmental impact of construction projects, environmental pollution control policies, measures and control effectiveness, emission reduction, and ordering the provincial department of environmental protection to lead the way in implementation.

2. Promptly disclosing major information on the monitoring of environmental protection

After the Fukushima nuclear accident occurred in 2011, the Department of Environmental Protection of Zhejiang Province promptly released environmental monitoring data, and opened up a special column for disseminating the knowledge about nuclear radiation and dispelling public panic. After the blood lead incident took place, the Department of Environmental Protection of Zhejiang Province also set up a special column for spreading the knowledge on protection against blood lead, stabilizing public sentiment, and it promptly updated the information on rectifying the storage battery industry to guarantee the public's right to know.

3. Building up information regarding the disclosure of information

The Department of Environmental Protection of Zhejiang Province has given great importance to information disclosure on its portal website, it has launched a special column on that website for the disclosure of information, and it promptly disclosed the results of the monitoring of environmental quality and the information on the examination and approval of projects subject to environmental impact assessment. To strengthen the disclosure of enterprises' environmental protection information, the Department of Environmental Protection of Zhejiang Province has built, on its portal

[14]Chen Yuanyuan, The Ranking of 120 Cities in Terms of the Pollution Information Transparency Index (PITI), *China Environment News*, June 10, 2014.

website, a unified platform for enterprises across the province to voluntarily disclose their monitoring information, and it has required the enterprises under national key monitoring to voluntarily disclose the monitoring scheme and the results of the monitoring on the platform.[15] Regarding water pollution prevention and control, the provincial environmental information center has built a platform for the disclosure of information on Zhejiang's water pollution and for public participation, with a view to promptly disclosing the information on the results of the monitoring of water pollution and environmental quality to the public. Besides the website, the Department of Environmental Protection of Zhejiang Province has actively developed other channels for the disclosure of information, and it launched an official microblog and the official WeChat "Zhejiang Environmental Protection" in 2013 and 2014, ensuring that the public understands and participates in environmental protection.

7.2.3.2 The Mechanism for Publicity and Education Regarding the Environment

With respect to ecological development, the people are the main actors, the people's line of thought and actions have a direct impact on the effect of ecological development.[16] Therefore, it is essential to strengthen the publicity and education about ecological development and foster an outlook on ecological development so that the philosophy of ecological development is deeply rooted among the people. The report delivered during the 18th National Congress of the Communist Party of China also vowed to strengthen the publicity and education about ecological development, to reinforce the awareness of the entire population about saving, environmental protection and ecology, to develop the social custom of rational consumption, and to foster a good conduct of love for the ecological environment.[17]

Zhejiang has given great importance to the publicity and education about ecological development and has made outstanding achievements. During the period of the 10th Five-Year Plan, the Zhejiang Environmental Publicity & Education Center, the Hangzhou Environmental Protection Publicity & Education Center, the Ningbo Environmental Protection Publicity & Education Center, and the Environmental Protection Bureau of Shaoxing County have been rated as the advanced national collective for environmental publicity and education by the State Environmental Protection Administration. Zhejiang's work experience and innovative practices regarding the publicity and education about ecological development are as follows.

---

[15]Wang Qi, Zhejiang's Enterprises Voluntarily Disclose Monitoring Information, *China Environment News*, February 11, 2014.

[16]Strengthening the Publicity and Education about Ecological Development – Eight Points on Studying and Implementing the Guiding Principles from the 18th National Congress of the Communist Party of China, *China Environment News*, November 28, 2012.

[17]Tao Detian, Strengthening the Publicity and Education about Ecological Development, Boosting the Building of a Beautiful China, *Environmental Education*, 2013(Z1).

1. Establishing a special body for the publicity and education regarding ecological development

Zhejiang has built the provincial environmental publicity and education center for the publicity and education regarding ecological development. The corresponding branches have also been established in some prefecture-level cities, such as Hangzhou and Ningbo. The Zhejiang Environmental Publicity & Education Center mainly performs the following functions: developing the provincial environmental publicity and education plan and annual plan, organizing efforts to carry them out; organizing, guiding, coordinating and supervising the environmental publicity and education across the province; organizing, guiding and coordinating the publicity relating to the building of an ecological province and Zhejiang's major environmental protection tasks and taking charge of the environmental news report.

2. Conducting publicity and education regarding ecological development in various ways

Zhejiang has given great importance to cooperating with mainstream media, it has launched a special column for disseminating the knowledge on ecological development and environmental protection. To popularize the knowledge on ecological development and environmental protection, the Department of Environmental Protection of Zhejiang Province has compiled easy-to-understand reading materials including the *Knowledge on the Circular Economy*. With the rise of mobile reading, the Department of Environmental Protection of Zhejiang Province has actively made breakthroughs, launched an e-book on the popularization of environmental protection science—a mobile application *Environmental Protection Knowledge for Urban Residents* in 2012, producing good effects.[18] In 2010, the Standing Committee of the People's Congress of Zhejiang Province adopted a resolution, identifying each June 30 as the Zhejiang Ecology Day. On each Zhejiang Ecology Day, the Department of Environmental Protection of Zhejiang Province organizes various types of theme publicity activities. In the meantime, the Department of Environmental Protection of Zhejiang Province has worked with the provincial civilization office, the provincial Women's Federation, the provincial department of education and the provincial department of health to carry out a series of tasks on green development and to energetically conduct environmental education. The Department of Environmental Protection of Zhejiang Province has also supported and worked with local authorities, Party schools and administration institutes at various levels to carry out special training on ecological development and environmental protection, and to improve the work on education and training.

3. Expanding the scope of the publicity and education about ecological development

First, in terms of content, Zhejiang has not only spread the knowledge on ecological development, but it has also released some cases involving violations of environmental laws and the environmental monitoring quality reports to accept wide public

[18] Shao Tian, Yan Liyang, Zhejiang Seeks Innovations and Breakthroughs in Environmental Publicity and Education, *China Environment News*, July 17, 2012.

supervision. Second, with respect to the platform, Zhejiang has released relevant information through multiple platforms including mainstream media, books, websites, mobile phone clients to carry out publicity and education. Finally, with regard to the scope of coverage, Zhejiang has conducted publicity and education at schools and training bodies, and has also gone deep into communities to guide families to consciously participate in ecological development.

7.2.3.3 The Mechanism of Public Participation in Environmental Protection

With increasing public awareness about environmental protection, more and more people have volunteered to participate in it. For ecological and environmental protection, it is also necessary to ensure that the public fully exercises supervision and participates in the decision-making. Therefore, ecological development needs to be promoted by improving the mechanism of public participation in environmental protection and promoting an environmental democratic process.

Zhejiang has been giving importance to improving the mechanism of public participation in environmental protection, and has affirmed the public's role in it. When delivering a speech, Xu Zhen, Director-General of the Department of Environmental Protection of Zhejiang Province, said, "Public support is essential for environmental protection, the public is an important ally and the greatest supporter of the environmental protection department in improving the work on environmental protection." Zhejiang's main experience, practices and institutional innovations are as follows.

1. Improving the mechanism of public participation in environmental protection, promoting public participation

With a series of institutional improvements and institutional innovations, Zhejiang has intensified its support for public participation in environmental protection. The *Administrative Measures of Zhejiang Province for Environmental Pollution Supervision* encourages and supports the public to participate in environmental protection; the *Measures of Zhejiang Province for the Environmental Management of Construction Projects* requires that environmental impact assessment on construction projects should be conducted by inviting social groups, research institutions, the units and individuals in relevant sensitive environmental areas to attend symposiums, feasibility study meetings and hearings and thus safeguard the public's right to participation. In order to further promote public participation, the Department of Environmental Protection of Zhejiang Province has proposed to issue the *Opinions on Practically Strengthening Public Participation in Environmental Protection*, establish the system of public opinion surveys on environmental decision-making and take the degree of public support as the important reference for decision-making.[19]

[19]Ye Hui, Shao Tian, From Spectators to the Main Forces – Zhejiang Has Made Institutional Improvement to Encourage and Guarantee Public Participation in Environmental Protection, *Zhejiang Today*, 2012(22).

2. Building a platform for public participation, adopting multiple modes to promote public participation

In order to strengthen public supervision over environmental protection, the Department of Environmental Protection of Zhejiang Province has established the "12369" environmental protection hotline and online reporting platform. The public has actively participated in this process. In 2011 alone, there were complaints and reporting about 53,587 environmental protection cases. The cities have energetically explored the mode of public participation in environmental protection. Jiaxing City has established such public participation modes as an environmental protection federation, a citizens' inspection group for environmental protection, a service group of experts and a group for the publicity on ecological development, with a view to reinforcing public supervision over environmental law enforcement.[20] Taizhou City has promoted the mode of "citizens' single-point" environmental law enforcement under which citizen representatives choose enterprises to conduct surprise inspection. Zhoushan City has trained non-governmental environmental supervisors for participation in environmental law enforcement.

3. Pushing forward the development of non-governmental organizations for environmental protection, making public participation specialized and organized

The non-governmental organizations for environmental protection are the non-governmental groups voluntarily organized by environmental protection personnel; they have played an important role in public participation. They leverage the advantages characterized by a high degree of specialization and a high degree of organization to actively organize various environmental protection activities, popularize environmental protection knowledge and strengthen environmental supervision.

7.3 Institutional Guarantee for Ecological Development

The *Decision of the Party Committee of Zhejiang Province on Building a Beautiful Zhejiang and Creating a Good Life*, adopted on May 23, 2014, systematically expounded the issue of an institutional guarantee for building a beautiful Zhejiang and stressed the improvement of the systems involving "strict control at the source, strict management of the process and severe punishment in case of bad consequences" for the purpose of promoting Zhejiang's ecological development and building a beautiful Zhejiang.

[20] Yu Wei, Focusing on Public Participation in Promoting Environmental Democracy – The Initiation of Zhejiang Public Participation Local-level Partnership Projects of the EU-China Environmental Governance Programme, *World Environment*, 2012(6).

7.3.1 *The System of Strict Control at the Source*

7.3.1.1 The System of Main Functional Areas

The system of main functional areas is an innovation in China. The 11th Five-Year Development Plan details the basic direction of and main tasks for establishing the main functional areas and divides the spaces of the national land into four types of main functional areas, including the area for optimized development, the area for focused development, the area where development is restricted and the area where development is prohibited. With the efforts of four years, the State Council issued the *National Plan for the Main Functional Areas*, marking the development of the main functional areas from philosophy towards reality.[21] In the 12th Five-Year Development Plan adopted in 2011, the plan of the main functional areas was elevated to a national strategy.[22]

Zhejiang has attached great importance to promoting the development of the main functional areas. In 2006, when delivering a speech at the 7th provincial conference on environmental protection, the then Secretary of the Party Committee of Zhejiang Province, Xi Jinping, stressed, "It is necessary to proceed from the environmental capacity to optimize the regional layout, give full scope to the levering effect of the resource environmental policies, adopt a differentiated policy of regional development; it is essential to identify the functional orientation and developmental direction for the different areas in light of the resource endowment, the environmental capacity and ecological conditions, to combine the regional economic plan with the environmental protection goal, and to proceed from environmental capacity and natural resource conditions to carry out optimized development, focused development, and to restrict and prohibit development; it is also necessary to strictly follow the requirement of ecological functional zoning to determine the main functions of different areas, clarify the orientation of the performance of the officials which reflects the Scientific Outlook on Development and develop a distinctive pattern of development."[23] In the same year, Zhejiang was identified as the pilot province for basic research on a provincial plan for the main functional areas by the National Development and Reform Commission. To promote the work on the main functional areas, the Zhejiang Provincial Development and Reform Commission also issued the *Circular Concerning the Pilot Work on the Main Functional Zoning*. In 2007, based on the pilot work in different areas, the People's Government of Zhejiang Province issued the *Circular of the General Office of the People's Government of Zhejiang*

---

[21] Yang Weimin, Yuan Xilu, and Zhang Gengtian, et al., Implementing the Strategy of the Main Functional Areas, Building an Efficient, Coordinated and Sustainable Beautiful Homeland – General Report on Researching the Strategy of the Main Functional Areas, *Management World*, 2012(10).

[22] Wu Dianyan, Wu Zhengzheng, An Exploration of Several Issues in the Implementation of the Plan for the Main Functional Areas, *People's Tribune*, 2011(24).

[23] Xi Jinping, *Carrying out Concrete Work to Stay Ahead – Line of Thought and Practice in Promoting New Development in Zhejiang*, The Party School of the CPC Central Committee Press, 2006, p. 201.

Province on the Formulation of the Provincial Plan for the Main Functional Areas. In 2012, based on completing the basic research on the main functional areas and the comprehensive evaluation of the spaces of the national land, the work on formulating Zhejiang's plan for the main functional areas, which started in 2006, was basically finished. Zhejiang's main practices and innovations in promoting the main functional areas are as follows.

1. Promoting the formulation of the plan for ecological functional areas

The plan for ecological functional areas is a special, comprehensive, basic and restrictive plan for ecological and environmental protection and the important foundation and basis for formulating the plan for the main functional areas.[24] As early as 2003, the *Plan for Turning Zhejiang into an Ecological Province* divided Zhejiang into 6 ecological areas and 15 ecological subareas. In 2006, the Department of Environmental Protection of Zhejiang Province issued the *Technical Guide of Zhejiang Province Concerning the Plan for Ecological Environmental Functional Areas in Counties (County-level Cities, Districts) (Trial)*, so the plans for ecological functional areas began its formulation for various pilot areas. In 2007, based on the formulation and verification of the plan, by experts, for ecological functional areas in 14 pilot cities (counties) across the province, in order to promote the formulation of the provincial plan for ecological functional areas, the Department of Environmental Protection of Zhejiang Province issued the *Circular on Further Strengthening the Formulation of the Plan for Ecological Functional Areas.* As of late 2007, the plans for ecological functional areas in the cities (counties) across the province were basically formulated and passed the acceptance check by experts. The successful formulation on these plans has laid a foundation for formulating the plans for the main functional areas in various areas.

2. Strengthening organizational guidance

In order to formulate Zhejiang's plan for the main functional areas and coordinate the work of various departments, Zhejiang has established a leading group for the formulation of the provincial plan for the main functional areas, in which the Executive Vice Governor serves as the leader, the deputy secretary general of the People's Government of Zhejiang Province and the director of the Zhejiang Provincial Development and Reform Commission serve as deputy leaders, the deputy director of the Zhejiang Provincial Development and Reform Commission, the chief accountant of the Department of Finance of Zhejiang Province, the deputy director of the Economic and Trade Commission of Zhejiang Province, the chief planner of the Department of Construction of Zhejiang Province, the deputy director of the Department of Science and Technology of Zhejiang Province, the deputy director of the Department of Land and Resources of Zhejiang Province, the deputy director of the Department of Agriculture of Zhejiang Province, and the deputy directors of other relevant departments directly under the provincial government are members. Under the direct guidance of

[24] Wang Zheming, An Exploration of Zhejiang's Practice in Ecological Functional Areas, *Environmental Protection*, 2010(14).

the leading group, in 2007, the *Work Program of Zhejiang Province for the Plan for the Main Functional Areas* was issued, calling for establishing a consulting group of experts for the provincial plan for the main functional areas and the leading group for formulating the provincial plan for the main functional areas with a view to promoting the formulation of the plan. With cooperation from the members of the leading group for formulating the provincial plan for the main functional areas, the *Plan of Zhejiang Province for the Main Functional Areas* was finalized in 2013.

3. Upholding the coordination among the provincial, municipal and county levels, emphasizing the connection among the existing plans

According to the *Work Program of Zhejiang Province for the Plan for the Main Functional Areas,* the provincial plan should be formulated in a top-down way through a combination at the higher and lower levels and coordination among the provincial, municipal and county levels; such formulation should be promoted by stressing the connection with the results of the existing plan, valuing basic survey and research, the supporting role of science and technology, correctly handling the relationship with the surrounding areas and guaranteeing the quality of the plan.

4. Adjusting measures to local conditions, adding more ecological economic areas

During the formulation of the plan, Zhejiang has followed the principle of acting according to local conditions and has given expression to Zhejiang's characteristics. Specifically, the functional areas under protection account for 70% of the total regional area; this tallies with Zhejiang's geographical feature that there are 70% mountains, 10% water and 20% farmland; more ecological economic areas are creatively added in light of the reality of the underdeveloped areas in Zhejiang, thus vigorously promoting the development of the underdeveloped areas.

5. Developing the supporting policies, guaranteeing the implementation of the main functional areas

The *Plan of Zhejiang Province for the Main Functional Areas* requires local governments to adjust the existing policies and systems according to the plan. To promote the implementation of the main functional areas, Zhejiang has established the system of supporting policies covering finance, industry, agriculture, investment, environmental protection, population, land and performance, and has stated that differentiated evaluation should be carried out for the different main functional areas; Zhejiang has also taken the completion of the goal of the main functional areas as the important basis for selecting and appointing cadres. In the meantime, the *Plan of Zhejiang Province for the Main Functional Areas* specifies the duties of the state organs directly under the provincial government and of the cities and counties for promoting the implementation of the plan for the main functional areas, calls for establishing a system for the management of dynamic monitoring of the spaces of national land covering the whole province, and supervising the implementation of the plan.

7.3.1.2 The System of Environmental Impact Assessment

The environmental impact assessment refers to a method and system for analyzing, forecasting and evaluating the possible environmental impact exerted by the implementation of plans and construction projects, putting forward the measures for preventing or mitigating the adverse environmental impact, and conducting tracking and monitoring. This system has the characteristics of both management and prevention. China has carried out the *Environmental Impact Assessment Law* as from 2003, but it has gradually found some drawbacks during implementation. Given these problems, Zhejiang has made some innovations and improvements to the existing environmental impact assessment system in light of the practical work experience.

1. Giving importance to public participation

In Zhejiang, except for the projects which should be kept confidential as required by laws, the environmental impact assessment of construction projects and the examination and approval based on environmental impact assessment are subject to information disclosure and solicitation of public opinions. The *Environmental Impact Assessment Law* only requires solicitation of public opinions on the projects which may cause a great impact on the environment, and has no requirement for other projects.

2. Guaranteeing the public's right to know

In Zhejiang, if the projects under environmental impact assessment have a direct impact on the major interests of others or directly involve a major interest relationship between the applicant and others, the stakeholders should be informed of that situation. The applicant and the stakeholders have the right to state their views, defend them or attend hearings. The provincial environmental protection bureau should listen to them, while the *Environmental Impact Assessment Law* has no requirement in this regard.

3. Strengthening the management of supervision

Zhejiang has given great importance to supervising and managing the projects under environmental impact assessment and has introduced multiple policies for controlling the unhealthy tendency in the industry. Industrial supervision is strengthened by holding the personnel of environmental bodies to account and giving criminal sanctions to the environmental impact assessment units which cause severe environmental pollution through the projects under environmental impact assessment due to the falsification of environmental impact assessment materials. Supervisors may provide a report on environmental supervision for the projects which have a great impact on the environment and regularly deliver those reports to the environmental protection department, thus promoting the implementation of various measures of ecological protection.

4. Simplifying the procedure of environmental impact assessment

In Zhejiang, since October 1, 2008, for the construction projects to be examined and approved by the environmental protection departments below the provincial level, the outline of environmental impact assessment has no longer needed to be prepared and submitted for examination and approval. Where the environmental impact assessment reports on the projects subject to environmental impact assessment really need to be examined and verified by the competent department in advance, the environmental protection department works with relevant departments to examine and verify them and does not separately examine and verify them. By contrast, the national procedure is more complicated. The outline of environmental impact assessment needs to be first prepared and is subject to examination and approval by the environmental protection department. The environmental impact assessment report needs to be preliminarily examined by the competent department before being delivered to the environmental protection department for final examination and approval.

7.3.1.3 The System of Total Quantity Control

The system of total quantity control is the general term for the pollution control modes and management regulations which allocate the quantity of pollutant emissions to pollution sources according to the economic, technical and social conditions within a particular period and thus control the quantity of pollutants caused by pollution sources in a certain spatial scope within the limit of environmental capacity.[25] Generally, in order to improve the quality of the environment, the allowable total quantity of pollutant emission will decrease with time. At present, China's system of total quantity control is mainly used to control air pollution and pollution of the water environment. However, in practice, with further implementation of the system, some problems have loomed large. These problems mainly include a lack of special legislation, incomplete legal basis, a small number of detailed rules for implementation, an incomplete basis, specification and accounting system for the concrete operation of the indicators of the total quantity, a lack of supporting systems, underdevelopment of the mechanism of supervision and responsibility.[26]

Zhejiang has issued the *Administrative Measures of Zhejiang Province for Environmental Pollution Supervision* and the *Administrative Regulations for the Basis of Lake Taihu,* bringing about certain innovations to the system of total quantity control in the practice of total quantity control over pollutants; its main institutional improvements are as follows.

[25] Liu Xu, Qin Nanqian, An Overview of China's System of Total Quantity Control in the Pollution Prevention and Control Law, *Legal System and Society*, 2011(4).

[26] Yan Haichao, What Is Lacking in the System of Total Quantity Control? *China Environment News*, May 21, 2012.

1. Identifying concrete indicators concerning total quantity control

In the case of carrying out the system of total quantity control, Zhejiang has worked out the total quantity of pollutants which enterprises and public institutions are permitted to emit, the quantity of pollutant emission which needs to be reduced and the time of emission reduction. Zhejiang has required actions to strengthen the dynamic management of the polluting units, establish accounts of management on the basis of one account for one factory, supervise the changes in the newly increased quantity of pollutant emissions, the quantity of reduced pollutant emissions and the quantity of excessive or secret emissions, and incorporate them into the dynamic management system, and carry out the dynamic monitoring of the polluting units.

2. Intensifying the supervision of environmental law enforcement

Multiple modes have been adopted to strengthen supervision over the pollutant emission behavior of the polluting units. Environmental law enforcement has been strengthened; the units where no environmental protection and pollution prevention facilities were built have been ordered to make rectification within the specified time. The units whose pollutant emission still fails to reach the relevant standard after expiration of the specified time will be resolutely shut down.

3. Strictly executing the project access system

Zhejiang has strictly executed national and provincial access policies and prohibited the launching of new projects which emit nitrogen and phosphorus pollutants. The quantity of emissions of other newly increased pollutants from new projects must be linked with reduction indicators according to the specified proportion. The projects where no environmental protection and pollution prevention facilities have been built or put into operation will be shut down.[27]

4. Improving various supporting systems

Various economic policies have been adopted to boost the total quantity control of pollutants. The credit policy has been adjusted to curb the implementation of the projects with high energy consumption and heavy pollution, encourage the implementation and development of various environmental protection projects and renewable energy projects, and promote environmental protection products. Zhejiang has attempted to establish the system of compensated use and trading of emission rights which is consistent with resource prices and is conducive to environmental protection, and has, based on strengthening supervision over the polluting units, adopted market means to enhance the efficiency of resource allocation and increase the level of environmental protection.

5. Strictly controlling the new total quantity

Zhejiang requires that the newly increased emission quantity from new projects in pilot areas must be verified by emission rights exchange and obtained through the

[27]Shen Manhong, Wei Chu, Cheng Hua, et al. 2012 Zhejiang Eco-economic Development Report. China Financial & Economic Publishing House, 2012, p. 303.

trading of emission rights, thus the newly increased pollutant emission quantity will be strictly controlled.

7.3.2 *The System of Strict Management of the Process*

7.3.2.1 The Rule of Law in the Environmental Field

Protecting the ecological environment is an important part of ecological development. At present, environmental pollution and ecological damage are severe in China, so it is essential to rely on the rule of law relating to the environment to regulate the people's behavior, strictly punish the acts of damaging the ecological environment and thus effectively protect the ecological environment.

Zhejiang has been promoting the rule of law to protect the environment and make ecological progress. When working in Zhejiang, Xi Jinping vowed to speed up local environmental legislation, improve the local environmental laws, regulations and system of standards, intensify the punishment of the acts which violate laws, intensively solve the problems concerning low costs for violating laws and high costs for observing laws; fully adopt judicial means to support and strengthen the work on environmental protection.[28] Since the goal of building an ecological province was put forward, Zhejiang has made many achievements in promoting the rule of law in the environmental field; the main experience is as follows.

1. Improving the system of legal laws and regulations

Zhejiang's environmental legislation is incomplete and has many problems. Since 2003, in order to promote the building of an ecological province and protect the ecological environment, Zhejiang has developed and revised more than 40 relevant local regulations and rules, including the *Regulations of Zhejiang Province on Air Pollution Prevention and Control*, the *Regulations of Zhejiang Province on Environmental Protection against Radiation from Nuclear Power Plants* and the *Regulations of Zhejiang Province for Forest Management*, thus the local system of laws and regulations has been preliminarily improved.[29]

2. Improving the system of supervision of environmental law enforcement

Zhejiang has established a mechanism of joint environmental law enforcement by the public security organs, procuratorial organs and people's courts; the liaison office of the public security organ for environmental protection has been established in the cities divided into districts and more than 80% of the counties (county-level cities, districts). Zhejiang has actively promoted environmental public interest litigations,

[28]Xi Jinping, *Carrying out Concrete Work to Stay Ahead – Line of Thought and Practice in Promoting New Development in Zhejiang*, The Party School of the CPC Central Committee Press, 2006, p. 202.

[29]Su Xiaoming, Zhejiang's Practice and Innovations in the Building of the System of Ecological Development, *Observation and Ponderation*, 2014(4).

and has made breakthroughs in overcoming difficulties in compulsory law enforcement involving non-litigation cases relating to environmental protection. Zhejiang has brought about innovations to supervision modes and methods, and improved six major environmental law enforcement mechanisms including inter-departmental joint law enforcement, public participation, enterprises' self-supervision, coordinated supervision at boundaries, information disclosure and grid-based law enforcement.

3. Strictly exercising the supervision of environmental law enforcement

Zhejiang has always taken a tough stand towards violations of environmental laws. Since 2008, Zhejiang has investigated and handled 56,000 cases of violations of environmental laws and imposed a penalty of 2.36 billion yuan. In particular, since 2011, Zhejiang has promoted joint environmental law enforcement by the public security organs, procuratorial organs and people's courts; as of late 2013, 1,052 people had been detained administratively and 429 people were detained criminally due to violations of environmental laws, thus effectively curbing the acts which violate environmental laws.

7.3.2.2 The System of Ecological Compensation

The mechanism for ecological compensation is the public system designed to adjust the interest relations among the stakeholders of ecological protection by adopting government and market means in light of the service value of the ecological system, ecological protection costs and the costs of developmental opportunity with a view to protecting the ecological environment and promoting the harmonious development between people and nature.[30] According to the report delivered during the 16th National Congress of the Communist Party of China, a mechanism for ecological compensation is built more rapidly under the principle that whosoever develops it should protect it and whosoever obtains benefits should make compensation.

Zhejiang has various types of ecological functional areas including areas for the conservation of the sources of drinking water, natural reserves, areas for the protection of forests and biodiversity. These ecological functional areas are mostly situated in the underdeveloped areas in southwestern Zhejiang. To take into account the overall situation and protect the environment, the economic development in these areas has been affected, and it is necessary to make ecological compensation for these underdeveloped areas. Ecological compensation was practiced early in various areas of Zhejiang. In particular, since the building of an ecological province, relevant practice and experiments have increased significantly in various areas, and some local governments have issued the corresponding regulations and policies. In the meantime, local authorities have also explored and developed multiple modes for

[30]Chen Jinqi, A Study of Zhejiang's Practice in the Mechanism for Ecological Compensation, Its Significance and the Improvement of Its Strategies, *Journal of the Party School of CPC Hangzhou*, 2010(6).

ecological compensation including the trading of emission rights, the trading of water rights and non-local development, thus greatly enriching the system of ecological compensation.

In 2005, Deqing became the first in Zhejiang to operate the system of ecological compensation and issued the *Implementation Opinions on the Establishment of the Mechanism for Ecological Compensation for Western Towns*, making ecological compensation for the three towns to the south of National Highway no.104 within the county. In 2010, the *Implementation Opinions on Enlarging and Improving the Mechanism for Ecological Compensation* was issued, expanding the scope of compensation and making the compensation area account for 1/3 of the county's total area. To define the usages of funds, Deqing County has established a special fiscal account for ecological compensation funds, and especially issued the *Administrative Measures for the Payment and Use of the Ecological Compensation Funds for Western Towns*. Deqing County has made significant achievements in ecological compensation, a large number of the polluting enterprises in western towns have been shut down, a domestic sewage treatment station with a daily handling capacity of 500t has been built, and the ecological environment and quality of the water in reservoirs have improved significantly.[31]

When working in Zhejiang, Xi Jinping gave great importance to the development of the system of ecological compensation. On April 5, 2005, during the meeting of the leading group for the building of an ecological province, he pointed out that most underdeveloped areas in Zhejiang were ecological functional areas. Recently, the China National Environmental Monitoring Center conducted a survey on the quality of the ecological environment in the cities and counties across the country, suggesting that Qingyuan County in Zhejiang ranked no.1, Jingning, Longquan, Taishun and Yunhe ranked no.5, no.8, no.9 and no.10, respectively. These counties are underdeveloped areas in southwestern Zhejiang, all of which are subject to the dual task of promoting ecological protection and accelerating development. Therefore, establishing and improving the mechanism for ecological compensation must be an important point to put on the agenda. There was also ecological compensation in the past, including payment transfer for the underdeveloped areas and improvement of the compensation standard for ecological forests, but it is essential to gradually intensify relevant work and improve the mechanism. It is necessary to establish a mechanism for ecological compensation, adopt the plan, legislation and market means to address the issues concerning interest compensation from the downstream areas for the upstream areas, from the developmental areas for protection areas, from the areas which obtain benefits for the damaged areas, from the industries at the end for the industries at the source. The focus of fiscal transfer payments should be shifted to regional ecological compensation; special policies have been adopted for the ecologically fragile areas and the ecological protection areas. It is also necessary to study and explore the adoption of a market means for ensuring that various classes in the society make ecological compensation. Of course, the mechanism for ecological

[31]Lai Huineng, *The Development Report on the Building of a Moderately Prosperous Society in China—Deqing Sample*, Red Flag Press, 2014, pp. 144–146.

compensation cannot be improved in a short time, but it is essential to move towards this direction, gradually intensify efforts, and ultimately give shape to a mechanism of virtuous social circulation which caters to the needs of various parties. This should be an important part of efforts in building a harmonious society.[32]

Subject to the requirement of the Party Committee and the People's Government of Zhejiang Province, Zhejiang Province organized experts and bodies to adopt the draft of the *Opinions of the People's Government of Zhejiang Province on Further Establishing and Improving Ecological Compensation* on the basis of an in-depth survey and research as well as the opinions from various parties. Based on solicitation of opinions from 14 provincial departments and preliminary revision of these opinions, the People's Government of Zhejiang Province issued the *Several Opinions on Further Improving the Mechanism for Ecological Compensation* in August, 2005, making Zhejiang the first province nationwide to develop the mechanism for ecological protection compensation. In 2006, in order to protect the areas at the source of the Qiantang River, the People's Government of Zhejiang Province issued the *Interim Measures for a Special Provincial Fiscal Subsidy for the Ecological Environmental Protection in the Areas at the Source of the Qiantang River*. With experiments of transfer payment for many years, the People's Government of Zhejiang Province issued the *Trial Measures of Zhejiang Province for Fiscal Transfer Payment for Ecological Environmental Protection* in 2008,[33] initiating ecological compensation for the whole river basin and fiscal transfer payment concerning ecological environmental protection for 45 cities, counties (districts) at the sources of eight major river basins.

Zhejiang's practices in and innovations to the mechanism for ecological compensation are as follows.

1. Making ecological compensation in multiple ways

Different compensation modes should be adopted for different areas. With many years of practice, Zhejiang has explored and developed a number of different ecological compensation modes. For the ecological functional areas at the upstream section of the river basin, non-local development has been fostered, preferential systems and policies have been introduced to build the poverty alleviation-oriented economic development zones for compensation. For instance, Jinhua City has built the Jinpan Poverty Alleviation-oriented Economic & Technological Development Zone as the development area for Pan'an County which is a source area, and has adopted a series of supporting policies. In the ecological protection areas and islands in the southwestern mountain land and forest land, the modes of "migration from mountains to shake off poverty" and "construction on large islands and migration from small islands" have mainly been adopted; a certain amount of economic compensation and relevant preferential policies have been provided to the people involved in ecological

[32]Xi Jinping, *Carrying out Concrete Work to Stay Ahead – Line of Thought and Practice in Promoting New Development in Zhejiang*, The Party School of the CPC Central Committee Press, 2006, p. 195.

[33]Deng Ai'lin, Zhao Jingjing, The Suggestions on China's Mechanism for Ecological Compensation – Take Zhejiang Province's Experience as an Example, *Management and Administration*, 2014(1).

migration; the project for quality training concerning the rural labor force of 10 million people has been widely carried out to strengthen employment training and make them overcome poverty and become rich. Regional ecological compensation is made at the upstream and downstream sections of some river basins and rivers, a certain amount of economic compensation is provided from the downstream section to the upstream section so as to guarantee the quality of water. For instance, Fucun Town at the downstream section of Jindong District, Jinhua City annually provides 50,000 yuan to Yuandong Town at the upstream section with a view to protecting the quality of the water at the upstream section.[34]

2. Holding ecological environmental destroyers to account

Zhejiang has strengthened its monitoring of pollutant emissions, and intensified the punishment for acts of excessive and secret pollutant emissions. Zhejiang has reinforced its evaluation of the quality of the water at the interface cross sections, and strictly followed the requirements of the *Measures of Zhejiang Province for the Management and Evaluation of the Quality of Water Protection at the Interface Cross Sections of the Rivers across Administrative Areas* to take effective measures to punish the parties responsible for the failure of the quality of the water to reach the relevant standard, and order them to make the quality of the water at the interface cross sections reach the relevant standard within the specified time.

3. Exploring the market-oriented system of ecological compensation

Zhejiang has combined the system of compensated use of resources with the system of ecological compensation, collected a compensation fee for mineral resources and the fee for water resources, built the system of reserve funds for improving the natural ecological environment at mines, which is used to protect the environment at mines; established the system of compensation funds for forest ecological benefits under which certain funds are allocated from provincial finances each year to make compensation for the projects of ecological public-welfare forests; Zhejiang has experimented a mechanism for the compensated use of water resources under which the win-win outcome—quota trading and water rights trading—is achieved; it has carried out the system for compensated use and the trading of emission rights, introduced a market mechanism to promote the reduction of pollutant emission costs and environmental protection.

4. Establishing a multi-channel financing mechanism

The ecological compensation funds cannot be merely based on fiscal support and should be made available by expanding financing channels and establishing a multichannel financing mechanism. Changxing County, Zhejiang has made attempts in this regard. The control over construction rights and operational rights is loosened to encourage social capital to enter the public utility field, address a shortage of funds in public utility, and make improvements in the quantity and quality of public services.

[34]Miao Kun, Jiang Ni, Harmoniously Promoting Policy Innovation to Lead Practice, Coordinate and Improve Mechanisms – A Survey on Zhejiang's Practice in Ecological Compensation, *Environmental Economy*, 2008(8).

7.3.3 The System of Severe Punishment of Bad Consequences

7.3.3.1 The Accountability System for Ecological Environmental Protection

When working in Zhejiang, Xi Jinping started practical experiments in the accountability system for ecological environmental protection. In 2004, the Party Committee and the People's Government of Zhejiang Province issued the *Measures of Zhejiang Province for Holding the Illegal Environmental Acts to Account (Trial)*, specifying the environmental protection responsibility and requiring serious investigation and handling of the cases involving violations of environmental laws. In 2014, Zhejiang put forward the lifelong accountability system for ecological environmental damage, under which all of the personnel responsible for severe ecological environmental damage caused by blind decision-making are held to account accordingly, whether or not they are at their posts.[35] Zhejiang has made new progress in the accountability system for ecological environmental protection. Zhejiang's main innovations over the years are as follows.

1. Holding to account the officials making mistakes

The *Measures of Zhejiang Province for Holding the Illegal Environmental Acts to Account (Trial)* specifies the environmental protection responsibility of the leaders of local Party committees and governments and identifies five main illegal environmental acts of officials: (1) violating national industrial policies or running counter to industrial developmental orientation, violating the environmental resource protection plans for areas or river basins, causing environmental deterioration or ecological damage; (2) letting illegal environmental acts go unchecked, covering up and indulging illegal environmental acts, or not taking actions to solve or improperly dealing with the environmental pollution problems which prompt strong reactions from the society; (3) not promptly and properly handling the major environmental pollution accidents at the local level, causing more losses, or hindering the competent department for environmental protection from reporting according to regulations; (4) obstructing, restricting the competent department for environmental protection from performing environmental management, investigating and handling the illegal environmental acts according to laws, resulting in the competent department for environmental protection failing to perform its duties according to laws; (5) inspiring, instigating, ordering the competent department for environmental protection to approve the unqualified construction projects in violation of laws, or releasing the unqualified pollution prevention and control facilities during the acceptance check. The main personnel responsible for these acts are held to account.

[35] Liao Xiaoqing, Lin Lixian, and Sheng Xiangliang, Zhejiang Will Establish a Lifelong Accountability System for Ecological Environmental Damage, *Zhejiang Daily*, August 1, 2014.

2. Holding to account the department of supervision for inadequate supervision

The *Measures of Zhejiang Province for Holding the Illegal Environmental Acts to Account (Trial)* requires the competent departments for environmental protection and the administrative departments for environmental supervision at various levels to strengthen their supervision, and hold the units responsible for the following seven weak supervision acts to account: approving the environmental impact assessment report (report form) in violation of laws, falsely reporting, concealing, refusing to report environmental pollution, ecological damage accidents or not promptly dealing with accidents, releasing, in an acceptance check, the unqualified pollution prevention and control facilities in construction projects or the projects designed to make improvements within the specified time but failing to finish the tasks according to regulations. Besides the circulation of a notice of criticism against units, the personnel in charge who assume the main leadership responsibility and other personnel who are directly responsible are subject to punishment at different levels according to the severity of the circumstances, including an inner-Party warning, removal from Party posts, administrative demotion or even removal, expulsion from the Party, and administrative dismissal.

3. Holding to account the examination and approval departments which are not strict with examination and approval

The *Measures of Zhejiang Province for Holding the Illegal Environmental Acts to Account (Trial)* requires the development and reform, construction, planning and other departments to perform their own functions. A notice of criticism is circulated against the responsible departments which approve, release, during an acceptance check, the projects which do not undergo environmental protection examination and approval in violation of regulations. The personnel in charge who assume the main leadership responsibility and other personnel who are directly responsible are subject to punishments at different levels according to the severity of the circumstances, including inner-Party recording of a serious demerit, inner-Party warning and administrative warning, removal from Party posts, administrative demotion or even removal.

4. Holding to account the state organs, groups, enterprises, public institutions which commit illegal environmental acts

The *Measures of Zhejiang Province for Holding the Illegal Environmental Acts to Account (Trial)* specifies the responsibility of state organs, groups, enterprises, public institutions for observing environmental regulations and laws and consciously protecting the environment. According to the measures, in the case of the following six illegal environmental acts, besides an administrative penalty imposed on the responsible units which commit illegal acts, the personnel in charge of the units are held to account: resisting, obstructing the environmental law enforcement activities of the competent department for environmental protection or other departments for the management of environmental supervision, including inspection, survey and acceptance checks, refusing to execute the decision about rectification involving environmental pollution or not finishing the task of making improvements within the specified time by exceeding the time limit without justified reasons. Different punishment measures are taken according to the severity of the circumstances.

5. Holding to account the leaders who indulge their subordinates with making mistakes

According to the *Measures of Zhejiang Province for Holding the Illegal Environmental Acts to Account (Trial)*, where the subordinates make mistakes, the leaders who indulge or cover up their acts are held to account and are subject to punishment at various levels, including inner-Party recording of a serious demerit, inner-Party warning and administrative warning, removal from Party posts, administrative demotion or even removal.

7.3.3.2 The System of Environmental Damage Compensation

Zhejiang has made beneficial experiments in marine ecological environmental damage compensation. Zhejiang carried out the *Administrative Regulations of Zhejiang Province for the Fishery Industry* as from April 1, 2006, introducing the system of compensation for pollution damage for the ecological environment in the fishery industry. With many years of carrying out surveys and studies, in 2013, the Zhejiang Province Oceana and Fisheries Bureau drafted the *Interim Measures of Zhejiang Province for the Administration of Marine Ecological Damage Compensation and Loss Compensation (Draft)*, stating solicitation of public opinions and proposing to introduce the system of compensation for ecological environmental damage across the sea following the fishery industry.

Zhejiang's main practices and innovations in marine ecological environmental damage compensation are as follows.

1. Specifying the principle that compensation must be made in case of ecological environmental damage

According to the *Administrative Regulations of Zhejiang Province for the Fishery Industry*, where the ecological environment in the fishery industry is destroyed and damaged and thus economic loss is incurred to others, the responsibility for making compensation should be assumed; where loss is incurred to national fishery resources, the competent department for the fishery industry above the county level orders compensation. According to the *Interim Measures of Zhejiang Province for the Administration of Marine Ecological Damage Compensation and Loss Compensation (Draft)*, where a marine pollution accident occurs, illegal development and utilization of marine resources causes marine ecological damage within the sea area under Zhejiang's jurisdiction, marine ecological damage compensation should be made; where marine projects cause changes in the marine ecological environment, marine ecological loss compensation should be made.[36]

[36]Both administrative regulations specify the principle that compensation must be made in case of ecological environmental damage.

2. Specifying the departments responsible for claiming compensation

According to the *Administrative Regulations of Zhejiang Province for the Fishery Industry*, the competent department for the fishery industry above the county level orders compensation. According to the *Interim Measures of Zhejiang Province for the Administration of Marine Ecological Damage Compensation and Loss Compensation (Draft)*, the competent department for the ocean and fisheries of the people's government above the county level claims marine ecological damage compensation and loss compensation from the units and individuals who have caused or are causing marine ecological damage and marine ecological loss on behalf of the state.

3. Specifying the limits of jurisdiction of the departments responsible for claiming compensation at different levels

The *Interim Measures of Zhejiang Province for the Administration of Marine Ecological Damage Compensation and Loss Compensation (Draft)* specifies five types of cases which are handled by the Zhejiang Province Oceana and Fisheries Bureau and in which compensation is claimed by the Zhejiang Province Oceana and Fisheries Bureau, including the cases in which damage involves more than two cities divided into districts and causes direct economic loss above 10 million yuan, the survival environment for precious and endangered aquatic wildlife is severely polluted, the ecological functions of the fishery industry partially disappear, and other major pollution accidents in which compensation should be claimed by the Zhejiang Province Oceana and Fisheries Bureau. Other types of cases are handled by the competent departments for the ocean and fisheries at the municipal and county levels.

4. Specifying the usage of compensation

According to the *Interim Measures of Zhejiang Province for the Administration of Marine Ecological Damage Compensation and Loss Compensation (Draft)*, marine ecological damage compensation and loss compensation will be incorporated into the special fiscal account and specially used for ecological environmental restoration, protection, improvement and management of the ocean and the fishery industry.

Chapter 8
Experience and Inspirations from the Building of a Beautiful Zhejiang

Ran Wang

The report delivered during the 18th National Congress of the Communist Party of China incorporates ecological development into the five-sphere integrated plan for building socialism with Chinese characteristics, and calls for vigorously promoting ecological development, building a beautiful China and achieving the sustainable development of the Chinese nation. This means that ecological development has been given a newer and richer connotation, and also suggests that ecological development has become the important means for economic development and environmental protection. Therefore, it is essential to adopt the new philosophy of development established according to the Scientific Outlook on Development, give greater importance to ecological development, environmental protection and coordinated and sustainable development among the economy, society and the population, resources, environment. The Party Committee and the People's Government of Zhejiang Province have followed the objective laws and sized up the situation to take a new path in building an ecological province and a green Zhejiang, thus they have accumulated the Zhejiang Experience which is unique and achieved a huge success. Summing up and promoting the Zhejiang Experience in ecological development is of important theoretical value and practical significance for practicing the Scientific Outlook on Development, vigorously promoting ecological development and achieving the great rejuvenation of the Chinese nation.

8.1 Carrying Through One Blueprint

The successive Secretaries of the Party Committee of Zhejiang Province have actively advocated ecological protection and have given great importance to environmental greening and comprehensive improvement, made great contributions to promoting

R. Wang (✉)
The Zhejiang Academy of Social Sciences, Hangzhou Shi, China

J. Pan and M. Shen (eds.), *Chinese Dream and Practice in Zhejiang – Ecology*, Research Series on the Chinese Dream and China's Development Path,
https://doi.org/10.1007/978-981-13-7209-4_8

Zhejiang's environmental protection and ecological development. The strategic goal of Zhejiang's ecological development can be broadly summarized as four stages: a green Zhejiang → an ecological province → an ecological Zhejiang → a beautiful Zhejiang. The evolution of that strategic goal presents a consistency of policy of carrying through one blueprint and also the progressiveness in connotation and level. This is a model of making innovations amid inheritance. The building of a green Zhejiang, an ecological province, an ecological Zhejiang and a beautiful Zhejiang epitomizes Zhejiang's goals of ecological development in different periods. These goals are expressed in different ways, but the main line is consistent: upholding the Scientific Outlook on Development and the Outlook on Ecological Civilization, properly dealing with the relationships between the "gold and silver mountain" and clean, clear waters and lush mountains, placing equal emphasis on the "gold and silver mountain" and clean, clear waters and lush mountains, achieving sustainable economic and social development, ensuring the people's happiness and well-being. Ecological development is a beneficial project which entails great efforts in unifying the line of thought and massive investments and slowly produces effects, so it is very difficult to carry out the project for a long time if actions are taken to seek instant benefits in a short-sighted way. The admirable spirit of the successive Party Committees and the People's Governments of Zhejiang Province consists in tenaciously persevering in keeping lush mountains, successively working hard and carrying through one blueprint. This is a valuable experience acquired from the building of a beautiful Zhejiang and deserves to be promoted in an all-round way.

8.1.1 Giving Importance to Environmental Protection: The Successive Party Committees and the People's Governments Have Worked Hard Year by Year

Environmental protection is the basic requirement for ecological development. The successive Party Committees and the People's Governments of Zhejiang Province have made great efforts in this regard year by year by proceeding from advocating afforestation to build an ecological province and a green Zhejiang. The successive leaders of Zhejiang Province, including Jiang Hua, Tie Ying, Wang Fang, Xue Ju, Li Zemin, and Zhang Dejiang, have showed great care for and given high attention to environmental protection, they have made contributions to Zhejiang's environmental protection and ecological development in different periods.

When serving as the Secretary of the Party Committee of Zhejiang Province, Xi Jinping developed a well-rounded philosophy of development for turning Zhejiang into an ecological province; he comprehensively expounded the theory of "two mountains"—clean, clear waters and lush mountains and the "gold and silver mountain"; Xi Jinping criticized the practice of only seeking the "gold and silver mountain" in disregard of clean, clear waters and lush mountains, he required that equal emphasis should be placed on the "gold and silver mountain" and clean, clear waters and lush mountains, he predicted that clean, clear waters and lush mountains was the "gold and silver mountain", and he advised that the people should prefer clean, clear waters and lush mountains to the "gold and silver mountain", and that criticism, requirement, prediction and advice constitute the complete theory of the "two mountains". The theory of the "two mountains" is the important guiding thought for ecological development with a beautiful China as the goal. In January, 2003, Zhejiang became the 5th pilot province for building an ecological province in China, suggesting the rapid promotion of the work on comprehensively turning Zhejiang into an ecological province. In March, the *Plan of Zhejiang Province for Building an Ecological Province* was adopted by experts through a feasibility study; Xi Jinping attended and delivered a speech during the meeting. In May, the Party Committee and the People's Government of Zhejiang Province set up Zhejiang's Leading Group for the Building of an Ecological Province, in which Xi Jinping served as the leader. In the same month, he chaired a meeting of the Standing Committee of the Party Committee of Zhejiang Province to discuss and adopt, in principle, the *Plan of Zhejiang Province for Building an Ecological Province*. In June, the Standing Committee of the People's Congress of Zhejiang Province adopted the *Decision on Building an Ecological Province*. In July, when delivering the report during the 4th Plenary Session of the 11th Party Committee of Zhejiang Province, Xi Jinping vowed to further leverage the "Eight Advantages" and push forward the "Eight Measures", and one important aspect consisted of further leveraging Zhejiang's ecological advantages to build an ecological province and a green Zhejiang. In the same month, the Party Committee and the People's Government of Zhejiang Province held the provincial mobilization meeting on building an ecological province, in which Xi Jinping delivered a speech titled "Comprehensively Initiating the Building of an Ecological Province, Striving to Build a Green Zhejiang". In August, the *Plan of Zhejiang Province for Building an Ecological Province*—a programmatic document for guiding Zhejiang to build an ecological province—was officially circulated. Thus, Zhejiang initiated the building of an ecological province. With a great deal of attention, care and guidance from Xi Jinping, Anji became the first nationwide to build an ecological county in China in June, 2006; so far, this is the only national ecological county in Zhejiang.

Afterwards, Zhejiang continued the building of an ecological province. In April, 2007, Zhao Hongzhu chaired the plenary meeting of the Leading Group for the Building of an Ecological Province. In each of several subsequent years, he chaired the plenary meeting of the Leading Group for the Building of an Ecological Province to summarize the previous year's work on building an ecological province and make arrangements for the current year's work on it. During the 3rd Plenary Session of the 12th Party Committee of Zhejiang Province in April, 2008, he vowed to take the perspective of ecological progress to regard "strengthening ecological development and environmental protection, optimizing the living environment" as the important part of efforts to comprehensively improve the people's well-being. Xia Baolong pointed out that it was necessary to follow the requirement of the Central Committee for ecological development and the building of a beautiful China, seize the present great opportunities for speeding up the building of an ecological province, uphold the strategy of building the province through ecological development, focus on the goal and task of building an ecological Zhejiang and a beautiful Zhejiang, observe the objective laws, make a systematic plan, intensify the reforms, strengthen security, be bold in taking responsibilities, make sure that Zhejiang carries out concrete work to stay ahead in building an ecological province.

8.1.2 Strengthening the Building up of Water Conservation: Concretely Carrying Out the Present Work, Stressing Long-Term Development More

Situated in the region of rivers and lakes south of the Yangtze River, Zhejiang has riverways with a total length of more than 60,000 km; Zhejiang is also a region with frequent occurrence of multiple disasters including floods, drought, typhoons and humidity, so the building up of water conservation is particularly important for Zhejiang's agricultural production. The successive Party Committees of Zhejiang Province have given importance to "concretely carrying out the present work, stressing long-term development more" and acted in response to the changing situation, proceeded from advocating the governance of mountains and waters to take water conservation as the important basic industry in the national economy, promote the sustainable utilization of water resources and sustainable economic and social development in Zhejiang.

Xi Jinping showed great care for improving eight main water systems. In the environmental improvement activity for building an ecological province, Xi Jinping conducted surveys and studies at the primary level and guided the work on improvements. In June, 2003, Xi Jinping inspected the work on water conservation and flood prevention in such cities (counties) as Cixi, Shangyu, Haining, Deqing and Yuhang District of Hangzhou City. He pointed out that the building up of water conservation, flood prevention and disaster resistance, the protection, development and utilization of water resources, the governance and effective improvement of the environment of water ecology were always the major tasks of the Party committees and governments at various levels and the crucial issues for speeding up Zhejiang's efforts in building a moderately prosperous society in all respects and basically achieving modernization in advance. In September, 2005, when surveying the ecological development and environmental protection in the Qiantang River basin in Hangzhou and Quzhou, he stressed that the Qiantang River was Zhejiang's mother river, the pollution prevention and control in the Qiantang River basin had a direct bearing on 14 million people in production and life within the basin and the realization of Zhejiang's goal of building an ecological province; the pollution prevention and control in the Qiantang River basin must be at the forefront across the province. In August, 2006, when inspecting the development and improvement of Southern Lake Taihu in Huzhou, he stressed that it was necessary to take the development and improvement of Southern Lake Taihu as a priority in promoting scientific development in relevant areas, as a breakthrough in building a resource-conserving and environmentally friendly society, an important aspect in fostering a correct outlook on officials' performance, persistently and solidly pushing forward relevant work to better develop and improve Southern Lake Taihu.

Zhao Hongzhu gave great importance to the prevention and reduction of disasters. In June, 2007, during the 12th Party Congress of Zhejiang Province, he emphasized that it was necessary to intensify the protection and development of the areas at the source of river water systems and important ecological functional areas. In case of a typhoon, he would stay at the provincial flood control and drought relief headquarters to directly command disaster relief efforts, express sympathy and solicitude and guide relevant work in disaster areas in order to protect the safety of the people's life and property. After the Super Typhoon Sepat struck Zhejiang in August, 2007, he vowed to carry out the Seawall, Bank and House Consolidation Project and gave a great deal of attention to promoting the work in this regard. In April, 2008, during the 3rd Plenary Session of the 12th Party Committee of Zhejiang Province, he stressed that it was necessary to focus on the Seawall, Bank and House Consolidation Project to further strengthen the construction of disaster prevention and reduction infrastructures and the shelters for averting disasters, thus enhancing the capability for resisting natural disasters.

Xia Baolong stressed the necessity for combining environmental protection with ecological development. He pointed out, "At present, the Central Committee is giving great importance to environmental protection and ecological development, the people have very high expectations about environmental development and Zhejiang is a southern coastal economically developed province; based on the great amount of attention from the successive Party Committees and the People's Governments of Zhejiang Province given to ecological development, the Party Committee and the People's Government of Zhejiang Province have introduced new measures for resolving the current problems and also eliminating the root causes, including "three renovations and one demolition" and "governing five waters in tandem"; these efforts have provided the favorable climatic, geographical and human conditions for us to speed up the building of an ecological province; as opportunities are fleeting, we must seize the favorable situation to comprehensively push forward the building of a beautiful Zhejiang, focus on governing waters, make comprehensive improvement in our water, air and land and really bring about lush mountains and clean, clear waters."

8.1.3 Developing a Green Economy: Valuing the "Gold and Silver Mountain", Placing More Emphasis on Clean, Clear Waters and Lush Mountains

A green economy is essential for ecological development. The successive Party Committees and the People's Governments of Zhejiang Province have valued the "gold and silver mountain" and placed more emphasis on clean, clear waters and lush mountains; they have proceeded from pollution control to develop the circular economy and take a path, with Zhejiang's characteristics, for developing a green economy. During the early stage of reform and opening-up, with rapid economic development, environmental problems became increasingly salient in Zhejiang. Besides pushing forward economic development, the Party Committee and the People's Government of Zhejiang Province gradually gave more importance to and increasingly strengthened the leadership over ecological and environmental protection, they developed such ideas as "enhancing the awareness about environmental protection, controlling environmental pollution, protecting and rationally utilizing natural resources to gradually improve the ecological environment" and "taking pollution prevention and control as part of efforts in promoting modernization and the strategy of sustainable development ", so the trend of environmental deterioration in Zhejiang has been basically controlled and the quality of the environment in some areas has improved to some extent.

Xi Jinping developed the developmental line of thought for a circular economy. In order to control pollution, it is necessary to repay the "old debts" and not incur "new debts", and developing a circular economy is the best choice for doing that. Therefore, under Xi Jinping's advocacy and organization, Zhejiang orderly pushed forward a circular economy and took it as a central part of the efforts in building an ecological province; in 2005, Zhejiang fully initiated the work on developing a circular economy and building a conservation-minded society. In April, 2005, when attending the meeting of Zhejiang's Leading Group for the Building of an Ecological Province, he stressed that it was necessary to more rapidly develop a circular economy and promote sustainable development of resources and environment. In May, he chaired the meeting of the Leading Group for Financial and Economic Affairs under the Party Committee of Zhejiang Province and especially listened to the work report on Zhejiang's development of a circular economy. In June, the Party Committee and the People's Government of Zhejiang Province held a special provincial conference on the circular economy, in which he delivered a speech entitled "Earnestly Developing a Circular Economy, Actively Exploring a New Path of Scientific Development". Immediately after the conference, the Party Committee and the People's Government of Zhejiang Province set up Zhejiang's Leading Group for Developing a Circular Economy and Building a Conservation-minded Society in which Xi Jinping served as the leader. In August, the People's Government of Zhejiang Province unveiled the *Plan of Zhejiang Province for the Development of a Circular Economy*, specifying clear requirements for various tasks relating to the development of a circular economy in Zhejiang within a certain period of time.

Zhao Hongzhu continued to give great importance to and solidly push forward the development of a circular economy, making new progress. In December, 2007, with the approval of the State Council, Zhejiang was included in the second batch of national pilot provinces for a circular economy by the National Development and Reform Commission. According to the *2010 National Ecological Environmental Assessment Report* released by the China National Environmental Monitoring Center, Zhejiang's Ecological Environment Index was excellent, Zhejiang stayed ahead nationwide for many consecutive years. In June, 2010, according to the strategic requirement put forward during the 17th National Congress of the Communist Party of China, for ecological development, the 7th Plenary Session of the 12th Party Committee of Zhejiang Province earnestly summarized the experience in building an ecological province and became the first nationwide to make the *Decision on Promoting Ecological Development*. The Secretary of the Party Committee of Zhejiang Province, Zhao Hongzhu, stressed that it was necessary to uphold the strategy of building an ecological province, take the path of building the province through ecological development, build an ecological Zhejiang which is affluent, beautiful, harmonious and peaceful, and turn Zhejiang into the national ecological development demonstration area.

Xia Baolong stressed that it was essential to place equal emphasis on environmental improvement and ecological development. He vowed to concurrently promote environmental protection, transformation and upgrading and turn ecological advantages into economic advantages; he also stressed that it was essential to combine the visible hand with the invisible hand, properly deal with the relations between government acts and market acts, lay equal stress on institutional improvement and cultural development, provide a guarantee through rigid institutions and external constraints, offer support through a high degree of ecological development and internal cultural self-consciousness to ensure actual effects in building an ecological province. The governments at various levels should enhance the sense of responsibility, duly take responsibilities for preserving the land and fully complete the tasks for building an ecological province.

8.1.4 Optimizing the Ecological Philosophy: If Ecological Development Is Achieved, Progress Will Be Made

The leading ecological philosophy is the key to ecological development. The successive Party Committees of Zhejiang Province have stressed that if ecological development is achieved, progress will be made; if ecological degradation occurs, retrogression will come, the "gold and silver mountain" should be valued and more emphasis should be placed on clean, clear waters and lush mountains. Zhejiang's ecological philosophies have been continuously optimized and developed in practice, from the planned "Oriental Geneva" to "eternal existence and sustainable utilization of lush mountains", "taking the path of comprehensive development, intensive operations, deep processing, multiple enhancements, ecological balance and benign circulation" to "eliminating barren mountains within five years, greening Zhejiang within ten years".

Based on the work of the successive Party Committees of Zhejiang Province, Xi Jinping earnestly carried out the guiding principles adopted by the Central Committee of the Communist Party of China, proceeded from Zhejiang's reality, fully respected the creative practice of the people of Zhejiang, drew upon domestic and foreign beneficial practices and experience in environmental protection and ecological development, gradually developed the philosophy of ecological development with Zhejiang's characteristics.

First, stressing that if ecological development is achieved, progress will be made; if ecological degradation occurs, retrogression will come. This was often mentioned by Xi Jinping when he emphasized the significance of ecological development. In his view, for Zhejiang, attaching great importance to sustainable development, resource and ecological issues, developing a circular economy and building an ecological province have a vital bearing on the overall situation, the future and the people's well-being.

Second, stressing that the "gold and silver mountain" should be valued and more emphasis should be placed on clean, clear waters and lush mountains. This was often mentioned by Xi Jinping when he explained the programs of ecological development. In his opinion, realizing that clean, clear waters and lush mountains are the "gold and silver mountain" and that the environment itself can bring wealth is a higher realm and reflects the essential requirement of the Scientific Outlook on Development.

Third, stressing that if early actions are taken in regard to environmental protection and ecological progress, more achievements will be made with less effort; otherwise, fewer achievements will be made with greater efforts; if actions are taken too late, the situation will become more passive. This was often mentioned by Xi Jinping when he commented on the attitude toward ecological development. In his view, we cannot allow the developmental mode which only takes into account short-term interests in disregard of long-term interests, making treatment after pollution and taking restoration actions after destruction.

Fourth, stressing that environmental protection is equivalent to the protection of wealth. This was often mentioned by Xi Jinping when he analyzed the benefits of ecological development. In his view, destroying the ecological environment disrupts the productive forces; protecting the ecological environment protects the productive forces; improving the ecological environment develops the productive forces.

Fifth, stressing that as the economy develops further, greater importance should be given to environmental protection and ecological development. This was often mentioned by Xi Jinping when he intensified efforts in persistently promoting the work on building an ecological province. In his view, a government which does not value ecology is a government which lacks vision; a leader who does not value ecology is an incompetent leader; an enterprise which does not value ecology is a futureless enterprise; a citizen who does not value ecology is a citizen who does not have any awareness regarding modern civilization.

Zhao Hongzhu continued giving importance to environmental protection and ecological development. He pointed out that a good ecological environment was the foundation for survival and development and the basic element of productive forces. He stressed that it was necessary to strictly enforce laws, strengthen environmental law enforcement, take a tough stance and apply strict measures where necessary, make sure that the people across the province can breathe fresh air, drink clean water, eat safe foods, and enjoy a beautiful natural landscape and a comfortable living environment. In order to put the philosophies of ecological development into action, Zhao Hongzhu conducted surveys in Huzhou City, Shangyu of Shaoxing City, Xinchang,

Fenghua of Ningbo City, and Yuyao. To chart the blueprint for Zhejiang's ecological development during the next stage, he took charge of the key research subject "A Study of Countermeasures and Lines of Thought for Zhejiang's Efforts in Promoting Ecological Development", and he also proposed holding a special study meeting of the Party Committee of Zhejiang Province to extensively carry out the guiding principles adopted by the Central Committee of the Communist Party of China, continuously expand the line of thought and measures for speeding up the transformation of the economic developmental mode and promoting economic transformation and upgrading, earnestly study the overall requirements, goals and tasks and safeguard the measures for ecological development.

Xia Baolong stressed that it was necessary to comprehensively expand ecological development, build a beautiful Zhejiang, place equal emphasis on environmental improvement and ecological development, take strong actions to tackle the problems, adopt the zero tolerance policy towards environmental pollution, seek good solutions to ensure continuous improvement and constantly increase the environmental capacity. He pointed out, "Once the local environment is improved, it is equivalent to the planting of phoenix trees, so golden phoenix will surely come; it is imperative to speed up transformation and upgrading, make improvements to force transformation, enhance the industrial quality; improving the quality of life is also aimed at seeking development better."

8.2 Harmonious Environmental, Economic and Social Development

According to the working policy of resolving the current problems and eliminating the root causes, maintaining stability and promoting adjustment, the Party Committee and the People's Government of Zhejiang Province have firmly kept to the important position of environmental protection in the transformation of the method of achieving economic development, they have effectively given scope to the guiding and driving role of environmental protection in macro control, vigorously promoted economic transformation and upgrading and introduced new lines of thought and measures for boosting the integrative development of the environment and the economy. The Party Committee and the People's Government of Zhejiang Province have identified the important positioning of environmental protection in maintaining growth and promoting transformation, conformed to the characteristics of the stages of of economic and social development, they have placed equal emphasis on environmental protection and economic development, actively explored and practiced the ways to achieve optimal development of environmental protection, continuously summarized and explored the line of thought and measures for pushing forward the integrative development of the environment and the economy, thus greatly promoting harmonious environmental, economic and social development.

8.2.1 *Adopting Institutional Means to Strengthen the System of Environmental Access*

The new "three-in-one" system of environmental access with a combination of two evaluations is a great innovation in Zhejiang's environmental protection; it is the important means for enabling the optimal development of environmental protection, participating in comprehensive decision-making involving economic, social and environmental issues; moreover, it has effectively played the role of regulation and optimization in practice, and greatly ensured that the environmental protection department has more say in and makes greater contributions to the transformation of manner of development. The system of environmental access covers spatial access, total quantity access and project access, while a combination of two evaluations means a combination of evaluation by experts and public evaluation.

The management of environmental space, that of the industrial environment and that of the construction projects have really promoted the optimization and upgrading of economic layout, industrial structure and production technology and have prevented and controlled environmental pollution at the source. A combination of two evaluations is a great innovation to the mechanism for the assessment of the environmental impact and is also an important guarantee for carrying out the "three-in-one" institutional system as well as an institutional guarantee for really putting civic rights in place. Only when both evaluations are combined can the fairness, openness and scientificity of macro decision-making be guaranteed. At the operational level, it is necessary to carry out continuous practical experimentations, give play to the guiding role of institutions, and actively build a scientific, well-functioning environmental management system which integrates spatial environmental access, planning and environmental impact assessment, project examination and approval, technical support and examination and approval supervision.

8.2.2 *Taking Science and Technology as Support to Promote Scientific Environmental and Social Development*

Scientific and technological advancement is the fundamental impetus for developing the environmental protection programs. Reinforcing environmental scientific and technological innovation is essential for improving the work on environmental protection in the new period, increasing the level of environmental management and promoting the historic transformation of the work on environmental protection; it is urgently needed for addressing the structural, composite and compressive environmental issues; it is the important guarantee for achieving the targets of constraint concerning emission reduction; it is also the important way to cope with resource and environmental factor constraints, the challenges from science and technology and market competition.

First, intensifying scientific research and technology application relating to environmental protection. Zhejiang has given full scope to the role of the public scientific and technological innovation service platform for environmental protection in resource integration, it has built a mechanism of interest and risk sharing, has better cooperated with and relied on supporting units of environmental protection technology, scientific research institutions and the institutions of higher learning to accelerate environmental scientific and technological innovation and it has effectively carried out technology promotion, technical consulting, technical training and demonstration projects. In the meantime, Zhejiang has built a number of characteristic environmental protection research and development centers with expertise and turned them into the main forces in scientific and technological innovation relating to environmental protection; it has supported a batch of key scientific and technological projects for environmental protection to make great technological breakthroughs in the control of water pollution, air pollution and soil pollution and sludge treatment; it has promoted a number of advanced appropriate technologies to strengthen connection with technical needs and to increase the level of the application of technology to environmental protection at the primary level.

Second, reinforcing the development of the system of environmental standards. Standards embody scientific and technological advances and represent a certain technical level; they reflect the level of environmental protection and ecological development in a country and territory. It is necessary to conform to the environmental protection requirements of regional economic and social development and the people, adopt the "stepped" environmental standards and improve environmental standards in good time. On the one hand, Zhejiang has proceeded from its current industrial agglomeration and developmental direction to make and revise emission standards and access conditions involving the common, sensitive toxic organic pollutants with regional characteristics in such industries as chemical and pharmaceutical industries; on the other hand, with a focus on consolidating the technical foundation for environmental protection, Zhejiang has intensified its efforts to make environmental protection standards which play an important technical supporting role, including the guide rules for the best practicable technologies, technical specifications for environmental protection engineering and clean production standards.

Third, upgrading and improving environmental protection equipment and pollution control facilities. The monitoring system is the "eyes" of the environmental protection system; Zhejiang has continuously increased the coverage and accuracy of monitoring, strengthened and improved the building of water, air and soil environmental monitoring systems, including conventional automatic surface water and air monitoring systems. In recent years, Zhejiang has carried out the building of monitoring systems, including the system of automatic monitoring of the quality of the water at the sources of drinking water, the network of three-dimensional monitoring and early warning of combined air pollution, the system of automatic monitoring of the quality of the water at interface cross sections of the rivers across administrative divisions; Zhejiang has also promoted the building of domestic garbage disposal, domestic sewage treatment and hazardous waste disposal facilities, improved the system of the network of urban sewage pipelines, it has conducted upgrading and renovation in sewage treatment and relied on contiguous improvement of the rural environment to more rapidly increase the level of rural sewage treatment and garbage disposal. Moreover, Zhejiang has also thoroughly carried out the *Plan of Zhejiang Province for Digitization and IT Application in Environmental Protection*, established and improved the three-in-one IT-based control system and environmental information security guarantee system covering environmental quality, pollution source and environmental management.

8.2.3 Taking the Scientific Outlook on Development as a Guide to Promote Harmonious Environmental and Social Development

As early as 2004, Xi Jinping pointed out, "It is necessary to take "whether the line of thought regarding development and the strategy of development are correct, the relationships between quality and quality, between speed and benefit are properly handled" as an important consideration in evaluating the leading cadres as to whether they have fostered a correct outlook on performance. In evaluation, emphasis is placed on economic, social, humanistic and environmental indicators; there is a shift from the mere pursuit of speed to a comprehensive evaluation of the speed of growth, the level of employment, educational input and environmental quality.[1]" According to this requirement, Zhejiang has optimized its philosophy of development and thoroughly practiced the Scientific Outlook on Development to achieve harmonious environmental and social development.

First, improving the system of the evaluation of cadres. Zhejiang has built a system of indicators for the evaluation of cadres with Zhejiang's characteristics which tallies with Zhejiang's reality, and it has developed different evaluation indicators in light of different regional characteristics, fully leveraged the local advantages, tailored evaluation measures to industrial, agricultural and ecological development, so that various areas in Zhejiang have been completely freed from the constraint of GDP-based evaluation, and thus have embarked on the new path for harmonious ecological economic development. Since 2006, the Department of Organization under the Party Committee of Zhejiang Province has issued the *Implementation Measures of Zhejiang Province for Comprehensive Evaluation of the Leading Groups and the Leading Cadres of the Party and the Government of Counties (County-level Cities, Districts) (Trial)*, further improving the system of indicators for the evaluation of cadres and guiding the formation of the Scientific Outlook on Development and the Outlook on the Performance of the Officials.

Second, incorporating the ecological environmental indicators into the cadre evaluation system. In order to protect the ecology and build an ecological Zhejiang and a beautiful Zhejiang, the Department of Organization under the Party Committee of Zhejiang Province has incorporated the ecological environmental indicators into the

[1] Xi Jinping, *Zhijiang Xinyu*, Zhejiang People's Publishing House, 2007

system of evaluation of cadres and has carried out the accountability system and the one-vote veto system. In the evaluation of the cities and counties, 11 prefecture-level cities in Zhejiang have been classified into three categories in light of the local reality and the level and speed of economic development. For the areas at a high level of economic development, evaluation focuses on economic structural adjustment, scientific and technological innovation and social services; for the areas unsuitable for developing an industrial economy, evaluation focuses on ecological development and environmental quality; for the areas which fail to meet the requirements of indicators of an ecological environment assigned by the Party Committee and the People's Government of Zhejiang Province, a notice of criticism is circulated, the granting of relevant honors and titles under evaluation is voted down; the projects which violate regulations or exceed relevant standards should be rectified within the specified time; if rectification is not duly put in place, relevant responsible personnel will be held to account. In the key ecological functional areas, GDP-based evaluation has been canceled, and evaluation focuses on ecological protection, ecological economy and the people's living standard, with a view to promoting ecological development in these areas.

8.2.4 Seeking Impetus from Innovations to Achieve Leapfrog Environmental and Social Development

The conventional mode of development has led to a craze for GDP growth and increasing contradictions between the environment and the economy, between people and nature; the innovative mode of development can promote rapid and steady economic and social development and ensure that the environment and the economy, the people and nature reinforce each other and develop together.

First, bringing about innovations to people's mindset. The innovation to the mindset must be guided by the Scientific Outlook on Development; regarding the innovation to the mindset concerning ecological development, it is essential to first answer these questions: For whom is development pursued?—The overwhelming majority of the people; What is the objective of development?—The happiness of the overwhelming majority of the people. The ecological environment is the basic condition for the people's survival; the ecological and environmental rights and interests are the people's basic rights and interests; the ecological and environmental quality has a direct bearing on the quality of the people's life. Jiaxing City has seized the opportunity from the mechanism involving the compensated use and the trading of emission rights, and has adopted the environmental protection system to "squeeze out" the investment projects with high input, high consumption, heavy pollution and low returns and it has introduced the investment projects with low consumption, less pollution, high technology and high returns, so Jiaxing City has preliminarily overcome the dilemma caused by an extensive developmental mode.

Second, scientific and technological innovation. Science and technology determines environmental costs and benefits, while environmental costs and benefits determine the method of economic development. The scenarios including "ecological but uneconomical", "circular but uneconomical" and "low-carbon but uneconomical" are caused by scientific and technological underdevelopment. The developmental mode which is both "ecological and economical", "circular and economical" and "low-carbon and economical" must rely on ecological scientific and technological innovation. The upgrading and transformation of the traditional industries is an important task, so it is essential to earnestly push forward the research on, development and promotion of ecological technology in the traditional industries. Zhejiang has energetically pushed forward the research, development and promotion of new and high technologies, occupied the commanding height in green, circular and low-carbon science and technology and made them become the new growth points for industrial development. From online monitoring to pollution emissions based on swiping cards, from single tests to the building of a system of testing, Zhejiang has stayed ahead nationwide.

Third, institutional innovation. Ecological development calls for synergy among the government, enterprises and the general public, and for coordination among the government mechanism, the market mechanism and the social mechanism, and for a scenario in which these three forces work together and complement each other's advantages to bring about a joint force. In improving the mechanism of evaluation, Zhejiang has adopted differentiated modes for the evaluation of the performance of the officials according to the functional orientation of different players in different areas so as to effectively stimulate the supply of public goods, including ecological environmental development. In the meantime, Zhejiang has introduced a mix of the compulsory policies including total quantity control, the selective policies for weighing the advantages and disadvantages including the trading of emission rights, moral education-focused guiding policies including environmental protection education. On optimizing the market mechanism, Zhejiang has accumulated successful experience in carrying out a system for the compensation of ecological protection, a system for the trading of water usage rights trading system, a system for the trading of emission rights, a system for a ban on fishing, a system involving the compensated use of mining rights and a mechanism of third-party governance. Regarding the promotion of innovation to the social mechanism, it is necessary to, based on public participation in supervision over environmental protection, build the most extensive united front in environmental protection, give full play to the role of scientific research institutions, the institutions of higher learning, experts and scholars as think tanks so that necessary intellectual and technical support is provided for environmental protection, give full scope to the roles of non-governmental organizations for environmental protection and volunteers to push forward the work on the safeguarding of rights in environmental protection, environmental protection publicity and education, environmental improvement and give full play to the roles of letters and visits relating to environmental complaints, prize-based reporting of offences, appraisal and review of industrial tendencies, thus ensuring better public participation in environmental protection and supervision over environmental protection, converting the people's willingness, passion and wisdom into concrete actions for joint ecological development and for the sharing of an ecological civilization.

8.2.5 Strengthening the Planning of Functional Areas, Optimizing the Pattern of Land Development

Xi Jinping pointed out, "land is the spatial carrier of ecological development; it is necessary to follow the principle of 'achieving equilibrium among the population, resources and the environment, concurrently stressing economic, social and ecological benefits' to holistically plan the development of land spaces, scientifically arrange production space, living space and ecological space to leave more space for national restoration; it is necessary to unswervingly speed up the implementation of the strategy for the main functional areas, determine and strictly keep the ecological red line according to the main functional orientation of optimized development, focused development, restricted development and prohibited development, build a pattern of urbanization that is scientific and rational, a pattern of agricultural development and one of ecological safety, guarantee national and regional ecological safety and improve ecological service functions. It is essential to foster an awareness of the ecological red line. On protecting the ecological environment, transgressing the bounds is not permitted; otherwise, punishment should be imposed.[2]" Since the reform and opening-up, Zhejiang has made huge achievements in economic and social development, but the lack of coordination in regional development, the extensive growth mode and the increasing pressure on resources and on the environment have become increasingly salient. Planning the main functional areas in a scientific, well-regulated and orderly way is of very important significance. The planning of the main functional areas embodies the philosophy of seeking people-oriented development, pursuing coordinated urban and rural development, harmonious development of people and nature, and it is an important measure for carrying out the Scientific Outlook on Development and building a harmonious socialist society. Planning the main functional areas, following the different requirements of optimized development, focused development, restricted development and prohibited development to identify the main regional functions and develop a differentiated regional policy of development is an effective way to optimize the regional layout and promote coordinated regional development. Planning the main functional areas, proceeding from the carrying capacity of resources and the environment to guide the economic layout and population distribution, reduce massive cross-regional allocation of water, energy and other resources, promote spatial equilibrium among the population, the economy, resources and the environment is an urgent need for safeguarding the natural ecosystem and improving the efficiency of resource utilization. Planning the main functional areas is conducive to shaping a regional pattern of development in which functional positioning and development orientation are clear, there is a good order of

[2]Xi Jinping, *Xi Jinping: The Governance of China*, Foreign Languages Press, 2014, p. 209.

development, the intensity of development is appropriate and there is coordination between economic and social development and the population, resources and the environment, and it is the main carrier for coordinating various spatial plans and special plans, thus enhancing the efficiency of the utilization of the spaces and the resources.

8.2.6 Earnestly Building a Conservation-Minded Society

Xi Jinping pointed out, "Saving resources is the fundamental policy for protecting the ecological environment; it is necessary to actively promote an economical and intensive utilization of resources, push forward the fundamental transformation of the manner of utilizing resources, strengthen the management of whole-process conservation, greatly reduce the intensify of energy, water and land consumption, energetically develop a circular economy, promote reduction, reuse and recycling during production, circulation and consumption.[3]" Zhejiang has taken the reduction of energy consumption and the conservation of resources as important indicators of economic and social development, and incorporated the building of a conservation-minded society into various fields of economic and social development. First, adjusting the economic structure. In the production field, it is necessary to build a resource-conserving agricultural production system focusing on land and water conservation, venture down a new path towards urbanization, develop an industrial structure conducive to the sustained utilization of resources and environmental protection and promote the virtuous circulation of economic and social development. Regarding consumption, it is necessary to earnestly advocate the concept and behavior of rational and moderate consumption, and make efforts to make sure that the people in the whole society consciously save energy, water, materials and grain, and that they classify and collect garbage, reduce the use of disposable goods, and that they thus gradually develop a civilized conservation-minded manner of consumption. In the field of urban construction, it is necessary to fully consider resource conditions and the environmental carrying capacity, promote economical and intensive utilization of land, freshwater, energy and other resources, strictly control construction land, build a conservation-minded system of residential building and transportation. Second, strengthening scientific and technological innovation. Scientific and technological innovation is the key to building a conservation-minded society; it is necessary to vigorously promote the existing technologies to make them really useful, and conduct systematic integration and comprehensive innovation of the introduced technologies. In the meantime, it is necessary to give full scope to the crucial role of talents in technical innovation, accelerate the transformation of scientific and technological achievements towards productive forces, and as early as possible make economic development really rely on scientific and technological advancements and the improvement of the quality of the workers. Third, intensifying reforms. Reform

[3] Xi Jinping, *Xi Jinping: The Governance of China*, Foreign Languages Press, 2014, p. 209.

is the impetus for building a conservation-minded society; it is necessary to give full play to the role of the market mechanism and economic lever, adopt the price, fiscal and tax and financial means to promote the conservation and effective utilization of resources, eliminate the institutional factors leading to an extensive mode of economic growth in a targeted way, intensify the reforms and make institutional innovations to convert resource conservation into an impetus for development, so that those who conserve resources will obtain more benefits and opportunities in market competition, those who waste resources will bear more costs and pay a higher price. Fourth, strengthening supervision. Supervision is the guarantee for building a conversation-minded society; it is necessary to make and improve the laws and regulations for promoting economical and effective utilization of resources, establish and improve various regulations and systems, uphold scientific and strict management, resolutely change various ways which cause the wasting of resources. It is necessary to develop stricter standards of conservation, build a compulsory elimination system, improve the market access system, and build a system for the evaluation of resources for new construction projects. It is also necessary to further intensify law enforcement relating to resource protection and conservation, seriously investigate and handle the illegal and irregular behaviors which destroy and waste resources. Fifth, making sure that government departments play an exemplary role. Government departments have a huge potential for conservation, they should consciously set a good example in conserving resources. The leading cadres at various levels and all of the civil servants should set an example, practice strict economy and combat waste; it is necessary to carry out a change to the state of energy saving in government buildings and office system, to boost the conservation of energy regarding official vehicles, promote the conservation of energy in government departments, build a scientific system for the evaluation of the performance of the officials, incorporate the responsibility for the conservation of resources and the actual results into the system of target responsibility and that of the evaluation of the cadres of the governments at various levels and produce a good demonstration effect on building a conservation-minded society.

8.3 Joint Governance by the Government, the Market and the Society

With a focus on the burdensome historical task of building a moderately prosperous society that is beneficial to the people across the province in all respects, Zhejiang has upheld the strategy of building an ecological province, undertaken the path of building the province through ecological development, continuously intensified, broadened and innovated the line of thought regarding and the measures for building an ecological province and actively pushed forward ecological development, thus the mechanism of ecological development and environmental protection has been further improved, and thus a good pattern of joint governance and management, co-building and sharing by the whole society has taken shape. Each year the leaders of the Party

Committee and the People's Government of Zhejiang Province chair a meeting on ecological and environmental protection to study major issues and arrange key tasks, while the People's Congress of Zhejiang Province strengthens legislation and law enforcement examination relating to ecological and environmental protection, and the Political Consultative Conference of Zhejiang Province actively exercises democratic supervision and offers advice and suggestions. Local authorities and departments have promoted ecological and environmental protection in various fields of economic and social development, and so an atmosphere in which the whole society pays attention to, participates in, supervises and supports environmental protection has gradually taken shape, thus providing the political guarantee and the basis among the masses for expanding the building of an ecological province and increasing the level of ecological development in the new situation, as well as shaping a harmonious pattern of joint governance and ecological and environmental protection by the government, the market and the society.

8.3.1 Intensifying Reforms, Strengthening the Guiding Role of the Government

Ecological development is a systematic project involving the economic, political, cultural and social fields; under the traditional system of government, there is severe fragmentation among different departments involving water conservation, environmental protection, shipping and gardening. In order to overcome the drawback by which governmental functions are excessively distributed among many departments, in May, 2003, Zhejiang established Zhejiang's Leading Group for the Building of an Ecological Province, in which the Secretary of the Party Committee of Zhejiang Province, Xi Jinping, served as the leader, the Governor of Zhejiang Province served as Executive Deputy Leader, members of the Standing Committee of the Party Committee of Zhejiang Province, the Secretary General of the Party Committee of Zhejiang Province, vice governors of Zhejiang Province served as deputy leaders, the people chiefly in charge of the Party Committee and the People's Government of Zhejiang Province and relevant departments of Zhejiang Province were members. The leading group had an office which was located within the provincial environmental protection bureau. Afterwards, this organization was established. The cities and counties also set up their own leading groups accordingly, with offices (ecology offices) of the leading groups at various levels located within the environmental protection departments, making important contributions to building Zhejiang into an ecological province.

As from 2009, Zhejiang's competent department for the environment was upgraded from the provincial environmental protection bureau to the provincial department of environmental protection; this has increased the administrative level of the department for environmental protection and also expanded the scope of the official duties and authority of the that department; the provincial department for environmental protection strengthens the functions of the former environmental protection bureau, and performs an additional function of coordinating major issues, including environmental supervision, planning and standards; it also strengthens the organization of the building of an ecological province, as well as guidance and coordination in promoting ecological development; moreover, it strengthens the guidance of, coordination in and supervision over environmental improvement and ecological protection and it intensifies efforts in achieving national and provincial emission reduction goals and environmental supervision. The establishment of the provincial department for environmental protection is conducive to performing the function of environmental protection, promoting the improvement of the ecological environment and ecological development.

Since 2013, Zhejiang has carried out the fourth round of the reform of the system of administrative examination and approval, with a focus on establishing a centralized system of examination and approval, more quickly improving the system of joint examination and approval, building a system involving early examination and approval and the standardized management of intermediary services, carrying out a system of an agency to deal with the whole process of examination and approval services concerning the projects whose entry into parks is sought and large projects, establishing the access system for the items subject to examination and approval and improving the system for the responsibility of examination and approval. Zhejiang has greatly reduced administrative functions and powers by releasing the list of governmental powers; it developed the list of the projects in the investment fields which enterprises are prohibited and restricted from entering by releasing the negative list of enterprise investment projects; it has granted no government approval to the enterprise investments outside the list. At present, administrative service centers and village (community)-level service centers have been built in all of the cities, counties, towns and villages in Zhejiang, the government service system has provided full coverage, and government service departments have offered services at the Home of Citizens in a centralized way, providing great convenience for the people.

8.3.2 *Making Active Innovations: Giving Scope to the Role of the Market Mechanism in Regulation*

For a long time, market failure has occurred in the field of resource environment due to the nature of environmental resources as public goods and the externality of the environment, so government intervention is necessary. With an increasing scarcity of environmental resources and decreeing costs for defining the property rights of resources, it is possible to introduce a market mechanism into the field of the resource environment. Therefore, it is essential to renovate the market mechanism and reform the system of environmental resource property rights so that the market mechanism plays a decisive role in resource allocation. Zhejiang's innovations to the market mechanism mainly include reform of forest rights, the mechanism for the trading of emission rights and water rights.

8.3.2.1 Promoting the Reform of Forest Rights, Putting the Assets from Forest Rights to Use

Zhejiang has actively promoted the renewal (issuance) of a certificate of forest rights to stabilize mountain rights and forest rights. With the renewal (issuance) of this certificate and the signing of contracts for the contractual mountains, Zhejiang has become one of the provinces being the first nationwide to basically complete the task of reforming forest rights. Since 2011, Zhejiang has actively pushed forward IT applications relating to forest rights so that all of the data on the attribution of the certificate of forest rights are entered into the system and digital management of forest rights is gradually achieved. Zhejiang issued a number of policy documents, including the *Administrative Measures of Zhejiang Province for the Circulation of Forests, Forest Trees and Forest Land*, and it established the Eastern China Forestry Exchange, set up forest rights management bodies and forest rights trading centers in many counties (county-level cities, districts) to facilitate the management and trading of forest rights. Zhejiang has become the first nationwide to offer forest right mortgage loans, it has explored a number of new loan modes, introduced multiple policies and systems, improved the incentive measures including fiscal interest subsidies, risk compensation and policy-oriented insurance, it has established a new mode of forest rights mortgage loans. Moreover, the People's Government of Zhejiang Province has issued the *Several Opinions on Accelerating the Development of Modern State Agricultural, Forest and Fish Farms*; Zhejiang has become the first nationwide to push forward the reform of state forest farms, and has identified many forest farms as public institutions. Under the guidance of local forestry departments, forest workers in various areas have established various types of forestry-related specialized cooperative organizations, thus effectively promoting the improvement of forestry management and the development of the forestry industry.

8.3.2.2 Building on the Market Mechanism to Promote Compensated Use and the Trading of Emission Rights

Zhejiang has experienced the process from local pilot work to province-wide promotion in practicing compensated use and the trading of emission rights.

First, taking laws as the yardstick. Zhejiang has issued a number of provincial policy documents, including the *Guiding Opinions Concerning the Pilot Work on the Compensated Use and Trading of Emission Rights*, the *Interim Measures of Zhejiang Province for Pilot Work on the Compensated Use and Trading of Emission Rights* and the *Detailed Rules for the Implementation of Interim Measures of Zhejiang Province for Pilot Work on the Compensated Use and Trading of Emission Rights*. These policy documents cover different contents and focus on different aspects, but they improve the framework for the system of policies, laws and regulations involving the compensated use and trading of emission rights from different perspectives.

Second, seeking support from technologies. Zhejiang has established accounts concerning the increment in the number of the sources of pollution, the decrement of that number, the amount of excessive emissions or the amount of secret emissions by installing online monitoring equipment and the systems for online monitoring of the environmental quality in key polluting enterprises so as to be able to monitor the emission behavior of the enterprises that emit pollutants. Zhejiang has made great efforts to build an online monitoring system and promote the networking of pollutant emitting enterprises. The emission system based on swiping cards has been set up. Under such a system, once the enterprises' emission quota has been used up, the valves at sewage pipelines will be automatically closed, thus effectively monitoring the emission behavior of enterprises each month. Zhejiang has built a networking system of the comprehensive management of environmental protection covering the trading of emission rights and the management of emission permits, the examination and approval of construction projects and environmental law enforcement, so data sharing has been achieved.

Third, deriving impetus from innovations. The cities in Zhejiang have made innovations in light of the reality to avoid or overcome institutional barriers or resistance, to solve practical problems and develop various modes. For instance, the Jiaxing Model for the initial allocation of emission rights helps circumvent the legislation issue by recognizing the emission rights obtained by enterprises before pilot work, thus guaranteeing fairness; the system of emission rights mortgage loans introduced by Shaoxing and the system of emission rights leasing adopted by Pinghu City effectively overcome the shortage of resources and the difficulties in obtaining loans for small and medium-sized enterprises. The People's Government of Zhejiang Province has, based on the successful experience from different areas, improved institutional innovations and promoted them across the province, and it has vigorously pushed forward the compensated use and trading of emission rights.

Fourth, building on respecting the market mechanism. In Zhejiang, the prices of emission rights were once set by the government, but these prices often deviated from the balanced prices on the market and hindered trading. Later, many local authorities attempted to conduct auctions of emission rights; as prices are set by the market, auction prices are not necessarily the balanced prices on the market, but they reflect the willingness of the enterprises to pay emission rights, so auction prices are representative to some extent. Nanhu District of Jiaxing City and Hangzhou City conducted auctions of emission rights on many occasions, many enterprises participated in them, and basically all of the emission rights were bought, thus promoting the trading and transfer of emission rights.

8.3.2.3 The Mechanism of the Trading of Water Rights

The trading of water rights is essentially the transfer of the right to use water resources among different traders The water involved in the cases involving the trading of

Zhejiang's water rights is basically reservoir water, it is easy to define its ownership and the difficulty in defining water rights caused by the mobility and inter-regional flow of water is prevented. Local governments represent local public interests. The trading of water rights among local governments can be conducted at lower costs on the one hand and can effectively guarantee local long-term interests on the other hand. Several cases of the trading of water rights in Zhejiang are based on surplus water, priority is given to guaranteeing water use in production, life and in the ecological field, thus effectively preventing a lack of water in the place from which water is supplied and the resulting contrast between the place from which water is supplied and the place where water is used. All of the cases of the trading of water rights in Zhejiang are spontaneously carried out by local governments in light of the reality; both sides in trading have actual needs. The trading of water rights can address the shortage of water resources on the one hand and exchange surplus water for capital on the other hand, both sides obtain benefits and a win-win outcome is achieved; this offers experience and serves as reference as follows.

First, overcoming the tradition that the allocation of water rights is administratively monopolized. For a long time, the allocation of water rights was administratively monopolized in China, mainly characterized by directive-based utilization and administrative allocation of water. In the management of river basins, generally, water in various areas of river basins is administratively allocated by the higher authorities, issues of the utilization of water or disputes over water are dealt with mainly by administrative means in the dry season. In the transfer of water across regions or river basins, water diversion projects are carried out mainly by the central or higher administrative departments to administratively allocate water resources among regions; water diversion projects are carried out solely by the state or provided with very high investment subsidies. Under the market economic conditions, whether in the management of water affairs in the upstream and downstream sections of river basins or the transfer of water across river basins, there are increasing difficulties in adopting administrative means, the effectiveness of coordinating interest conflicts has been on the decrease. In the trading of water rights, administrative coordination is slow and unreliable, economic strength is very high, so the party which suffers a water shortage elects to directly purchase water for obtaining water rights through the market mechanism; this is different from the previous method of transferring water across river basins and breaks the tradition that water rights are allocated by administrative means.

Second, marking the official emergence of a market for water rights in China. Water resources are owned by the state, thus the primary allocation of water rights must be made by government departments, but the secondary allocation of water rights is not necessarily made by administrative means; if it is conducted through the market, the market for the trading of water rights ("the water rights market") emerges. Likewise, if water goods are allocated through the market, a market for water goods emerges. In practice, we generally call the water rights market and the water goods market "the water market". On the market for tap water, currently, ladder-like water prices have been adopted, pricing varies with different sectors in many cities, in which, although prices have not yet been completely set by the market, the market mechanism has played a crucial role, and prices are fully determined by supply and demand. Water shortage brings huge business opportunities to enterprises, so the water goods market develops rapidly. With the trading in water rights, Zhejiang has filled the gap in the water right market and has become the first to initiate trading through consultation based on equality and voluntariness, and an inter-city market for the transfer of water rights has taken shape for the first time.

Third, proving that the market mechanism is the effective means for allocating water resources. As water rights are traded by means of a market mechanism, the interests for both sides increase. The water supplier obtains surplus water sources through water saving projects and the broadening of new source projects, the costs are generally low thanks to an economy of scale and natural resource advantages; the water purchaser is generally disadvantageous in natural resources, self-construction of the source of water usually entails higher costs. With trading in water rights, both sides will be forced to save more water and protect water resources; the market plays the role of optimizing resource allocation. If both sides address issues through administrative means, the contrasts between the two sides will certainly increase and may even become disputes, the market mechanism essentially plays the role of coordinating conflicts of local interests.

8.3.3 Ensuring Multi-party Cooperation, Achieving Synergy Among Social Forces

The development of the ecological environment has a bearing on a large number of households and everyone's food, clothing, housing and transportation. During ecological development, Zhejiang has given full scope to the roles of the government and the market, and has also actively mobilized social forces to participate in the process and provide support, thus effectively promoting ecological development.

8.3.3.1 Improving the Mechanism of Public Participation in Environmental Protection

With a series of institutional improvements and institutional innovations, Zhejiang has encouraged and supported public participation in environmental protection. According to the *Measures of Zhejiang Province for Environmental Management of Construction Projects*, the assessment of the environmental impact on construction projects should be conducted by inviting social groups, research institutions, the units and individuals in relevant sensitive environmental areas to attend symposiums, feasibility study meetings and hearings and thus safeguard the public's right to participation.

8.3.3.2 Improving the Platform for Public Participation in Environmental Protection

To strengthen public supervision over environmental protection, the Department of Environmental Protection of Zhejiang Province has built the environmental protection hotline and online reporting platform to actively mobilize the people to participate in the supervision of environmental protection. The cities have actively experimented with such modes of public participation in environmental protection as environmental protection federations, citizens' inspection groups for environmental protection, service groups of experts, groups for the publicity regarding ecological development, non-governmental environmental supervisors with a view to reinforcing public supervision over environmental law enforcement.

8.3.3.3 Affirming and Giving Play to the Role of Non-governmental Organizations

The non-governmental organizations for environmental protection are the non-governmental groups voluntarily organized by environmental protection personnel; they have played an important role in public participation. They leverage the advantages characterized by high specialization and a high degree of organization to actively organize various environmental protection activities, popularize environmental protection knowledge and strengthen environmental supervision. Zhejiang has given great importance to the development of non-governmental organizations, and various non-governmental environmental protection organizations are very active, thus greatly promoting Zhejiang's ecological development.

8.4 Summary

The Party Committee and the People's Government of Zhejiang Province have achieved a great deal of success in building an ecological province and a green Zhejiang, giving rise to the Zhejiang Experience, which is unique. Summarizing and promoting the Zhejiang Experience in ecological development is of important theoretical value and practical significance for practicing the Scientific Outlook on Development, earnestly promoting ecological development and achieving the great rejuvenation of the Chinese nation. The Zhejiang Experience in ecological development is mainly reflected in the following three aspects.

First, gaining the strategic insight of carrying through one blueprint. A good ecological environment is the fairest public goods and the most universal benefit for the people's well-being. Ecological development is a great cause which entails massive investments, produces effects slowly, requires hard work at the present time, but benefits generations to come. The admirable spirit of the successive Party Committees and the People's Governments of Zhejiang Province consists in tenaciously persevering in keeping lush mountains, successively working hard and carrying through one blueprint year by year. This is a valuable experience from the building of a beautiful Zhejiang.

Second, ensuring harmonious environmental, economic and social development. Xi Jinping pointed out, "It is necessary to correctly handle the relationship between economic development and ecological environmental protection, firmly uphold the philosophy that the protection of the ecological environment is the protection of productive forces, the improvement of the ecological environment is the development of productive forces, more consciously promote green, circular and low-carbon development, not unduly seek economic growth at the expense of the environment.[4]" The Party Committee and the People's Government of Zhejiang Province have adopted institutional means to strengthen the environmental access system; it has relied on science and technology to promote environmental and social development, it has taken the Scientific Outlook on Development as a guide to boost harmonious environmental and social development, it has derived impetus from innovations to achieve leapfrog environmental and social development, intensified the planning of functional areas, optimized the pattern of land development and it has actively built a conservation-minded society to make clean, clear waters and lush mountains deliver economic and social benefits, concurrently enhance economic, social and ecological benefits, make the people rich and ecology beautiful. This is another valuable experience in building a beautiful Zhejiang.

Third, achieving joint governance by the government, the market and the society. The Party Committee and the People's Government of Zhejiang Province have regularly held a working conference on ecological and environmental protection to study major matters and arrange key tasks, while the People's Congress of Zhejiang Province strengthens legislation and law enforcement examination relating to ecological and environmental protection, and the Political Consultative Conference of Zhejiang Province actively exercises democratic supervision and offers advice and suggestions. Local authorities and departments have promoted ecological and environmental protection in various fields of economic and social development, the atmosphere in which the whole society pays attention to, participates in, supervises and supports environmental protection has gradually taken shape, thus providing a political guarantee and a basis among the masses for expanding the building of an ecological province and increasing the level of ecological development in the new situation, as well as shaping a harmonious pattern of joint governance and ecological and environmental protection by the government, the market and the society. This is the third valuable experience in building a beautiful Zhejiang.

[4]Xi Jinping, *Xi Jinping: The Governance of China*, Foreign Languages Press, 2014, p. 209.

Postscript

Zhejiang is reputed as a region of rivers and lakes south of the Yangtze River and a paradise on earth. As a province which met with growing pains during the earliest stage of industrialization, Zhejiang has boldly made ecological innovations. Such great projects as the building of a green Zhejiang, the building of an ecological province, the building of an ecological Zhejiang and the building of a beautiful Zhejiang have brought enormous ecological welfare to the people of Zhejiang, and have also created a beautiful experience for the rest of the country. The Chinese Dream and Zhejiang's Practice—Ecology Volume ("Ecology Volume") is aimed at summing up Zhejiang's experience in building a beautiful China.

Based on Xi Jinping's series of important speeches on ecological development, the "Ecology Volume" systematically summarizes the theoretical basis, strategic line of thought, major actions, construction effects and basic experience in Zhejiang's ecological development. The whole book consists of eight chapters, which are "The Outlook on Ecological Civilization Leads the Building of a Beautiful Zhejiang", "The Development of the Conservation-oriented and Environmentally Friendly Ecological Economy", "The Protection of the Ecological Environment with Beautiful Mountains and Rivers", "Inheritance of the Ecological Culture with Harmony between People and Nature", "The Construction of a Beautiful Countryside with Harmony between People and Nature", "The Green and Low-carbon Development of Ecological Cities", "Explorations in the Institutions for Ecological Development with Checks and Balances" and "Experience and Inspirations from the Building of a Beautiful Zhejiang".

The "Ecology Volume" is the result of cooperation among the experts and scholars engaged in research on ecological development from the Chinese Academy of Social Sciences, Ningbo University, Zhejiang Sci-Tech University and the Zhejiang Academy of Social Sciences. The outline for the writing of this book was drafted by the Deputy Leader of the Research Group, Shen Manhong, it was deliberated and improved by all the members of the Research Group organized by the Leader of the Research Group, Pan Jiahua, and finalized by the Department of Publicity under the Party Committee of Zhejiang Province and the Zhejiang

J. Pan and M. Shen (eds.), *Chinese Dream and Practice in Zhejiang – Ecology*,
Research Series on the Chinese Dream and China's Development Path,
https://doi.org/10.1007/978-981-13-7209-4

Academy of Social Sciences. The authors of the various chapters carried out the writing according to the outline. After the first draft was finished, Shen Manhong fully checked and reviewed the entire book and put forward detailed revision suggestions. The authors of the various chapters improved the manuscript according to those suggestions. After review by experts, the manuscript was revised again, and then Shen Manhong conducted a final compilation and editing of the manuscript, Pan Jiahua reviewed and determined it. During the examination of the manuscript, associate professor Zhang Lei assisted Shen Manhong in carrying out a great deal of the work. The authors of the "Ecology Volume" carried out the following work:

Chapter 1, Shen Manhong (Ningbo University, The Zhejiang Research Center for Ecological Development);
Chapter 2, Li Zhibin, Lin Kaiwen, and Tian Hongyan (Zhejiang Sci-Tech University, The Zhejiang Research Center for Ecological Development);
Chapter 3, Li Zhibin, Su Qingkai, and Tian Hongyan (Zhejiang Sci-Tech University, The Zhejiang Research Center for Ecological Development);
Chapter 4, Li Hongyu (The Chinese Academy of Social Sciences);
Chapter 5, Zhang Lei (Zhejiang Sci-Tech University, The Zhejiang Research Center for Ecological Development);
Chapter 6, Zhu Shouxian (The Chinese Academy of Social Sciences);
Chapter 7, Liu Jian (The Zhejiang Academy of Social Sciences);
Chapter 8, Wang Ran (The Zhejiang Academy of Social Sciences).

The reviewing experts made great efforts and gave valuable advice in developing the outline and improving the manuscript. We hereby express our heartfelt thanks to them!

The authors of the "Ecology Volume", are experts and scholars engaged in research on ecological development. Even if there were other very arduous tasks, the authors still remarkably undertook the writing of this book and earnestly revised it according to the suggestions of the Leader of the Research Group and the reviewing experts, thus guaranteeing the quality of the "Ecology Volume". We highly appreciate the scientific cooperative and dedicated spirit of the authors!

The Research Group of The Chinese Dream and Zhejiang's Practice, Ecology Volume.

December 1, 2014

MIX
Papier aus verantwortungsvollen Quellen
Paper from responsible sources
FSC® C105338

If you have any concerns about our products,
you can contact us on
ProductSafety@springernature.com

In case Publisher is established outside the EU,
the EU authorized representative is:
Springer Nature Customer Service Center GmbH
Europaplatz 3, 69115 Heidelberg, Germany

Printed by Libri Plureos GmbH
in Hamburg, Germany